OIL PIPELINES
AND PUBLIC POLICY

A Conference Sponsored by the
American Enterprise Institute for Public Policy Research

OIL PIPELINES AND PUBLIC POLICY

Analysis of Proposals for Industry Reform and Reorganization

Edited by Edward J. Mitchell

American Enterprise Institute for Public Policy Research
Washington, D.C.

Library of Congress Cataloging in Publication Data

Main entry under title:
Oil pipelines and public policy.

    1.   Petroleum—United States—Pipe lines.
2. Pipe lines—Economic aspects—United States.
I. Mitchell, Edward John, 1937-   II. American
Enterprise Institute for Public Policy Research.
HD9580.U5037   338.2′7′2820973   79-13728
ISBN 0-8447-2157-3
ISBN 0-8447-2158-1 pbk.

AEI Symposia 79-E

Printed in the United States of America

CONTRIBUTORS

Donald I. Baker
Partner in law firm of
Jones, Day, Reavis and Pogue (D.C.)

Henry M. Banta
Counsel, Senate Subcommittee on Antitrust and Monopoly

Stanley Boyle
Visiting Professor of Economics
University of South Carolina

David Brown
Attorney, Energy Section, Antitrust Division
U.S. Department of Justice

Michael E. Canes
Policy Analysis Director
American Petroleum Institute

Edward W. Erickson
Professor of Economics and Business
North Carolina State University

Donald L. Flexner
Deputy Assistant Attorney General

Scott Harvey
Economist, Bureau of Economics
Federal Trade Commission

George A. Hay
Director of Economics, Economic Policy Office
Antitrust Division, U.S. Department of Justice

Donald Kaplan
Chief, Energy Section, Antitrust Division,
U.S. Department of Justice

Edmund W. Kitch
Professor of Law
University of Chicago Law School
Adjunct Scholar
American Enterprise Institute

Ulyesse J. LeGrange
Controller, Exxon Corporation

Lucinda M. Lewis
Associate Professor of Economics
University of Pennsylvania

Gayle C. Linder
Consultant, Erickson Associates

Morris Livingston
Consultant, Standard Oil of Indiana

Walter S. Measday
Chief Economist
Senate Subcommittee on Antitrust and Monopoly

Edward J. Mitchell
Professor of Economics
University of Michigan Graduate School of Business
Director, Energy Policy Studies
American Enterprise Institute

Donald A. Norman
Senior Economist, Policy Analysis
American Petroleum Institute

Howard O'Leary
Partner in law firm of
Dykema, Gossett, Spencer, Goodnow and Trigg (D.C.)

William L. Peters
Consultant, Erickson Associates

Michael J. Piette
Assistant Professor of Economics
University of Hartford

Robert J. Reynolds
Senior Economist, Economic Policy Office
Antitrust Division, U.S. Department of Justice

Calvin Roush
Economist, Bureau of Economics
Federal Trade Commission

James Shamas
Director of Transportation
Getty Oil Company, Houston

Thomas C. Spavins
Economist, Economic Policy Office
Antitrust Division, U.S. Department of Justice

Shyam Sunder
Associate Professor of Accounting
University of Chicago Graduate School of Business

David J. Teece
Associate Professor of Business Economics
Stanford University Graduate School of Business

George S. Wolbert
Attorney in private practice, Houston

CONTENTS

PREFACE
Edward J. Mitchell

PART ONE
STRUCTURAL ISSUES

Oil Pipelines: The Case for Divestiture 3
Donald L. Flexner

The Pipeline Undersizing Argument and the Record of Access
and Expansion in the Oil Pipeline Industry 15
Edward W. Erickson, Gayle C. Linder, and
William L. Peters

Commentaries
George S. Wolbert 57
Walter S. Measday 60

Discussion 63

PART TWO
EFFECTIVENESS OF GOVERNMENT REGULATION

The Regulation of Oil Pipelines 77
Thomas C. Spavins

Effectiveness of Government Regulation 107
Ulyesse J. LeGrange

Commentaries
Shyam Sunder 113
David J. Teece 116

Discussion 119

PART THREE
Alternatives to the Present System

Appraising Alternatives to Regulation for Natural Monopolies 135
Lucinda M. Lewis and Robert J. Reynolds

Pipelines and Public Policy 141
Michael E. Canes and Donald A. Norman

Commentaries
Michael J. Piette   165
Stanley Boyle   167

Discussion 171

SUPPLEMENTS

Supplement 1: Testimony of John H. Shenefield, Assistant Attorney General, Antitrust Division, Department of Justice, before the Subcommittee on Antitrust and Monopoly of the Committee on the Judiciary of the United States Senate Concerning Oil Company Ownership of Pipelines 191

Supplement 2: Report of the Antitrust Division, Department of Justice, on the Competitive Implications of the Ownership and Operation by Standard Oil Company of Ohio of a Long Beach, California–Midland, Texas, Crude Oil Pipeline 217

Supplement 3: United States of America Federal Energy Regulatory Commission: Treatment of Certain Production-Related Costs for Natural Gas to be Sold and Transported through the Alaska Natural Gas Transportation System 253

Supplement 4: An Analysis of the Rates of Return on Petroleum Pipeline Investments 261
Exxon Pipeline Company/Exxon Company, U.S.A.

Supplement 5: Oil Pipelines: Industry Structure 317
S. Morris Livingston

PREFACE

In 1864, a group of teamsters attacked the Van Syckel pipeline in western Pennsylvania with pickaxes and attempted to pull it apart with their teams. This raucous reception for America's first oil pipeline seems to have set the tone of the public discussion that followed. Emotionally charged, criticism of the pipeline industry heard before the U.S. Industrial Commission in 1899 differed little from that heard before congressional committees in the 1970s. Fundamentally, the charge was discrimination by pipeline owners among shippers, an act illegal now for more than seventy years, and one that should lead to a plea for law enforcement, not a revolutionary restructuring of the industry.

Happily, the quality of discussion has radically improved as the Department of Justice today attempts to base its proposals for reform on serious economic analysis of the pipeline industry. It is this more intellectual approach to the issue by the Department that enabled our conference to be successful. Because the arguments involved primarily questions of economic logic and fact the discussion could be maintained at a high level. Industry, government, and academic participants all received the idea of the conference enthusiastically. My perception was that a considerable amount of learning took place among the participants.

For the reader, this volume appears to represent the state of the debate in 1979. The Justice Department's "undersizing" thesis and "competitive rules" were analyzed in depth. The issue of profitability of pipelines was brought into sharper focus. While it can hardly be said that the issues have been resolved or that anyone was converted, the agenda for research is more clear, and the need for further empirical research more obvious.

EDWARD J. MITCHELL

PART ONE

STRUCTURAL ISSUES

Oil Pipelines: The Case for Divestiture

Donald L. Flexner

Oil pipelines have in recent years become one of the Antitrust Division's top priorities, and dialogue of the sort offered by this conference provides a valuable opportunity to sharpen and clarify thinking on all sides of this issue.

It is not my intention to recite here the details of our current efforts on several fronts to achieve pipeline rate reform. However, as a representative of the Antitrust Division, I do think it is important to place our overall pipeline program—and in particular the possibility of divestiture—in the context of our institutional preference for competition rather than regulation. The Antitrust Division has previously expressed its alarm at the rise of regulation in energy markets.[1] We believe that in many energy industries where competition can and should work, the nation is being ill served by unnecessary regulation. I do not mean to suggest, of course, that competition is a panacea for all our energy problems. Our view, however, is that in the current climate—when the nation exhibits a heightened inclination to regulate energy markets—it is particularly important that we explore whether a competitive industry structure can serve as a viable alternative to excessive regulation.

I would like to set forth the means by which we determine whether shipper-ownership of pipelines creates competitive problems that require solutions—regulatory or otherwise. To answer this question, it is necessary to decide first whether market power exists in particular pipelines. Basic to this inquiry are two generally well-understood tenets of pipeline economics. First, compared with other modes of transportation, pipelines are the most efficient overland transport mode for the move-

---

AUTHOR'S NOTE: I wish to thank Richard A. Feinstein for his help with this paper.
[1] See, for example, remarks by John H. Shenefield before the Natural Resources Law, Administrative Law, and Public Utility Law Sections of the American Bar Association, August 8, 1978.

ment of significant amounts of petroleum between any two given points.[2] Indeed, even where waterborne competition exists, the large pipelines, which include most interstate lines constructed in recent years, are generally more efficient than all but the largest tankers. The second basic principle of pipeline economics concerns competition from other pipelines. As pipeline capacity varies with the square of the radius of the pipe, pipelines exhibit considerable economies of scale, and the average cost of a unit of throughput declines as pipeline size increases. This phenomenon of declining unit costs occurs throughout the range of technologically feasible pipeline sizes. Consequently, if a decision to build a pipeline capacity is warranted, it is more efficient to build a single pipeline of sufficient size to satisfy the entire demand for transportation between two points than to build several smaller pipelines at the same time.

These two economic facts—cost efficiency and economies of scale—constitute the natural monopoly characteristics of pipelines. A pipeline's "monopoly" in this sense is the natural result of ordinary economic and technological forces—forces that are distinguishable from monopolistic or predatory behavior. Pipelines' natural monopoly characteristics, along with the concomitant abuse of market power, were recognized by Congress as early as 1906, when the Hepburn Act[3] placed pipelines within the jurisdiction of the Interstate Commerce Commission. However, these characteristics convey monopoly power only in the market for transportation between two points; they do not convey monopoly power to any particular pipeline owner for the products the pipeline transports. To have such monopoly power, the pipeline must also exercise market power.

Market power can be expressed as a pipeline's cost advantage over its competitors, measured at the upstream or downstream end of the pipeline, or both. In general, a pipeline will derive market power from downstream conditions when:

- its throughput comprises a significant share of the downstream market
- it can supply the downstream market with less expensive petroleum than can other suppliers, because of lower transportation costs than alternative modes (including other pipelines), because of access to lower-cost crude (or products) than are available to

---

[2] U.S. Senate, Committee on Energy and Natural Resources, National Energy Transportation Report, vol. 1, no. 95-15, 95th Congress, 1st session (1976), pp. 182-184.

[3] 34 Stat. 584 (1906).

other suppliers, or because local producers (or refiners) cannot expand output without increasing costs

- alternative supply is less than perfectly elastic, that is, any increase in the supply required from alternative sources can be attained only at a higher price.

Each of these conditions is measured by determining the prices and quantities that would prevail if the market were competitive.

Similarly, a pipeline will generally derive market power from upstream conditions when:

- its throughput comprises a significant share of the upstream market
- increased supplies of petroleum can be absorbed upstream only at a lower price
- upstream suppliers will be willing to accept lower prices to sell their crude or products, either because of a lack of transportation alternatives at costs as low as the pipeline's, or because contraction of output would be difficult or unprofitable.

This type of market power is ordinarily termed monopsony power.

A concrete illustration of this analytical process can be seen in the Antitrust Division's report on the competitive implications of Sohio's pipeline from Long Beach to Midland, Texas.[4] There, we concluded that the proposed pipeline would not possess market power in the midcontinent area—PAD districts II and III.[5] It would be useful to set out the basis for that conclusion in some detail.

The terminal point of the proposed Sohio pipeline is Midland, Texas, the center of a pipeline network that transports crude oil from West Texas to PADs II and III. Oil from West Texas, in turn, becomes part of a larger stream that includes petroleum from other domestic producing areas and the considerable volume of imports entering the United States through the Gulf Coast. There are no subdivisions of this area that appear to be uniquely or specially served by the Sohio pipeline. If the Sohio pipeline were expanded to its

---

[4] See Antitrust Division, Department of Justice, "Report on the Competitive Implications of the Ownership and Operation by Standard Oil Company of Ohio of a Long Beach, California–Midland, Texas, Crude Oil Pipeline," June 1978; hereafter referred to as Sohio Report.

[5] Petroleum Administration for Defense districts. PAD II includes Ohio, Kentucky, Tennessee, Indiana, Illinois, Michigan, Wisconsin, Missouri, Iowa, Minnesota, North Dakota, South Dakota, Nebraska, Kansas, and Oklahoma. PAD III consists of Arkansas, Alabama, Louisiana, Mississippi, New Mexico, and Texas.

largest capacity planned, Sohio's share of the crude oil market in PADs II and III would be about 10 percent.

The incremental supply of oil to this market is high-cost imported oil, which determines the market price of all oil entering PADs II and III through the Gulf Coast region. Because such imports will remain the incremental supply for the entire range of throughput possible for the Sohio pipeline, Sohio could not, by restricting throughput to West Texas, increase the price of additional delivered oil to PADs II and III above the world price. Any such throughput restriction on the part of Sohio would merely invite greater imports of foreign crude at the world price—a result which could not benefit Sohio in the relevant market.

Standing in marked contrast to our conclusions about the Sohio pipeline is our 1976 analysis of the proposed Gulf Coast deepwater ports.[6] Because deepwater ports reflect the same natural monopoly characteristics of pipelines, the process by which we analyze the competitive implications of such ports is essentially identical to that employed for pipelines.

In assessing the competitive implications of the LOOP deepwater port, for example, we found that the amount of oil it would provide to Gulf Coast and midcontinent crude markets would be substantial, that it would have significant cost efficiencies over other modes of transportation, and that downstream refiners would likely be willing to pay more for delivered foreign oil if the capacity of the ports were reduced. In short, as I will explain more fully later, we concluded that LOOP would have downstream market power.

A pipeline's market power resulting from its natural monopoly attributes in transportation could be restrained through adroit regulation. Unfortunately, however, as we have stated repeatedly in other forums, pipeline rate regulation has historically been extremely ineffective. Moreover, we believe that the present structure of the petroleum industry renders regulation incapable of preventing evasion of the effects of such regulation—and hence the exploitation of market power—by integrated oil companies. This can be seen by comparing the different incentives facing independent and integrated pipelines. Because an independent pipeline company is involved only in the transportation business, its natural incentive is to maximize transportation profits. Where such a pipeline has monopoly power, it can be expected to

---

[6] See "Report of the Attorney General Pursuant to Section 7 of the Deepwater Port Act of 1974 on the Applications of LOOP, Inc., and Seadock, Inc., for Deepwater Port Licenses," November 5, 1976, p. 28; hereafter referred to as Deepwater Port Report.

exercise it. But, if its ability to charge monopoly rates is restrained through effective regulation, it has the incentive to serve the largest possible number of customers in order to maximize its profit by maximizing its throughput.

The incentives of an independent pipeline company thus differ from those of a vertically integrated pipeline company, which seeks to maximize overall profits, not just transportation profits. If a vertically integrated pipeline has market power, its incentive can be to limit throughput below the level at which an independent owner would size or operate the line, in the presence of effective rate regulation. It can thus use its structure to circumvent rate regulation, taking its monopoly profits in upstream or downstream markets.

Consequently, if the vertically integrated pipeline owner is also a significant seller in the downstream market, and if the pipeline has market power downstream, the pipeline owner may have an incentive to limit throughput so that the downstream market price reflects the cost of delivery by more expensive transport modes. The difference between this inflated market price and the cost of pipeline transportation is profit to the pipeline's shippers. To the extent that those shippers are affiliated with the pipeline company, the vertically integrated firm has circumvented regulation and thereby succeeded in capturing downstream profits initially denied by the rate regulator. The magnitude of the gains derived in this manner depends upon the vertically integrated firm's share of the downstream market and its ability to use the pipeline, rather than another more expensive mode, for transportation.

Similarly, if a vertically integrated pipeline owner is a significant buyer in the upstream market, and if the pipeline has market power upstream, the owner may have an incentive to limit throughput to depress the upstream market price. Again, the extent to which the vertically integrated pipeline shipper-owner gains such monopsony profits depends upon the position the firm holds in the upstream market.

I have already alluded to the analysis by which we concluded that Sohio lacked market power and, as a consequence, also lacked the incentive to undersize its proposed Long Beach to Midland pipeline, and to the very different conclusions reached in the Deepwater Port Report. The demand for deepwater ports is linked directly to the increased dependence of the United States on imported oil, and there is no evidence to suggest that this dependence is likely to decline significantly in the near future. The ports would serve the nation's most significant refining area, and, by permitting use of the largest tankers, they would provide the least expensive means of delivering imported crude to Gulf

7

Coast refineries. For example, Exxon projected that in 1980 the ports could transport 34 percent of all crude imports, and by 1990, when the ports would reach their full capacity, they could transport 63 percent of all crude imports. In short, the ports have the potential for transporting one-third of the nation's crude oil requirements each day.[7]

Because the profitability of the ports themselves is at least theoretically limited by rate regulation, the owners—generally integrated oil companies—would, in our view, have had the incentives I have already described to restrict port throughput. Furthermore, in examining the owners' plans for the design and operation of the ports, we found that those incentives had, in fact, been acted upon in relation both to the capacity of the ports and to access to the ports by nonowner-shippers.[8]

There was, of course, an understandable desire to design systems that would reflect optimal economic and engineering characteristics. At the same time, however, the systems were clearly designed in reasonable conformity with the particular desires of the owners—who, as I mentioned, for the most part have large, integrated operations. A system that reflects the needs and capabilities of such owners does not and perhaps cannot adequately reflect the interests of relatively small shippers. For example, the owners can effectively control which shippers have access to the ports simply by establishing design and operational criteria that small shippers cannot meet. If this strategy forecloses access for the small shipper, the owners increase their opportunity to capture downstream the monopoly profits that they may be denied at the port itself by rate regulation. However, by imposing uniform criteria[9] on all shippers, the owners avoid the charge that they are discriminatorily denying access to some.

For reasons set out in the Deepwater Port Report,[10] we did not recommend that this competitive problem be remedied by preventing integrated oil company ownership of the ports. However, the conclusions in that report are consistent with our understanding of pipelines—an understanding which leads us to believe that these competitive problems inhere in the structure of the industry.

The effect of joint ventures has received relatively little attention in the Antitrust Division's previous public expressions of its pipeline

---

[7] Deepwater Port Report, p. 51.

[8] See the discussion of the sizing decisions concerning the deepwater ports, later in this paper.

[9] For example, port capacity, storage, minimum tanker sizes, and pumping rates; see Deepwater Port Report, pp. 68-69.

[10] Ibid., pp. 103-109.

policy. Because most pipelines built in recent years are owned by more than one company, joint ventures are obviously an important element of the structure of the industry and are correspondingly significant in our analysis of that structure.

In many areas of the country, where workable competition may exist within other sectors of the petroleum industry, the downstream market shares of individual companies are often not large. However, the combined downstream market shares of the joint owners of a pipeline can present a very different picture.[11] Decision making by a joint venture, especially where unanimous or near unanimous consent of the members is required, is necessarily a collective process. That collectivity, in turn, reflects the fact that the individual interests of participants in the joint venture are maximized in accordance with the cumulative downstream market share of the members. In short, the joint venture will operate as though the sum of the downstream market shares of all the individual firms is affected by each collective decision. Accordingly, the market shares and resulting market effects are much more significant than the corresponding effects of a decision made by one firm with a relatively small market share.

The proposals for LOOP and Seadock, which were analyzed in the Antitrust Division's 1976 report to the secretary of transportation, provide some insight into the means by which a joint venture ensures that its decision reflects collective interests. The shareholders' agreements for both LOOP and Seadock indicated that the major participating owners retained tight control over decisions to authorize significant corporate expenditures. In the case of Seadock, an early draft of the shareholders' agreement, which merely required majority approval for such expenditures, was rejected in favor of approval by votes representing at least 70 percent of the ownership shares. Similarly, LOOP's shareholders' agreement required approval by owners of not less than 75 percent of the voting stock.[12] If, as we have argued, vertically integrated pipeline owners have the incentive to restrict throughput

---

[11] For example, in 1973, in any one of the states of Virginia, Maryland, Delaware, New Jersey, and in the District of Columbia, no single company among the group of integrated oil companies which jointly own Colonial Pipeline Company appeared to have a retail market share greater than 19 percent (Lundberg Survey, published by *National Petroleum News*). The sum of the retail market shares of all ten owners is much greater. In addition, in 1976 the owners of Colonial, taken collectively, owned about 49 percent of the refining capacity in those same states (*Oil & Gas Journal,* 1976). In those states, then, it would be appropriate to take into account their collective share of the supply, including the impact of their ownership of refineries, in determining the downstream effects of the joint ownership of Colonial.

[12] Deepwater Port Report, pp. 98-99.

where there is market power downstream, then reserving a collective power to veto pipeline expansion would further a throughput limitation strategy.

A conceptually similar means by which a joint venture can effectuate a noncompetitive result can be seen in the planning for the capacity of the deepwater ports. Our review of Seadock, for example, found among the owners a consistent pattern of conservative volume nominations—a pattern which results, of course, in a conservative capacity for the entire port. If the owners understate the volumes which they expect to ship and actually do ship through the port, and if the port's capacity reflects that systematic understatement, there is less capacity available for nonowners. Even if common carrier obligations preclude the outright denial of space to nonowners, there would likely be partial denials in the form of prorationing, and, as I shall explain shortly, prorationing does not effectively eliminate the shipper-owners' incentive to undersize the pipeline. As a result, throughput would nonetheless be restricted.[13]

No overt collusion is necessary for the participants in the joint venture to achieve this result. The capacity decision by the joint venture is a collective determination based on the demand forecasts of each member. Since each firm knows that all stand to gain if capacity is limited, by simply understating volume forecasts, individual owners attain collectively the desired result. We addressed this problem in the Deepwater Port Report by proposing what we termed "competitive rules."[14] Such rules alter the symbiotic relationship among the joint venture participants by providing each with independent incentives to increase throughput, and, ultimately, the capacity of the port.

It has been asserted that even in the absence of competitive rules, competition among the members of a joint venture will limit the anticompetitive effects of a pipeline. In responding to that assertion, it is appropriate to question how competition can occur when capacity choices are made interdependently in the manner just described. Under the appropriate conditions, incentives to undersize would exist regardless of whether a given project was structured as a joint venture stock company or as an undivided joint interest pipeline.

In addition, where a pipeline is organized as a joint venture and is exploiting its monopoly power through a single excessive tariff, a shipper who is a 10 percent owner will, at the margin, receive only

---

[13] Ibid., pp. 70-73.

[14] Ibid., pp. 103 ff. The paper by Robert J. Reynolds and Lucinda M. Lewis in Part III of this volume provides a discussion of this novel form of structural relief for the problems of natural monopoly.

one-tenth of the excessive profits resulting from the tariff. To the extent that the tariff is too high, that shipper essentially pays 90 percent of the excess to other owners. But such a shipper still captures a portion of the excessive profits and thus has no incentive to seek a lower tariff. This result is, of course, quite different from the suggestion that competition among the owners of a joint venture pipeline will reduce costs to consumers.

Up to this point I have attempted to explain our perception of the market power that can exist because of the structure of the pipeline industry. What remains is the question of remedies. As indicated at the outset of these remarks, the Antitrust Division has an institutional preference for avoiding regulatory interference with market forces. Because the vertically integrated structure of the oil pipeline industry creates competitive problems which cannot be solved by rate regulation alone, a regulatory response to the exploitation of market power by pipelines would require a layer of regulation in addition to that which seeks to prevent excessive tariffs. However, as our analysis indicates, the structure of the pipeline industry raises very complicated questions of access and capacity—questions that we believe are best addressed by private parties in a competitive climate. For example, in the deep-water port context, we believe that questions concerning minimum tenders, hose sizes, pumping rates, and the like are more appropriately resolved by independent business judgment than by regulation.

As suggested previously, the inadequacy of much current pipeline rate regulation seems evident.[15] The same can be said of simple prorationing rules—a regulatory tool often cited as an antidote to access restrictions. There is, in fact, little indication that prorationing rules have ever been effectively regulated. During the time that pipelines were within its jurisdiction, the Interstate Commerce Commission never developed any explicit guidelines for prorationing—equitable or otherwise. Moreover, even effective prorationing would not alter basic throughput restriction incentives. Rather, such prorationing would merely require the owners of a pipeline with market power to dilute their monopoly return by sharing it with others. In short, shipper-owners of an undersized pipeline receive benefits regardless of prorationing rules. Furthermore, the wide use of prorationing rules based on historical throughput shares has the effect of perpetuating such benefits.[16] An "owners last" prorationing rule, giving nonowners pri-

---

[15] See, for example, the paper on the present system of oil pipeline regulation by Thomas C. Spavins, in Part II of this volume.

[16] See Exhibit I accompanying the testimony of Fred F. Steingraber of Colonial Pipeline Company before the Subcommittee on Antitrust and Monopoly, Com-

ority over owners, would reduce the incentive of the owners to under-size. But such a rule would not eliminate that incentive, because owners would nevertheless receive the benefits of restricted capacity on that portion of the throughput which they do ship. An "owners last" rule would create an additional burden of ownership and would, in fact, unfairly penalize owners if prorationing became necessary because of an emergency.

Prorationing over all sources of supply has also been offered as a solution to the access problem.[17] Such prorationing makes access to the prorated pipeline dependent on each shipper's share of the total supply available to the market served by the prorated portion of the pipeline. However, there are several important disadvantages asso-ciated with such a system of prorationing. It would necessarily be extremely complex and would have to take into account all alternative sources of supply.[18] Such a system also would require knowledge of costs of supply from all alternative sources and would amount to industrywide average cost pricing. This would be economically in-efficient. In addition, prorationing over all sources of supply would require a vast information exchange among competitors—a phenome-non that courts interpreting section 1 of the Sherman Act have often found troubling. Even apart from the potentially anticompetitive im-plications of this information exchange, a regulatory scheme of such complexity would be likely to offer the regulated carriers opportunities to make the process ineffective. In sum, consistent with the predilection of the Antitrust Division, I am inclined to believe that our economy in general—and in particular the energy sector, already replete with complex and often anticompetitive regulations—would not be well served by the imposition of such a prorationing system on pipelines.

It may be that behavioral remedies such as prorationing rules are inherently unsuitable for an industry, where, if we are correct, the competitive problem is essentially structural. The Department of

---

mittee on the Judiciary, U.S. Senate, January 30, 1975. See also, the statement of Allen E. Murray, president, U.S. Marketing and Refining, Mobil Oil Corpo-ration, before the House Merchant Marine and Fisheries Committee, July 26, 1977.

[17] See, for example, Michael E. Canes and Donald A. Norman, "Pipelines and Antitrust," Critique no. 005, American Petroleum Institute, November 1978, p. 15.

[18] Prorationing is much more complex for all alternatives than for a single alternative mode. For example, if the capacity of the Alaskan Natural Gas Pipeline were artificially restricted, the alternative source of supply mode would be gas supplied from the U.S. Midwest. See Report of the Attorney General Pursuant to Section 19 of the Alaska Natural Gas Transportation Act of 1976, July 1977, pp. 30-43.

Justice prefers a structural remedy so that the natural monopoly characteristics of pipelines do not lead, through vertical integration, to the exploitation of market power. The exercise of market power is not, in our view, so much the result of specific acts that can be proscribed as it is a function of expected business incentives acted upon in the context of monopoly power.

In this context, behavioral remedies such as prorationing rules or some adjustment to a pipeline's capacity would, in effect, amount to regulation and would result in an undesirable confusion of antitrust enforcement with the regulatory process. In addition, complex regulatory remedies offer the opportunity for those regulated to manipulate the regulatory process.[19] By contrast, we perceive structural relief not as a penalty, but as an approach to creating markets where competition can work. Indeed, our confidence that competition will work in such markets is grounded in no small part in the fact that divestiture would create an astute class of nonowner-shippers who would have every reason to police the activities of independent pipelines.

---

[19] See generally, Bruce M. Owen and Ronald Braeutigam, *The Regulation Game* (Cambridge, Mass.: Ballinger, 1978).

The Pipeline Undersizing Argument and the Record of Access and Expansion in the Oil Pipeline Industry

Edward W. Erickson, Gayle C. Linder,
and William L. Peters

Questions about the organization, ownership, and operation of petroleum pipelines have been raised by Department of Justice (DOJ) contentions concerning deliberate undersizing of oil and gas pipelines and such transshipment facilities as rail shipment of coal and deepwater port terminals.[1] The issues raised include such general problem areas as: (1) competition among crude oil or products pipelines, across crude oil and products pipelines, and between pipelines and alternative modes of transport; (2) the effectiveness of traditional rate of return regulation; (3) the access of various categories of shippers to pipeline services where pipeline organization varies; and (4) the economic organization of increments to pipeline capacity. Here we focus upon some of the record concerning access and the economic organization of increments to pipeline capacity, including:

- the access of nonaffiliated shippers to transportation services on joint venture pipelines of which they are not part of the ownership,
- the conditions of entry of new pipeline entities into the petroleum transportation industry,
- the composition of that entry with regard to the relative importance of (a) individual or joint ownership and solo-operation (as in undivided joint interest situations) as compared to (b) joint ownership and operation by a jointly owned pipeline company, and

---

The research on which this paper is based was supported by the American Petroleum Institute. However, all opinions and conclusions are those of the authors alone.

[1] The full Department of Justice analysis and position is contained in a number of documents. These include "The Report of the Attorney General Pursuant to Section 7 of the Deepwater Port Act of 1974 on the Applications of LOOP, Inc., and Seadock, Inc., for Deepwater Port Licenses," November 5, 1976; "The Report of the Attorney General Pursuant to Section 19 of the Alaska Natural Gas Transportation Act," July 1977; "Competition in the Coal Industry," May 1978; and proposed testimony of John H. Shenefield, assistant attorney general, Antitrust Division, before the Subcommittee on Antitrust and Monopoly of the Committee on the Judiciary, U.S. Senate, June 28, 1978.

- the relative incidence among new pipeline entities of pipelines—whether jointly owned and operated or otherwise—which include ownership participation by companies not included among the eighteen largest major oil companies.

The patterns of answers to questions in these areas have certain implications with regard to the empirical relevance of the DOJ assertions and will be further discussed below. The empirical review which is presented subsequent to this discussion focuses upon a number of specific comparisons. With regard to access of nonaffiliated and non-affiliated/nonmajor shippers to pipeline service, we examine an arbitrary sample of five interesting pipelines—Colonial, Plantation, Explorer, Buckeye, and Williams. The period examined for each pipeline begins with the first year of operation (Colonial, 1963; Williams, 1966; Explorer, 1972) or with 1960 (Buckeye, Plantation) and extends through 1976. The analysis of the empirical record of the entry of new pipeline entities covers the period 1947–1977.[2] Both products and crude oil pipelines are examined,[3] and both the value of carrier property and barrel-miles of throughput are used as measures of pipeline size. In all of the empirical review of entry of new pipeline entities, attention is focused upon the type of ownership (joint venture versus other) and the composition of ownership (major oil company versus other).

The Shell Game Theory of Monopoly

The U.S. petroleum industry is in general vertically integrated. The largest firms in the industry—commonly called the majors—are a heterogeneous set of economic organizations. This heterogeneity operates along and across dimensions or axes which may variously be defined in terms of principal refining facilities (by size, type, and location); international versus domestic specialization in crude oil production; degree of diversification into various alternative non-oil or gas energy sources; crude oil self-sufficiency; representation in various kinds of marketing and geographic sub-markets; and activities in petrochemicals and non-energy manufacturing, mining, or commerce, et cetera. But the standard perception of the industry's firms often seems to be that of a uniform set of homogeneous entities all composed of corporate building blocks of

---

[2] There is also a supplemental appendix which updates a previous, more highly aggregated, analysis of entry by type and organization of pipeline capacity and throughput. See Appendix Tables A–1 through A–4.

[3] In Appendix Tables B–1 through B–4, the identities and some relevant operating details of the new pipeline entities are presented. It will be noted upon examination of these data that there is flexibility in the use of pipeline capacity and migration from (and to again) initial crude or other products classifications.

16

equal relative importance, corresponding to the vertical stages of the oil industry.[4] This standard perception, together with the indisputable fact that the firms have, to varying degrees, organized themselves as a sequential set of integrated activities across the adjacent vertical stages of the industry, has led various industry critics to hypothesize that one of the private advantages of such organization—if not necessarily the principal rationale—is that it allows monopoly profits earned through the exercise of market power at one stage of the industry to be transferred to, and presumably hidden in, other stages of the industry. This conception is analogous to the economics of the midway in which the vertical stages of the industry are alleged to be the walnuts in a sophisticated shell game among and under which the pea of monopoly profit is passed and is hidden, regardless of the appearance of effective competition or effective regulation at any or all stages.[5]

Access and Expansion: Vulnerabilities of the Pipeline Undersizing Argument. Logical problems are already beginning to appear in the various DOJ versions of the pipeline undersizing argument. This has best

[4] These vertical stages include: (1) crude oil and natural gas exploration, development, and production, (2) crude oil transportation, (3) refining, (4) refined products transportation, and (5) refined products marketing. Pipelines, the subject of examination here, are one of the several modes of crude oil and refined products transportation.

[5] In addition to various problems of economic logic, this conception must ultimately bear the burden of identifying where the monopoly profits are hidden and specifying their effect upon overall petroleum company rates of return. Petroleum company rates of return are, however, generally regarded to be within the range consistent with effective competition. Were the monopoly profits, however, earned in a relatively small stage of the industry—such as pipelining—and generally transferred to the much larger aggregate which is the sum of the other stages of the industry, then substantial fruits of considerable market power could be hidden in the "noise" of profit and rate of return data. There are two problems with such an accommodation to the shell game theory of monopoly profits as applied to pipeline undersizing. First, the DOJ posits effective regulation of the pipeline sector so that there is no monopoly profit pea to be shunted from this stage to other stages. Second, according to the DOJ hypothesis, the exercise of market power through undersizing of pipelines should result in a recorded profit enhancement downstream in marketing. To the extent that profit data by stage of the industry are available and reliable, the marketing activities of the industry are, as a class, the least profitable component among a set of activities whose overall returns are only normal. Thus, by DOJ assumption, there is no pea, and were there a pea, the shell under which it is alleged to be hidden is, upon inspection, empty. For a discussion of the economic rationale for differences in rates of return across stages of the industry, see Edward W. Erickson and Robert M. Spann, "Vertical Integration and Cross-Subsidization in the U.S. Petroleum Industry: A Critical Review," Economics Special Report, Department of Economics and Business, North Carolina State University, December 1977. One of the classic versions of this argument is the assertion concerning the effects of the depletion allowance advanced by de Chazeau and Kahn. See Appendix C for a reprise of the historical arguments concerning percentage depletion.

been put in summary by Canes and Norman when they observe, "That something is amiss in the Justice analysis is easily seen by recalling that the Department assumes workable competition in non-pipeline stages *and* asserts that monopoly profits are earned there."[6] Although the DOJ assumptions may only be for purposes of argument, the pipeline undersizing argument does suffer from a line of specific weakness because of the potential dissipation of the alleged gains to other firms. Access to pipelines by nonowner shippers and widespread participation in incremental expansion of pipeline system capacity create serious problems for the pipeline undersizing argument.

Grant for the purpose of argument that there is an economically sound pipeline undersizing argument in which the assumptions concerning competitive structure and performance are consistent with observed results in adjacent stages.[7] Then any gains from undersizing are dissipated in the short run to the extent that nonowner shippers have access to the pipeline, that other pipelines (either products or crude oil) compete in serving the market which is intended to be exploited by undersizing, or that alternative transportation modes provide effective competition. In the long run, the practicability of organizing and maintaining collusive pipeline undersizing schemes depends upon the observed record of entry into pipelining, and the diversity of firms who enter joint ventures to construct and operate pipeline facilities.

If a pipeline is deliberately undersized as part of a strategy to transfer market power and monopoly profits to an adjacent stage of the industry, then to retain the fruits of that strategy the pipeline owner(s) must be able to restrict access to use of the pipeline. Otherwise, non-owner-shippers will take advantage of the price differential in the exploited market, ship over the pipeline at regulated rates, and dissipate the downstream market gains the owners of the pipeline anticipated when the pipeline was undersized.

This process will displace owner shipments originally targeted for the market intended to be monopolized by pipeline undersizing—presumably to competitive markets. As a result, unless access to the undersized pipeline is restricted, the pipeline owners will find that non-owner shippers are capturing a significant portion of the anticipated

---

[6] Michael E. Canes and Donald A. Norman, "Pipelines and Antitrust," Critique #005, November 30, 1978, p. 14. See also, pp. 1-13 and 15-28 for a more complete logical critique of the DOJ undersizing argument. In addition, see Paul Kobrin, "A Formal Critique of the Justice Department's Pipeline Undersizing Assertion," American Petroleum Institute, Critique #004, August 21, 1978, for a complementary analysis.

[7] It must be emphasized that this is a willful suspension of disbelief. See Canes and Norman, "Pipelines and Antitrust," and Kobrin, "A Formal Critique."

TABLE 1
COLONIAL PIPELINE:
ACCESS AND ACTIVITIES OF NONAFFILIATED SHIPPERS AS ESTIMATED FROM YEAR END ACCOUNTS RECEIVABLE, 1963–1977

	1963	1964	1965	1966	1967	1968	1969	1970	1971	1972	1973	1974	1975	1976	1977
Number of nonaffiliated shippers	0	4	2	2	5	2	5	7	9	12	17	14	15	20	22
Number of nonaffiliated shippers as estimated from minor accounts	6	6	3	0	3	4	3	5	6	4	1	4	4	4	3
Total nonaffiliated shippers	6	10	5	2	8	6	8	12	15	16	18	18	19	24	25
Change from previous year	—	4	−5	−3	6	−2	2	4	3	1	2	0	1	5	1
Number of nonmajor, nonaffiliated shippers	0	2	0	0	1	0	2	3	5	9	11	9	9	15	16
Change from previous year	—	2	−2	0	1	−1	2	1	2	4	2	−2	0	6	1
Number of nonmajor, nonaffiliated shippers assuming minor accounts are nonmajor	6	8	3	0	4	4	5	8	11	13	12	13	13	19	19
Change from previous year	—	2	−5	−3	4	0	1	3	3	2	−1	1	0	6	0
Number of identifiable nonaffiliated shippers who were completely new shippers this year	0	4	0	0	3	0	2	4	3	3	5	1	3	6	3
Number of identifiable nonaffiliated shippers who were "new" shippers this year, but who shipped in previous years	—	0	0	0	0	0	1	0	0	0	1	0	2	2	0

NOTE: Dash (—) indicates not applicable.

TABLE 2

PLANTATION PIPELINE:
ACCESS AND ACTIVITIES OF NONAFFILIATED SHIPPERS AS ESTIMATED
FROM YEAR END ACCOUNTS RECEIVABLE, 1960–1977

	1960	1961	1962	1963	1964	1965
Number of nonaffiliated shippers	1	3	3	5	16	13
Number of nonaffiliated shippers as estimated from minor accounts	8	6	6	5	10	10
Total nonaffiliated shippers	9	9	9	10	26	23
Change from previous year		0	0	1	16	(3)
Number of nonmajor, nonaffiliated shippers	0	0	0	4	7	6
Change from previous year		0	0	4	3	(1)
Number of nonmajor, nonaffiliated shippers assuming minor accounts are nonmajor	8	6	6	9	17	16
Change from previous year		(2)	0	3	8	(1)
Number of identifiable non-affiliated shippers who were completely new shippers this year		1	2	5	12	0
Number of identifiable non-affiliated shippers who were "new" shippers this year, but who shipped in previous years				1	4	0

monopoly returns in the downstream market while their own expected profits are diluted by an increased incidence of sales in competitive markets.

In addition, the owner-shippers will be forgoing the regulated returns they would otherwise have earned by building an optimally sized pipeline. Moreover, to the extent that owner-shippers must now supplement pipeline movements by higher-cost alternative transport, they will be incurring higher weighted average unit costs which could have been avoided by constructing a larger pipeline. For all of these reasons, restriction of access to an undersized pipeline is a critical component for the success of any strategy which might lead owner-shippers to plan, construct, and persist in operation of an undersized line.

For well-known reasons, the larger the number and more diverse the parties to a collusive arrangement, the more difficult it is to agree upon a common course of action. This principle applies to strategies of

1966	1967	1968	1969	1970	1971	1972	1973	1974	1975	1976	1977
14	16	6	7	8	13	9	7	9	9	10	10
13	9	10	12	10	9	15	12	10	10	10	11
27	25	16	19	18	22	24	19	19	19	20	21
4	(2)	(9)	3	(1)	4	2	(5)	0	0	1	1
5	7	2	3	4	6	5	4	4	4	5	6
(1)	2	(5)	1	1	2	(1)	(1)	0	0	1	1
18	16	12	15	14	15	20	16	14	14	15	17
2	(2)	(4)	3	(1)	1	5	(4)	(2)	0	1	2
3	3	0	4	3	6	0	4	3	2	3	4
2	1	0	4	2	4	0	2	3	1	3	3

pipeline undersizing. In the short run, once an undersized pipeline exists, it is in the interest of the owner-shippers to restrict access to it in order to earn the full gain anticipated from the strategy.[8] In the long run, if the undersizing of pipelines is a general problem, one would expect those who expect to gain from such a strategy to limit entry into pipeline ownership.

Such restrictions on entry into incremental expansion of pipeline system capacity would serve three purposes. First, they might facilitate restriction of access to shipment over the pipeline itself in the event that it were more feasible to deny access to nonowner-shippers than to owner-shippers. Second, they limit the number of parties among which

---

[8] Recall that such gains accrue only in the event that adjacent market conditions permit. Solely for purposes of argument, we are here assuming that whatever these conditions must be, they are. The DOJ explicitly assumes effective competition in adjacent markets.

TABLE 3

EXPLORER PIPELINE:
ACCESS AND ACTIVITIES OF NONAFFILIATED SHIPPERS AS
ESTIMATED FROM YEAR END ACCOUNTS RECEIVABLE, 1972–1977

	1972	1973	1974	1975	1976	1977
Number of nonaffiliated shippers	0		1		1	4
Number of nonaffiliated shippers as estimated from minor accounts	1	4	2	2	3	3
Total nonaffiliated shippers	1	4	3	2	4	7
Change from previous year		3	(1)	(1)	2	3
Number of nonmajor, nonaffiliated shippers	0	0	0	0	1	2
Change from previous year		0	0	0	1	1
Number of nonmajor, nonaffiliated shippers assuming minor accounts are nonmajor	1	4	2	2	4	5
Change from previous year		3	(2)	0	2	1
Number of identifiable nonaffiliated shippers who were completely new shippers this year		0	1	0	1	3
Number of identifiable nonaffiliated shippers who were "new" shippers this year, but who shipped in previous years			0	0	0	2

the potential gains to undersizing must be shared and hence increase the returns to each individual pipeline owner. Third, restrictions on entry into pipeline ownership would also facilitate a solution to the formidable task of reconciling disparate desires among numerous and diverse potential participants with regard to such specific questions as, for example, pipeline size and routing.

We have, then, two general hypotheses. First, to the extent that pipeline undersizing is a genuine (rather than an arguendo) problem, one would expect to see evidence of more restricted access upon pipelines where owner-shippers have an incentive to retain the gains to undersizing than on pipelines where such incentives do not exist. Second, if pipeline undersizing is a general and persistent tendency in the planning and operation of pipelines, one would expect to see patterns of ownership of increments to pipeline system capacity which are consistent with facilitating the implementation of such a strategy. In addition, to the extent that various pipelines—both product and crude oil—compete with each other,[9] the general record of entry and expansion in pipeline system capacity is also of interest.

Empirical Results

The empirical review presented here focuses upon the questions of access and the composition of ownership of incremental expansion of pipeline system capacity. For purposes of comparability with respect to a sufficient period of time, the access comparisons are for Colonial, Plantation, Buckeye, and Williams—although some data for Explorer are also presented. The examination of entry into pipeline capacity includes both products and crude oil pipeline capacity, considers all capacity additions which are regulated by and report data to the ICC, and covers the period 1947–1976.

Access. In Tables 1 through 5, various data pertaining to the access of nonaffiliated shippers to pipeline transport over the Colonial, Plantation, Explorer, Buckeye, and Williams pipeline systems are presented. The basic source of these tabulations are year-end accounts receivable data as reported to the ICC. These accounts receivable data are broken down into three basic categories: (1) affiliated shippers, (2) nonaffiliated shippers, and (3) minor accounts. Nonaffiliated shippers are nonowner-shippers which are individually identified. Minor accounts are non-owner-shippers with individual accounts receivable below some arbitrary cutoff such as $100,000 which are aggregated into a single sum. An approximation of the minimum number of minor accounts may be estimated by the value of the sum and the level of the cutoff. Use of both the individually identified nonaffiliated shipper and the minor accounts data on year-end accounts receivable provide an estimate of the actual number of nonaffiliated shippers who actually used the pipeline over the course of the year which has a downward bias because some firms have accounts payable which are less than the cutoff value. However, there may also be some double counting of smaller subsidiaries of larger firms. In addition to the problem that the estimation procedure for the number of individual firms in the minor accounts aggregate sum may omit some firms and double count others, the overall procedure

---

[9] It is to be noted that such competition may be enhanced by the general practice of crude oil and product exchanges which are a prevalent business practice in the petroleum industry. A detailed examination of this possibility is beyond the scope of this analysis. It does appear plausible, however, that exchanges which operate as a substitute for pipeline capacity are inconsistent with the pipeline undersizing argument.

TABLE 4

BUCKEYE PIPELINE:
ACCESS AND ACTIVITIES OF NONAFFILIATED SHIPPERS AS ESTIMATED
FROM YEAR END ACCOUNTS RECEIVABLE, 1960–1977

	1960	1961	1962	1963	1964	1965
Number of nonaffiliated shippers	0	0	0	0	33	32
Number of nonaffiliated shippers as estimated from minor accounts	2	4	5	9	11	10
Total nonaffiliated shippers	2	4	5	9	44	42
Change from previous year	—	2	1	4	35	−2
Number of nonmajor, nonaffiliated shippers	0	0	0	0	19	18
Change from previous year	—	0	0	0	19	−1
Number of nonmajor, nonaffiliated shippers assuming minor accounts are nonmajor	2	4	5	9	30	28
Change from previous year	—	2	1	4	21	−2
Number of identifiable non-affiliated shippers who were completely new shippers this year	0	0	0	0	33	1
Number of identifiable non-affiliated shippers who were "new" shippers this year, but who shipped in previous years	—	0	0	0	0	0

NOTE: Dash (—) indicates not applicable.

yields a downward biased estimate of the number of nonaffiliated shippers because there may be some nonaffiliated shippers who used the pipeline over the course of the year but do not owe accounts receivable at year-end.

In our opinion, a lower bound property of the estimate of non-affiliated shippers is in general likely to operate because there is evidence of intermittent shipping. The last line in each of the tables is the number of shippers who were "new" shippers in any given year (that is, they did not ship in the immediately previous year), but who had been shippers in some still earlier year. If there is intermittent shipping on a year-to-year basis as measured by year-end accounts receivable data, then it is very likely that some intermittent shipping accounts (as well, perhaps, as some continuous shipping accounts) are on a paid-up basis at any given year-end.

1966	1967	1968	1969	1970	1971	1972	1973	1974	1975	1976	1977
30	30	17	21	20	19	18	20	24	22	26	23
14	33	14	20	15	19	21	21	4	25	18	22
44	63	31	41	35	38	39	41	28	47	44	45
2	19	−32	10	−6	3	1	2	−13	19	−3	1
18	18	8	10	9	8	8	8	14	10	10	9
0	0	−10	2	−1	−1	0	0	6	−4	0	−1
32	51	22	30	24	27	29	29	18	35	28	31
4	19	−29	8	−6	3	2	0	−11	17	−7	3
6	4	0	5	3	3	2	4	7	3	6	1
2	0	0	4	2	1	2	4	5	1	5	0

Because divestiture measures of various kinds have focused upon the largest integrated major oil companies, we here concentrate on nonaffiliated shippers who are not among the largest eighteen major oil companies.[10] An estimate of such shippers is derived by inspecting the individually identified nonaffiliated shippers (some nonaffiliated shippers are majors) and by assuming that all firms included in the estimate of the minimum number of firms composing the minor accounts sum are nonmajors. The year-by-year value of this estimate is given in line 7 of the tables, "Number of nonmajor, nonaffiliated shippers assuming minor accounts are nonmajor."

The intermittent shipment data in the last line of Tables 1 through

---

[10] We define these companies as Exxon, Texaco, Shell, Standard of Indiana, Gulf, Mobil, Standard of California, Arco, Getty, Union, Sun, Phillips, Continental, Cities Service, Marathon, Standard of Ohio, Amerada-Hess, and Ashland.

TABLE 5
WILLIAMS PIPELINE:
ACCESS AND ACTIVITIES OF NONAFFILIATED SHIPPERS AS ESTIMATED FROM
YEAR END ACCOUNTS RECEIVABLE, 1966–1977

	1966	1967	1968	1969	1970	1971	1972	1973	1974	1975	1976	1977
Number of nonaffiliated shippers	13	1	4	4	3	3	5	4	4	6	4	7
Number of nonaffiliated shippers as estimated from minor accounts	10	2	18	16	19	12	12	14	15	15	25	23
Total nonaffiliated shippers	23	3	22	20	22	15	17	18	19	21	29	30
Change from previous year	—	−20	19	−2	2	−7	2	1	1	2	8	1
Number of nonmajor, nonaffiliated shippers	8	1	1	3	2	1	1	2	2	4	2	5
Change from previous year	—	−7	0	2	−1	−1	0	1	0	2	−2	3
Number of nonmajor, nonaffiliated shippers assuming minor accounts are nonmajor	18	3	19	19	21	13	13	16	17	19	27	28
Change from previous year	—	−15	19	0	2	−8	0	3	1	2	8	1
Number of identifiable nonaffiliated shippers who were completely new shippers this year	13	0	3	2	0	1	2	1	0	3	1	4
Number of identifiable nonaffiliated shippers who were "new" shippers this year, but who shipped in previous years	—	0	3	0	0	1	1	0	0	1	1	3

NOTE: Dash (—) indicates not applicable.

TABLE 6

COMPARISON OF NONMAJOR, NONAFFILIATED SHIPPER PATTERNS ACROSS FOUR PIPELINES, 1966–1977

	Colonial	Plantation	Buckeye	Williams
Number of nonmajor, nonaffiliated shippers				
1966	5 [a]	18	32	18
1977	19	17	31	28
1977 minus 1966	14	(1)	(1)	10
percent change	280	(6)	(3)	56
Incidence of intermittent shipping [b]	6	25	26	10

NOTE: Explorer Pipeline is not included because it started operation in 1972 and the span of data is not comparable.

[a] Colonial technically had zero nonaffiliated, nonmajor shippers in 1966, but had six, eight, and three, respectively, in 1963, 1964, and 1965. Rather than show an infinite percent change for the increase in such shippers for Colonial over 1966–1977, the average number of nonaffiliated, nonmajor shippers for Colonial for 1963–1965 is used for 1966. 1966 is the first year of operation of Williams and for that reason is the starting point for this tabulation.

[b] Horizontal sum of all the values in the last line of each Tables 1, 2, 4, and 5.

SOURCE: Tables 1, 2, 4, and 5.

5 are of interest with regard to questions of access for an additional reason. To the extent that access to a pipeline is costly or somehow artificially restricted, and to the extent that access is somehow differentially valuable (that is, as a result of the operation of exploitive undersizing), then once a firm had gained access to the pipeline, one would expect that it would be reluctant to risk giving it up by ceasing, even temporarily, to be a shipper. However, if access is not restricted, then the calculus of the costs and benefits of being an intermittent shipper would be done on the basis of ordinary commercial considerations. Thus, the presence of intermittent shipping might be regarded as evidence for an absence of restrictions of access.

On the other hand, it might be argued that the presence of intermittent shipping is evidence that those nonaffiliated shippers with the least leverage, for whatever reason, upon the pipeline have the tenure of their access to the pipeline constantly in peril and are periodically somehow forced off the pipeline. Were this the case, however, and were there a differential value to access to the pipeline due to exploitive undersizing, one would expect considerable litigation to result. In the

27

absence of such litigation, the presence of intermittent shipping appears to be additional clear evidence for the absence of restriction.

In Table 6, summary statistics on the number and trend of non-affiliated, nonmajor shippers and the incidence of intermittent shipping are presented for Colonial, Plantation, Buckeye, and Williams for the period 1966–1977. Colonial and Plantation are joint venture pipelines with a high incidence of owner-shippers, and Buckeye and Williams are independent pipeline companies.

Were the exploitive undersizing strategy to have any meaning, one would expect to see evidence of restriction of access for Colonial and Plantation relative to Buckeye and Williams. There are differences in the details of the pattern of the numbers across pipelines, but the overall pattern is not consistent with restriction. For example, Colonial has the lowest incidence of intermittent shipment, but the highest absolute and relative growth in nonmajor, nonaffiliated shippers. Moreover, there is an incidence of intermittent shipping on Colonial, as there is on all the other pipelines. On the basis of the data examined and compared in Tables 1 through 6, there appears to be no pattern of restriction of access to pipelines upon which owner-shippers would have an incentive to practice such restriction were there exploitive pipeline undersizing.

Expansion. The interstate pipeline industry regulated by the ICC has expanded considerably since World War II. For product pipelines, 48.7 percent of the total 1976 value of carrier property represented capacity added since 1947, and 79.2 percent of total 1976 barrel-miles of products shipped were carried in that capacity. For crude oil pipelines, 28.9 percent of the total 1976 value of carrier property represented capacity added since 1947, and 50.6 percent of total 1976 barrel-miles of crude oil shipments were carried in that capacity. The barrel-mile measures exceed the value of carrier property measures because of technological improvements which increased the efficiency of larger diameter lines. There were sixty-three new interstate, ICC regulated, product pipeline entities which appeared over 1947–1976, and fifty-five new crude oil pipeline entities. (See Tables 7 through 12.)

With this record of entry and expansion, the composition of the participation in the ownership of the incremental pipeline system capacity is particularly relevant to a critical evaluation of the validity of the pipeline undersizing argument. For reasons discussed above, were general systematic, exploitive pipeline undersizing an operative strategy, one would expect to see a pattern of ownership of incremental pipeline system capacity which reflects an attempt to minimize the number and diversity of owners.

TABLE 7
PRODUCT PIPELINES:
THE INCIDENCE OF NEW ENTITIES AS MEASURED BY VALUE OF CARRIER PROPERTY, 1947–1976

	1947–1951	1952–1956	1957–1961	1962–1966	1967–1971	1972–1976
Number of new entities transporting products in first full year of operation	7	16	13	9	9	9
Value of carrier property for first full year of operation[a]	40,095	415,064	388,609	1,385,072	324,975	341,390
Value of carrier property for fifth full year of operation[a]	318,415	686,463	337,981	1,374,626	316,332	274,978[b]
Fifth year value of carrier property relative to first year value	7.942	1.654	.8697	.9925	.9587	.8055
Average total pipeline industry value of carrier property[a]	724,091	1,160,437	1,608,644	2,854,491	3,869,255	4,448,168
Fifth year value of carrier property relative to total industry	.4397	.5916	.2101	.4816	.0818	.0618

[a] Thousands of dollars.
[b] Represents 1976 data.

TABLE 8

PRODUCT PIPELINES:

THE INCIDENCE OF NEW ENTITIES AS MEASURED BY BARREL-MILE SHIPMENTS, 1947–1976

	1947–1951	1952–1956	1957–1961	1962–1966	1967–1971	1972–1976
Number of new entities transporting products	7	16	13	9	9	9
Number of barrel-miles for first full year of operation	3,301,207	22,672,007	14,510,152	179,909,411	20,661,566	57,415,919
Number of barrel-miles for fifth full year of operation	31,225,247	50,656,750	31,398,631	549,424,513	35,446,992	42,584,352[a]
Fifth year number of barrel-miles relative to first year	9.4587	2.2343	2.1639	3.0539	1.7156	.7417
Average total number of barrel-miles for pipeline industry[b]	109,783	206,119	302,584	617,988	1,036,031	1,023,402
Fifth year number of barrel-miles relative to total industry	.2844	.2458	.1038	.8891	.0342	.0416

[a] Represents 1976 figures.
[b] Thousands of barrel-miles.

TABLE 9

PRODUCT PIPELINES:
THE RELATIVE SIGNIFICANCE IN 1976 OF 63 NEW ENTITIES
APPEARING DURING 1947–1976

	Value of Carrier Property	Barrel-Miles
1976 total for all new entities since 1947	$3,166,907,000	1,035,802,000
1976 industry total	$6,501,337,000	1,307,449,941
Ratio of new entities relative to industry total	0.487	0.792

We here focus upon the diversity of owners and especially those owners outside the largest eight and eighteen major oil companies. We also focus particularly upon joint venture pipelines, although data are presented for sole ownership pipelines. This is for several reasons. First, the *FTC* v. *Exxon* et al. suit alleges that the eight largest oil companies have a special community of interest, so the observed relationship among them and various other categories of firms are of special interest. Second, the various divestiture measures which have been introduced from time to time in Congress generally focus upon some approximation of the largest eighteen major oil companies, so these additional ten firms—in combination with the largest eight, and as a separate group of "other majors"—are also of special interest. Third, there are a large number of other firms engaged in crude oil production and transportation and refined products transportation and marketing. We designate any firm which is not one of the largest eighteen major oil companies as "other." The relative importance of combinations of eight plus others, ten plus others, and eight plus ten plus others then are measures of both the diversity and, less directly, the size of the groups which have participated in the ownership of incremental expansions of pipeline capacity. The larger the number and the more diverse are the owners of incremental capacity, the less force does the exploitive pipeline undersizing argument have.

Increasing the number of participants multiplies the difficulty of coordinating a restrictive policy. In addition, the more diverse are pipeline owners in terms of cost structures and demand conditions, the greater is the likelihood that effective coordination would require large, noncommercial side payments among the participants to equalize the relative benefits of restriction. Such side payments should be easily observable, but, to our knowledge, they are not part of the DOJ allegations.

31

TABLE 10

CRUDE OIL PIPELINES:
THE INCIDENCE OF NEW ENTITIES AS MEASURED BY VALUE OF CARRIER PROPERTY, 1947–1976

	1947–1951	1952–1956	1957–1961	1962–1966	1967–1971	1972–1976
Number of new entities transporting crude	9	11	12	4	7	12
Value of carrier property for first full year of operation[a]	350,708	451,374	377,078	82,732	187,895	422,965
Value of carrier property for fifth full year of operation[a]	507,704	862,525	370,585	88,069	178,590	492,543[b]
Fifth year value of carrier property relative to first year	1.448	1.911	.9828	1.065	.9505	1.1645
Average total pipeline industry value of carrier property[a]	3,638,214	4,800,744	5,162,509	5,196,248	5,217,956	6,967,592
Fifth year value of carrier property relative to total industry	.1395	.1797	.0718	.0169	.0342	.0707

[a] Thousands of dollars.
[b] Represents 1976 figures.

TABLE 11

CRUDE OIL PIPELINES:
THE INCIDENCE OF NEW ENTITIES AS MEASURED BY BARREL-MILE SHIPMENTS, 1947–1976

	1947–1951	1952–1956	1957–1961	1962–1966	1967–1971	1972–1976
Number of new entities transporting crude	9	11	12	4	7	12
Number of barrel-miles for first full year of operation	30,860,032	106,755,000	43,145,000	7,937,000	27,976,000	61,547,000
Number of barrel-miles for fifth full year of operation	101,711,490	119,051,000	44,360,000	3,318,000	64,613,000	69,086,000[a]
Fifth year number of barrel-miles relative to first year	3.30	1.12	1.03	.418	2.31	1.12
Average total number of barrel-miles for pipeline industry[b]	595,435	852,717	971,062	1,124,971	1,397,159	1,603,054
Fifth year number of barrel-miles relative to total industry	.171	.140	.046	.003	.046	.043

[a] Represents 1976 figures.
[b] Thousands of barrel-miles.

33

TABLE 12

CRUDE OIL PIPELINES:
THE RELATIVE SIGNIFICANCE IN 1976 OF 55 NEW ENTITIES
APPEARING DURING 1947–1976

	Value of Carrier Property	Barrel-Mile Shipments
1976 total for all new entities since 1947	$2,078,110,000	814,471,000
1976 industry total	$7,202,567,000	1,609,173,784
Ratio of new entities relative to industry total	.289	.506

In Tables 13 and 14, the data for product pipelines are presented. In Table 13, incremental pipeline system capacity is measured by the value of carrier property. In Table 14, incremental pipeline system capacity is measured by barrel-miles of throughput. Of the twenty-one new entity joint control product pipelines, sixteen (that is, eight plus one plus two plus five equals sixteen) had participation by firms other than the eighteen largest oil companies. Of the forty-two new entity sole ownership product pipelines, thirty-eight were built by firms other than the eighteen largest oil companies.

In Tables 15 and 16, the data for crude oil pipelines are presented. In Table 15, incremental pipeline system capacity is measured by the value of carrier property. In Table 16, incremental pipeline system capacity is measured by barrel-miles of throughput. Of the twenty-six new entity joint control crude oil pipelines, twenty-one had participation by firms other than the eighteen largest oil companies. Of the twenty-nine new entity sole ownership crude oil pipelines, twenty-one were built by firms other than the largest eighteen oil companies.

Table 17 summarizes the comparisons in Tables 13 through 16 which focus upon the relative importance among new pipeline entities of pipelines which had ownership participation by firms not included among the eighteen largest major oil companies. For jointly owned products pipelines, 88.1 percent of the incremental system capacity as measured by value of carrier property and 95.8 percent as measured by barrel-miles of thoughput had participation by companies which were not major oil companies. For jointly owned crude oil pipelines, 83.2 percent of the incremental system capacity as measured by value of carrier property and 63.9 percent as measured by barrel-mile throughput had participation by companies which were not major oil companies.

TABLE 13

PRODUCT PIPELINES: PARTICIPATION IN NEW PIPELINE ENTITIES
BY TYPE OF OWNERSHIP AND IDENTITY OF OWNERS AND AS
MEASURED BY VALUE OF CARRIER PROPERTY, 1947–1976

Ownership Classification	Number of New Entities, 1947–1976	Product Carrier Property	
		1976 value for 1947–1976 new entities	Percent of 1976 total of all pipelines
Joint control [a]			
Top 8 majors	2	$ 82,366,000	1.34
Next 10 majors	2	7,767,000	.13
All others	8	57,261,000	.93
Combination 8 + 10	1	69,996,000	1.14
Combination 8 + others	1	60,051,000	.98
Combination 10 + others	2	28,721,000	.47
Combination 8 + 10 + others	5	1,037,011,000	16.88
Total joint control new entities	21	$1,343,173,000	21.87
Sole ownership			
Top 8 majors	0[b]	$ 84,299,000[b]	1.37
Next 10 majors	4	25,514,000	.42
All others: Oil	22	971,037,000	15.81
Non-oil	16	735,094,000	11.97
Total sole ownership new entities	42	$ 815,944,000	29.56
Total new entities	63	$2,159,117,000	35.15

Total 1976 value of carrier property for *all* product pipelines including
pre-1947 entities: $6,142,319,000.

[a] Each of the ownership categories is mutually exclusive.

[b] This curious result is simply a definitional problem. For example, if a new pipeline entity originally specialized in crude oil shipment and subsequently diversified into products shipments as well, it was classified as a new crude oil pipeline entity but not a new product line entity—even though it would have some fraction of its 1976 value of carrier property allocated to products and would have had some 1976 product barrel-mile shipments. Explorer Pipeline Co. would be such an example. There were four new entity pipelines which originated as crude lines and diversified into products. There were five new entity pipelines which originated as product lines and diversified into crude oil. There were two combined crude and product new entities which exited from crude oil shipping but remained in products shipping. There was one combined crude and product new entity which exited from products shipping but remained in crude oil shipping.

SOURCE: Interstate Commerce Commission, *Transport Statistics in the United States for the Year Ended December 31, 1976, Part 6, Pipelines*; and annual Form P data for prior years.

TABLE 14

PRODUCT PIPELINES: PARTICIPATION IN NEW PIPELINE ENTITIES
BY TYPE OF OWNERSHIP AND IDENTITY OF OWNERS AND AS
MEASURED BY BARREL-MILE SHIPMENTS, 1947–1976

| | | Product Barrel-Miles | |
Ownership Classification	Number of New Entities, 1947–1976	1976 number for 1947–1976 new entities (thousands)	Percent of 1976 total for all pipelines
Joint control [a]			
Top 8 majors	2	16,738,000	1.28
Next 10 majors	2	258,000	.02
All others	8	13,260,000	1.01
Combination 8 + 10	1	13,009,000	.99
Combination 8 + others	1	8,663,000	.66
Combination 10 + others	2	8,918,000	.68
Combination 8 + 10 + others	5	655,804,000	50.16
Total joint control new entities	21	716,650,000	54.81
Sole ownership			
Top 8 majors	0 [b]	49,292,000 [b]	3.77
Next 10 majors	4	6,609,000	.51
All others: Oil	22	120,620,000	9.23
Non-oil	16	142,629,000	10.91
Total sole ownership new entities	42	319,150,000	24.41
Total new entities	63	1,035,800,000	79.22

Total 1976 product barrel-mile shipments for *all* pipelines including
pre-1947 entities: 1,307,449,941.

[a] Each of the ownership categories is mutually exclusive.

[b] See note b to Table 13 for a discussion of some definitional problems.

SOURCE: Interstate Commerce Commission, *Transport Statistics in the United States for the Year Ended December 31, 1976, Part 6, Pipelines*; and annual Form P data for prior years.

As with the data on access to pipelines, this pattern of diverse, nonexclusionary ownership is inconsistent with the behavioral implications of the exploitive pipeline undersizing argument. Not only does the pipeline undersizing argument have serious problems of internal eco-

TABLE 15

CRUDE OIL PIPELINES: PARTICIPATION IN NEW PIPELINE ENTITIES
BY TYPE OF OWNERSHIP AND IDENTITY OF OWNERS AND AS
MEASURED BY VALUE OF CARRIER PROPERTY, 1947–1976

Ownership Classification	Number of New Entities,[b] 1947–1976	Carrier Property 1976 value for 1947–1976	Percent of 1976 total for all crude oil pipelines
Joint control [a]			
Top 8 majors	1	$ 13,614,000	.18
Next 10 majors	2	102,541,000	1.36
All others	8	125,997,000	1.67
Combination 8 + 10	2	44,643,000	.59
Combination 8 + others	3	27,978,000	.37
Combination 10 + others	4	142,412,000	1.88
Combination 8 + 10 + others	6	503,959,000	6.66
Total joint control new entities	26	$ 961,144,000	12.71
Sole ownership			
Top 8 majors	3	$ 117,350,000	1.55
Next 10 majors	5	208,451,000	2.76
All others: Oil	19	757,061,000	10.01
Non-oil	2	28,669,000	.38
Total sole ownership new entities	29	$1,111,531,000	14.70
Total new entities	55	$2,072,675,000	27.41

Total 1976 value of carrier property for *all* crude oil pipelines including pre-1947 entities: $7,561,585,000.

[a] Each of the ownership categories is mutually exclusive.

[b] See note b to Table 13 for a discussion of some definitional problems.

SOURCE: Interstate Commerce Commission, *Transport Statistics in the United States for the Year Ended December 31, 1976, Part 6, Pipelines*; and annual Form P data for prior years.

nomic consistency,[11] not only does the exploitive pipeline undersizing argument not square with what we otherwise know about the competitive structure and performance of adjacent stages of the industry, but—if one

[11] See Canes and Norman, "Pipelines and Antitrust," and Kobrin, "A Formal Critique."

TABLE 16

CRUDE OIL PIPELINES: PARTICIPATION IN NEW PIPELINE ENTITIES
BY TYPE OF OWNERSHIP AND IDENTITY OF OWNERS AND AS
MEASURED BY BARREL-MILE SHIPMENTS, 1947–1976

Ownership Classification	Number of New Entities,[b] 1947–1976	Barrel-Mile Shipments	
		1976 number for 1947–1976 new entities (thousands)	Percent of 1976 total for all crude oil pipelines
Joint control [a]			
Top 8 majors	1	187,000	.01
Next 10 majors	2	107,986,000	6.71
All others	8	27,741,000	1.72
Combination 8 + 10	2	1,359,000	.08
Combination 8 + others	3	3,681,000	.23
Combination 10 + others	4	36,768,000	2.28
Combination 8 + 10 + others	6	121,585,000	7.56
Total joint control new entities	26	299,307,000	18.60
Sole ownership			
Top 8 majors	3	38,172,000	2.37
Next 10 majors	5	123,524,000	7.68
All others: Oil	19	387,490,000	24.08
Non-oil	2	4,038,000	.25
Total sole ownership new entities	29	553,224,000	34.38
Total new entities	55	852,531,000	52.98

Total 1976 crude oil barrel-mile shipments for *all* pipelines including
pre-1947 entities: 1,609,173,784.

[a] Each of the ownership categories is mutually exclusive.

[b] See note b to Table 13 for a discussion of some definitional problems.

SOURCE: Interstate Commerce Commission, *Transport Statistics in the United States
for the Year Ended December 31, 1976, Part 6, Pipelines*; and annual Form P data
for prior years.

grants for purposes of argument whatever assumptions are necessary to
neutralize these two objections—the exploitive pipeline undersizing
argument is not consistent with the facts and behavior we observe concerning access to pipelines and incremental expansion of the pipeline
system.

TABLE 17

RELATIVE IMPORTANCE AMONG NEW PIPELINE ENTITIES OF PIPELINES
HAVING OWNERSHIP PARTICIPATION BY FIRMS NOT INCLUDED AMONG
THE LARGEST 18 MAJOR OIL COMPANIES, 1947–1976
(percent)

Ownership Classification	Product Pipelines		Crude Oil Pipelines	
	Value of carrier property	Barrel-mile shipments	Value of carrier property	Barrel-mile shipments
Joint control	88.1	95.8	83.2	63.9
Sole ownership	93.9	82.8	70.7 [a]	70.7 [a]
All new entities	91.5	91.8	76.5	68.3

[a] This is a numerical coincidence.
SOURCE: Tables 13, 14, 15, and 16.

Conclusions

In addition to the conclusion that the DOJ exploitive pipeline under-sizing hypothesis is neither internally consistent nor supported by the facts, there are two more general conclusions we advance. First, the shell game theory of vertically integrated monopoly behavior is neither a conceptually appealing nor empirically likely proposition. And second, if the intense scrutiny of the domestic petroleum industry concerning competitive conditions continues, we may be nearing a time when all the shells have simultaneously been picked up and demonstrated to be empty. In such a case, it might then be generally realized that the competitive and efficient, large scale, vertically integrated and horizontally diversified firms which typify the U.S. energy industry are not only the necessary but, together with the many smaller energy firms, the desired social instruments through which U.S. energy policy may best be implemented.

APPENDIX TABLE A–1

Trunk Line Mileage for Crude Oil Pipelines

	Trunk Line Mileage			Percent of Total Miles		
	1951	*1977*	*Change*	*1951*	*1977*	*Change*
Individually owned pipelines	56,016	44,537	(11,479)	86.2	74.6	(11.6)
Top 8 majors	43,812	31,400	(12,412)	67.4	52.6	(14.8)
Next 10 majors	4,683	8,031	3,348	7.2	13.4	6.2
All others	7,521	5,106	(2,415)	11.6	8.5	(3.1)
Joint venture pipelines [a]	8,976	15,202	6,226	13.8	25.4	11.6
Among top 8	0	24	24	0	*	—
Among next 10	21	1,004	983	*	1.7	1.7
Among all others	1,665	1,097	(568)	2.6	1.8	(0.8)
Combination 8 + 10	3,224	6,900	3,676	5.0	11.6	6.6
Combination 8 + others	1,663	2,890	1,227	2.6	4.8	2.2
Combination 10 + others	1,260	880	(380)	1.9	1.5	(0.4)
Combination 8 + 10 + others	1,143	2,407	1,264	1.8	4.0	2.2
Total	64,992	59,739	(5,253)	100.0	100.0	—

NOTE: Asterisk (*) indicates less than 0.1 percent. Dash (—) indicates not applicable.

[a] Each of the ownership categories is mutually exclusive.

APPENDIX TABLE A–2

Trunk Line Shipments of Crude Oil

	Billions of Barrel-Miles			Percent of Total Barrel-Miles		
	1951	*1977*	*Change*	*1951*	*1977*	*Change*
Individually owned pipelines	612.1	1,050.3	438.2	88.1	57.6	(30.5)
Top 8 majors	495.1	717.7	222.6	71.3	39.4	(31.9)
Next 10 majors	70.6	271.8	201.2	10.2	14.9	4.7
All others	46.4	60.8	14.4	6.7	3.3	(3.4)
Joint venture pipelines [a]	82.6	772.3	689.7	11.9	42.4	30.5
Among top 8	0	0.2	0.2	0	*	—
Among next 10	0.1	110.8	110.7	*	6.1	6.1
Among others	1.5	5.7	4.2	0.2	0.3	0.1
Combination 8 + 10	43.1	125.4	82.3	6.2	6.9	0.7
Combination 8 + others	18.0	339.2	321.2	2.6	18.6	16.0
Combination 10 + others	3.8	91.6	87.8	0.5	5.0	4.5
Combination 8 + 10 + others	16.1	99.4	83.3	2.3	5.5	3.2
Total	694.7	1,822.6	1,127.9	100.0	100.0	—

NOTE: Asterisk (*) indicates less than 0.1 percent. Dash (—) indicates not applicable.

[a] Each of the ownership categories is mutually exclusive.

APPENDIX TABLE A–3

TRUNK LINE MILEAGE FOR PRODUCT PIPELINES

	Product Line Mileage			Percent of Total Miles		
	1951	1977	Change	1951	1977	Change
Individually owned pipelines	11,116	44,639	33,523	59.0	74.3	15.3
Top 8 majors	6,563	12,098	5,535	34.8	20.1	(14.7)
Next 10 majors	3,476	6,345	2,869	18.5	10.6	(7.9)
All others	1,077	26,196	25,119	5.7	43.6	37.9
Joint venture pipelines [a]	7,720	15,460	7,740	41.0	25.7	(15.3)
Among top 8	2,236	4,889	2,653	11.9	8.1	(3.8)
Among next 10	0	12	12	0	*	—
Among others	10	344	334	*	0.6	0.6
Combination 8 + 10	479	5,058	4,579	2.5	8.4	5.9
Combination 8 + others	446	0	(466)	2.5	0	(2.5)
Combination 10 + others	0	781	781	0	1.3	1.3
Combination 8 + 10 + others	4,529	4,376	(153)	24.0	7.3	(16.7)
Total	18,836	60,099	41,263	100.0	100.0	—

NOTE: Asterisk (*) indicates less than 0.1 percent. Dash (—) indicates not applicable.

[a] Each of the ownership categories is mutually exclusive.

APPENDIX TABLE A–4

Trunk Line Shipments of Petroleum Products

	Millions of Barrel-Miles			Percent of Total Barrel-Miles		
	1951	1977	Change	1951	1977	Change
Individually owned pipelines	53.2	736.5	683.3	43.5	45.2	1.7
Top 8 majors	29.9	63.1	33.2	24.5	3.9	(20.6)
Next 10 majors	19.5	396.6	366.7	16.0	24.4	8.4
All others	3.8	276.8	273.0	3.1	17.0	13.9
Joint venture pipelines [a]	69.0	891.5	822.5	56.5	54.8	(1.7)
Among top 8	24.5	128.9	104.4	20.0	7.9	(12.1)
Among next 10	0	0.3	0.3	0	*	—
Among others	0	6.1	6.1	0	0.4	0.4
Combination 8 + 10	4.7	626.1	621.4	3.8	38.5	34.7
Combination 8 + others	3.7	0	(3.7)	3.0	0	(3.0)
Combination 10 + others	0	8.6	8.6	0	0.5	0.5
Combination 8 + 10 + others	36.1	121.5	85.4	29.5	7.5	(22.0)
Total	122.2	1,628.0	1,505.8	100.0	100.0	—

NOTE: Asterisk (*) indicates less than 0.1 percent. Dash (—) indicates not applicable.

[a] Each of the ownership categories is mutually exclusive.

APPENDIX TABLE B-1

JOINTLY OWNED PRODUCTS PIPELINES: THE IDENTITY OF NEW ENTITIES AND SOME DETAILS OF THEIR ACTIVITIES, 1947–1977

Joint Ownership[a]	First Year	Fifth Year	1976	Miscellaneous
Top 8				
Wyco P/L Co.	1948 products	products	products	
Olympic P/L Co.	1965 products	products	products	
Next 10				
Cherokee P/L Co.[b]	1954 products & crude	products & crude	—	1975—merged with Continental P/L Co.
Lake Charles P/L Co.	1967 products	products	products	
Others				
Oklahoma Mississippi River Products Line, Inc.	1954 products	products	no 1976 part 6 listing	1967—merged into Sunray D X Oil Co.
Wyoming Nebraska P/L Co.	1954 products	products	products & crude	1966—name changed to Cheyenne P/L Co.
Okan P/L Co.	1955 products	products	products	
Teche P/L Co.	1956 products	products	—	1963—sold to Texas P/L Co.
Calnev P/L Co.	1961 products	products	products	
Chase Transp.	1974 products	new in 1974; no 5 yr. figure	products	

Navajo Refining Co.	1969 products	no 5 yr. figure	—	
Collins P/L Co.	1970 products	products	products	1971—terminated
8 + 10				
Wolverine P/L Co.	1953 products	products	products	
8 + Others				
Laurel P/L Co.	1959 products	products	products	New company formed as result of merger of two other new companies
10 + Others				
Yellowstone P/L Co.	1953 products	products	products	
Wabash P/L Co.	1958 products	crude only	no 1976 part 6 listing	1973—merged with Marathon
8 + 10 + Others				
Badger P/L Co.	1955 products	products	products	
West Shore P/L Co.	1961 products	products	products	
Dixie P/L Co.	1962 products	products	products	
Colonial P/L Co.	1963 products	products	products	
Explorer P/L Co.	1972 products	products & crude	products & crude	

a Not including undivided joint interest pipelines.
b Transporting products and crude in first full year of operation.

45

APPENDIX TABLE B-2

JOINTLY OWNED CRUDE OIL PIPELINES: THE IDENTITY OF NEW ENTITIES AND SOME DETAILS OF THEIR ACTIVITY, 1947–1977

Joint Ownership[a]	First Year	Fifth Year	1976	Miscellaneous
Top 8				
Kenai P/L	1960 crude	crude	crude	
Next 10				
Mid-Valley P/L Co.	1949 crude	crude	crude	
Cherokee P/L Co.[b]	1954 crude & products	crude & products	—	Also transports products; 1975— merged with Continental P/L Co.
Others				
Arapahoe P/L Co.	1954 crude	crude	crude	
Minnesota P/L Co.	1955 crude	crude	crude	
Wheat Belt	1959 crude	no 5 yr. figures	—	1961—purchased by Sinclair
Jayhawk P/L Corp.	1961 crude	crude	crude	
Portal P/L Co.	1963 crude	crude	crude	
Black Lake P/L Co.	1967 crude	crude	crude	
Paloma P/L Co.	1968 crude	crude	crude	
White Shore P/L Corp.	1968 crude	crude	crude	
8 + 10				
Glacier P/L Co.	1961 crude	submitted final report 12/31/63	—	Acquired by Continental P/L Co. 12/31/64
Cook Inlet	1967 crude	crude	crude	

			crude & products	
8 + *Others*				
Michigan-Ohio P/L Corp.	1950 crude	crude	—	
Muskegon P/L Corp.	1957 crude	no 5 yr. figures		Dissolved into Marathon P/L Co. 8/10/60
Osage P/L Co.	1975 crude	new in 1975; no 5 yr. figures	crude	
10 + *Others*				
Plains P/L Co.	1951 crude	no 5 yr. figures	crude	Liquidated 8/1/55. Assets & liabilities assumed by Phillips P/L & Cities Service P/L Co., ½ each.
Platte P/L Co.	1952 crude	crude	crude	
Tecumseh P/L Co.	1957 crude	crude	crude	
Southcap P/L Co.	1968 crude	crude	no activity	
8 + 10 + *Others*				
West Texas Gulf P/L	1953 crude	crude	crude	
Butte P/L Co.	1955 crude	crude	crude	
Four Corners P/L	1958 crude	crude	crude	
Chicap P/L Co.	1968 crude	crude	crude	
Texoma P/L Co.	1975 crude	new in 1975; no 5 yr. figures	crude	
Seaway P/L Inc.	1976 crude	new in 1976; no 5 yr. figures	crude	

[a] Not including undivided joint interest pipelines.
[b] Transporting both crude and product in first full year of operation.

47

APPENDIX TABLE B-3

SOLO PRODUCTS PIPELINES: THE IDENTITIES OF NEW ENTITIES AND SOME DETAILS OF THEIR ACTIVITIES, 1947–1977

Sole Ownership[a]	First Year	Fifth Year	1976	Miscellaneous
Top 8				
—				
Next 10				
Sun Oil Co. of Michigan[b]	1953 products & crude	products & crude	products	
Pioneer P/L Co.	1953 products	products	products	
Shelly P/L Co.	1963 products	products & crude	products & crude	
Ponder River Corp.	1971 products	products	products	
Others: Oil				
Sunray P/L Co.	1949 products	no 5 yr. figures	—	1953—commercial properties sold to Oklahoma, Mississippi River Products Line, Inc., and company ceased to be active; 1957—name changed to American Oil P/L Co.
Fairfax P/L Co.	1951 products	products		
Shamrock P/L Corp.[b]	1954 products & crude	products & crude	products & crude	
APCO Products P/L	1955 products	no 5 yr. figures		
Emerald P/L Co.[b]	1957 products & crude	products	products	1958—tariffs cancelled

Plymouth P/L Co.	1958 products	no 5 yr. figures	—	1962—purchased by Marathon
West Emerald Corp.	1959 products	products	products	
Marathon P/L Co.[b]	1960 products & crude	products & crude	products & crude	
Allegheny P/L Co.	1962 products	products	products	
Trans-Ohio P/L Co.	1962 products	products	products	
American Petrofina Co. of Texas	1963 products	products	products	
Williams Bros. P/L Co.	1966 products	products & crude	products & crude	
Hydrocarbon Transport Inc.	1967 products	products	products	
OMR P/L Co.	1967 products	no 5 yr. figures	—	1972—merged with Sun P/L Co.
Santa Fe P/L	1968 products	products	products	
Gulf Central P/L Corp.	1970 products	products	products	
Rome P/L Corp.[b]	1972 products & crude	products & crude	products & crude	
Okie P/L Co.	1973 products	new in 1973; no 5 yr. figures	products	
Pasco P/L Co.[b]	1973 products & crude	new in 1973; no 5 yr. figures	products & crude	
American Petrofina P/L Co.[b]	1974 products & crude	new in 1974; no 5 yr. figures	crude	
Amdel P/L Inc.[b]	1974 products & crude	new in 1974; no 5 yr. figures	products & crude	
Minden P/L Co.	1976 products	new in 1976; no 5 yr. figures	products	
Others: Non-oil				
Southern Pacific P/L Inc.	1956 products	products	products	
Cenex P/L Co.	1960 products	products	no 1976 part 6 listing	1971—terminated

(*Table continues on the next page.*)

49

APPENDIX TABLE B–3 (continued)

Sole Ownership[a]	First Year	Fifth Year	1976	Miscellaneous
Shoshone P/L Co.[b]	1961 products & crude	products & crude	no 1976 part 6 listing	1967—terminated
Airforce P/L Inc.	1965 products	products	products	
UCAR P/L Inc.	1969 products	products	products	
Pinto P/L Co.	1976 products	new in 1976; no 5 yr. figures	products	
Sunray Oil Corp.	1947 products	products	no 1976 part 6 listing	1953—carries properties sold to Oklahoma, Mississippi River Products Line, Inc.
Claiborne P/L Co.	1948 products	no 5 yr. figures	no 1976 part 6 listing	1951—acquired by Triangle P/L Co.
Shamrock Oil & Gas Corp.	1948 products	products	no 1976 part 6 listing	1976—merged into Diamond Shamrock
Triangle P/L Co.	1948 products	products	no 1976 part 6 listing	1955—company dissolved; P/L facilities merged into Texas Eastern Transmission Corp.
Augusta P/L Co.	1953 products	products	no 1976 part 6 listing	1960—merged into Kaneb P/L
Kaneb P/L Co.	1953 products	products	products	
Texas Eastern Trans. Corp.	1955 products	products	products & crude	
Cosden Petro Corp.	1959 products	no 5 yr. figures	no 1976 part 6 listing	1963—liquidated assets sold to American Petrofina Co. of Texas

				1968—name changed to MAPCO
Mid-America P/L Co.	1960 products	products	products	
Jet Lines Inc.	1961 products	products	products	

[a] Including undivided joint interest pipelines.
[b] Transporting both crude and product in first full year of operation.

APPENDIX TABLE B–4

Solo Crude Oil Pipelines: The Identities of New Entities and Some Details of their Activities, 1947–1977

Sole Ownership[a]	First Year	Fifth Year	1976	Miscellaneous
Top 8				
Phillips P/L Co.	1948 crude only	transporting products 5th yr.	transporting crude & products	
Salt Lake P/L Co.	1948 crude only	transporting crude & products	transporting crude & products	1966—name changed to Chevron
Mobil Eugene Island	1976 new in 1977 no 5 yr. figures	no 5 yr. figures	no activity	
Next 10				
Cities Service P/L Co.	1949 crude	crude	crude & products	
Tri-State Refining Co.	1950 crude	no 5 yr. figures	—	1953—merged into Freedom-Valvoline Oil Co. (Ohio)
Ashland P/L Co.	1951 crude	crude	crude	
Sun Oil Co. of Mich.[b]	1953 crude & products	crude & products	products only	
Hess P/L Co.	1960 crude	crude	crude	
Owensboro-Ashland	1974 no activity	new in 1974; no 5 yr. figures	crude only	

Others: Oil

Company				Notes
Lakehead P/L Co. Inc.	1950 crude	crude	crude	
Crown Rancho P/L	1953 crude	crude & products	crude	
Nantucket P/L Co.	1953 crude	crude	crude	1971—merged into Acorn
Trans-Mountain Oil P/L Corp.	1954 crude	crude	crude	no co. 1976 listing in part 6
Shamrock P/L Corp.[b]	1954 crude & products	crude & products	crude & products	
Emerald P/L Corp.[b]	1957 crude & products	products	products	
Marathon P/L Corp.[b]	1960 crude & products	crude & products	crude & products	
Ute P/L Co.	1962 crude	no 5 yr. figures	—	1964—merged into Penn. Transp. Co.
Cal-Ky P/L Co.	1963 crude	crude	no 1976 part 6 listing	
Panotex	1965 crude	crude	no 1976 part 6 listing	
Bell Creek P/L Co.	1968 no activity			⎱ 1972—merger of both companies
Fairview P/L Co.	1968 no activity	crude	crude	⎰ with name changed to Wesco
Kerr-McGee P/L Corp.	1968 no activity	crude	crude	
Acorn P/L Co.	1971 crude	crude	crude	

(Table continues on the next page.)

APPENDIX TABLE B–4 (continued)

Sole Ownership[a]	First Year	Fifth Year	1976	Miscellaneous
Dome P/L Co.[b]	1971 crude & products	crude & products	crude & products	
Kiantone P/L Co.	1972 crude	crude	crude	
Pasco P/L Co.[b]	1973 crude & products	new in 1973; no 5 yr. figures	crude & products	
American Petrofina P/L Co.[b]	1974 crude & products	new in 1975; no 5 yr. figures	crude	
Hando P/L Co.	1976 crude	new in 1976; no 5 yr. figures	crude	
Plains P/L Co.	1976 crude	new in 1976; no 5 yr. figures	crude	
Vickers P/L Co.	1976 crude	new in 1976; no 5 yr. figures	crude	
Amdel P/L Inc.[b]	1976 crude & products	new in 1974; no 5 yr. figures	crude & products	
Others: Non-oil Great Northern	1960 crude	no 5 yr. figures; terminated in 1962	—	1962—liquidated assets & liabilities purchased by Portal P/L Co.

[a] Not including undivided joint interest pipelines.
[b] Transporting both crude and product in first full year of operation.

Appendix C: Percentage Depletion—A Precursor Assertion. The basic allegation concerning percentage depletion advanced by de Chazeau and Kahn was that a vertically integrated firm had an incentive to pay itself inflated transfer prices for oil it produced and refined in its own refineries.[12] They alleged that, at a percentage depletion rate upon crude oil production of 27.5 percent, a 10 cent per barrel increase in the crude oil transfer price would result in an after-tax gain to the integrated operation of 2.7 cents per barrel. The empirical relevance of this proposition has been questioned on the basis of crude oil self-sufficiency ratios which are too low and severance tax and royalty rates which are too high to make it a profitability strategy.[13] The general consensus appears to be that the conditions necessary to make the depletion induced transfer-pricing distortion a valid general proposition never held.

Nevertheless, it is of interest to us here for two reasons. First, it is the classic example of the vertically integrated sleight-of-hand school of thought which ultimately foundered upon empirical relationships (as we feel the pipeline undersizing argument is sure to do). Second, as the depletion allowance argument was woven into successive critiques of petroleum industry structure, behavior, and performance, the weight of its asserted implications became too much for it to bear and it gradually became logically contradictory. For example, although the initial thrust of the depletion allowance assertion was that downstream refining and marketing profits were passed upstream within the integrated operation to more favorably taxed crude oil production, it subsequently was used to argue that upstream crude oil profits were used to subsidize downstream refining and marketing operations.[14]

---

[12] Melvin G. de Chazeau and Alfred E. Kahn, *Integration in the Petroleum Industry* (New Haven: Yale, 1959), pp. 221-222.

[13] See, for example, Stephen L. McDonald, *Petroleum Conservation in the United States: An Economic Analysis* (Baltimore: Johns Hopkins, 1971), p. 192; Edward W. Erickson, Stephen W. Millsaps, and Robert M. Spann, "Oil Supply and Tax Incentives," *Brookings Papers on Economic Activity*, 2(1974), pp. 454-457; Richard B. Mancke, *U.S. Energy Policy* (New York: Columbia, 1975), pp. 102-104; and Erickson and Spann, "Cross-Subsidization." In addition, institutional and legal restrictions in crude oil production such as posted prices and ratable take provisions prevented a company which was a purchaser of crude oil from a given field from raising the price for only its own production. In addition, as a result of changes in the tax laws, percentage depletion is no longer a tax option for major oil companies.

[14] See, for example, Federal Trade Commission Staff, "Preliminary Federal Trade Commission Staff Report on its Investigation of the Petroleum Industry," released as Senate Committee Print, *Investigation of the Petroleum Industry,* by the Permanent Subcommittee on Investigations of the Committee on Government Operations, 93rd Congress, 1st sess. (1973).

Commentaries

George S. Wolbert

In reading Flexner's paper I was reminded of the Department's Deepwater Port Report, and some of the speeches John Shenefield has made since the report; they have the same amorphous quality. Flexner's premise is that pipelines are the most efficient overland transportation for significant amounts of petroleum between any two given points. There is a definite kernel of truth in that, but perhaps it ought to be phrased a little more carefully than saying "any two given points." Obviously, if I were going to ship oil from here to Accotink, or some other place in the Washington, D.C., metropolitan area, I would not build a pipeline to do it.

Pipelines do have considerable economies of scale, but going from economies of scale to a natural monopoly in one jump assumes too much and ignores facts that are inconsistent with the conclusion. I have a few questions here: How does one build a single pipeline that satisfies the entire demand for transportation between two points? There is no way of knowing in advance what the demand is going to be, but it has to be measured in order to determine the size of the line.

When Plantation built its Baton Rouge to Greensboro, North Carolina line in World War II they had no idea how the territory was going to grow. With wisdom born of the event we now consider it an "undersized" line, but it was not undersized when it was first built. At that time it would not have been technologically feasible to build a thirty-six-inch line—the size of the largest segment of the line Colonial built years later. The traffic was not there, and of course no one thought there would be such an increase in consumption that the market would prove to be great, and tankers looked as though they had everything cornered.

57

When the Congress passed a Full Employment Act, they did not think the economy was going to pick up. If I had constructed a thirty-six-inch line at that time and financed it, I think several things would have happened. First, the Department of Justice would have been after me as it has done with IBM for market foreclosure, for keeping others from coming into the market. Second, my shareholders would have sued for imprudent management—and this would have to be a double derivative suit, because we have shipper-owners. And third, the trustees for the lenders would now own the line and a big chunk of the shipper-owners.

If I read Scherer's and Kahn's definition of a natural monopoly correctly, it means decreasing costs throughout the entire range of the market, not the entire range of the pipeline. Take a look at the cost curve aggregates and notice how they flatten out for pipelines around thirty-two or thirty-four inches in size. But certainly, the concept of a line that could serve the entire New York metropolitan market by itself boggles the mind; it would make the Trans-Alaska Pipeline System (TAPS) look like a garden hose. The argument presupposes that the market does not have economic supply alternatives. But more and more crude oil is coming into New York harbor and being refined there in large capacity refineries, which makes me wonder whether there really is any kind of market power down there.

Nor is there any reason in the world that refined products will not be coming into New York harbor in large quantities—if only at the insistence of New England senators, looking for more and cheaper fuel oil for their constituents. When voters get cold, they vote for somebody else.

When Colonial built its line it took a big risk because it did not know that grass-roots refineries would not be built and refined products would not be imported at the end of the line. After a risk is taken, it always looks easy.

Another point is deliberate undersizing. I think Michael Canes and Donald Norman pointed out that there cannot be a transfer of monopoly profits into a competitive marketplace.[1] I think market power is properly modified, but I would like to make such modification more definite and certain. There are internal inconsistencies between the so-called Sohio Report, which said Sohio did not have market power because of alternative sources of imported crude into PAD II, and the Colonial situation, which faces the same kind of competition at destination from imported crude and products.

If I understand the undersizing theory, it applies both upstream and downstream. I think Flexner said that monopsony would depress

[1] See Part III in this volume.

the crude oil prices upstream. In the *Federal Trade Commission* v. *Exxon,* however, a key charge is that pipelines are being used to prop up domestic crude prices so as to squeeze independent refiners, as Edward Mitchell testified before the Antitrust and Monopoly Subcommittee. But in the real world, I have trouble seeing how even a nasty major could simultaneously hold domestic crude prices up and drive them down. That is difficult to do.

The theory ignores a number of inconsistent facts, the first of which is competition among pipelines. There are thirteen common carrier lines going from West Texas to the Texas Gulf Coast; nine going from East Texas to the Texas Gulf Coast; ten from St. James, Louisiana, which carry indigenously produced crude to Toledo. There is also Wyoming and Rocky Mountain competition; Arapahoe, Platte, and what used to be Service pipeline are all beating their brains out over this same trail.

Explorer was going to be a great venture—it was even investigated by the Justice Department—but a couple of things happened to it along the way. Texas Eastern expanded its line from Houston to Seymore and then extended it on to Chicago. In addition, Seaway and Texoma are crude lines that penetrate right into that destination market, bringing crude oil to be refined. As a result of this area competition there were cash calls of almost $45 million. That does not look to me like undersizing.

Where are the monopoly profits? Mitchell has testified that he did not find any. He said that over the long pull, domestic and international producers and refiners are less rewarding to investors than the average corporation represented by S&P 500. Except for the banner year of 1973 when there were so-called inventory profits, the Big Eight were less profitable than the S&P 500.

Erickson has said that nonaccess is absent, and in 1957 Senator Dirksen filed similar views with the O'Mahony Report, although there are no hard data that I know of that support them.

Historically, 40 percent of all crude movements were nonowner shipments. In 1974 when Senator Stevenson was holding hearings, he asked the Association of Oil Pipelines (AOPL) to survey the industry and get a better idea of the facts. Fifty-one product lines reported 909 shippers, 806 of which (89 percent) were nonowners; 573 of the nonowners (71 percent) were independents. Somewhat later, the Interstate Commerce Commission asked for similar information from all pipelines, and out of 1,759 shippers there were 1,549 (88 percent) nonowners, of which 1,011 (65 percent) were independents. More recently, AOPL made another survey of 74 crude and products lines and 2,194 shippers;

1,958 shippers (89 percent) were nonowners; of these, 1,408 (64 percent) were independents.

These figures have been entered into the record. The Consumer Energy Act hearings contain some of them, and some were in the Petroleum Industry hearings.

I think it is against the self-interest of the shipper-owner to undersize. The Department of Justice is assuming in the FERC Valuation of Common Carriers (RM 78-2) that profits are high. If so, why would a shipper-owner willingly forgo a profit? Of course, one might answer that he can get it downstream. But why not take it where he can get it?

A shipper-owner backs his own shipments out of the line if he undersizes. Colonial is a beautiful example of that. Since 1969 the shipper-owners' shipments have decreased and nonshipper-owners and independents have increased. According to the Lundberg survey in the Petroleum Industry hearings, somewhere around p. 779, shipper-owners have lost market along the line.

The conclusions of the Deepwater Port Report are being carried over (mutatis mutandis) to the pipelines. Even if everything that is said about deepwater ports is true, what might be found in a single facility of that kind cannot be carried over to a pipeline running from A to B that has all kinds of competition. For example, Williams alone has ten competitors. Its situation is not the same as that of a deepwater port.

I have come to the conclusion that what we have here is a remedy in search of a problem.

Walter S. Measday

I find myself in a good bit of agreement with what Flexner has said. As an old institutional economist—or noneconomist—I welcome an intellectually, logically coherent theory of undersizing and monopolization in the pipeline industry for oil company ownership.

On the other hand, Erickson and I see things differently. He says that the more people there are as parties to a collusive agreement, the harder it is to agree on a course of action. As a statistical probability, this is quite correct, but it certainly is not a universal law. I think it depends on who is doing it. When I look at the oil industry I see a lengthy history of careful noncompetition in a number of areas.

Erickson says that if undersizing exists, we would expect to find restricted access. Then he shows that the number of shippers on Plantation and Colonial has grown considerably more than that on Williams or Buckeye in terms of the percent of change. What happened was that

Colonial and Plantation started with very few nonowner-shippers, and under what I think was pressure from the Justice Department and Congress they permitted nonowners on their lines, which resulted in an enormous rate of growth in the number of shippers. On Williams and Buckeye, where all shippers from the beginning were nonowners, it is hard to get comparable expansion. I am therefore not completely happy with this example as a demonstration of ease of access.

I will agree that access for these lines has increased considerably in recent years. The real question is probably whether the earnings record of Colonial is such that the nonowners have an economic disadvantage with respect to the owners, even though it is still possible for them to get on the line.

On the other hand, I have not seen any growth of nonowner-shippers on crude lines. On several of these lines the producers generally sell their crude to the pipeline, which seems to prefer to carry crude under the title of the line, or the title of the owners of the line, rather than under outside title. There may be an agreement to buy the crude oil at the origin point and sell it back to the producer at the destination point, but this is a tortuous procedure and is probably not true common carriage.

To support his argument that entry into pipeline ownership is not restricted, Erickson has prepared a number of tables that he offers as evidence that almost anybody who wants to can become a pipeline owner. This raises an issue that the Department of Justice commented on in the LOOP and Seadock applications: Is it necessary to become an owner at the very moment that the line starts? What happens if a shipper later decides to become an owner?

On Colonial, I think the only two new owners have been Arco and BP, who split the Sinclair share as part of an antitrust settlement after Arco acquired Sinclair. Given Colonial's earnings record, there is probably not an investment in the economy which would be as attractive. But the agreement still says that stock cannot be transferred to anybody outside without the permission of the owners, so it is pretty hard to get in.

In Table 13, Erickson finds a very sizable shift toward sole ownership of crude lines. I think part of the problem here is that he uses data from the Interstate Commerce Commission. Neither Capline nor TAPS shows up there. In other words, with an undivided interest form of operation each share is reported as a sole ownership line, but the whole line is still a joint operation.

Erickson shows all sorts of entry and then concludes, again, that entry is not restricted. Certainly, independents, in some cases, can and do become pipeline owners. Many lines have independents. But how much influence do the independents have?

The first example I could think of is Four Corners. Arco has sole ownership now, but the original ownership pattern was that Shell owned 25 percent; SOCAL, 25 percent; Gulf, 20 percent; and then Conoco, Arco, and Superior each 10 percent—Superior was the outside independent. The owners assumed that there was going to be enough crude oil production in the Four Corners area to justify the line. When it turned out there was not—and of course the Alaska oil came in, which meant there was no market for Four Corners oil on the West Coast anyway—they had to abandon the line. This is therefore a terrible example. Nevertheless, if, at the time of the agreement on expansion, the other owners had wanted to undersize the line, is there any way that Superior could force an expansion? I am willing to bet that the original agreement said something like 75 percent of the owners had to agree before there could be an expansion. These are the sort of restrictions that I think the Department of Justice has in mind. Even though an independent may have a piece of a line, it is not necessarily going to have the deciding voice on expansion of the line.

Other lines have other types of restrictions. The MCN line off-shore will carry oil only for owners, but anybody who wants to ship oil can buy some stock in the line, providing holders of at least 90 percent of the stock agree to let him on. Mobil and Continental each have more than enough stock to blackball anybody. Whether that power has ever been exercised is another question, but this type of restriction does not necessarily show up in the statistics.

The real issue is whether a shipper on the line can earn an owner-ship share in any way? And, even if he does, does he have anything to say about the sizing or the many other operating characteristics of the line? We have to look at the agreements themselves. Erickson's paper is enormously interesting, but I'm not sure that he has answered all of the basic problems here on the possibilities of monopoly power accruing to the owners of a pipeline.

Discussion

DONALD I. BAKER: Clearly, the whole undersizing theory depends on what might be called long-range foresight. These are big capital-intensive projects with at least a thirty-year life.

In the course of the earlier discussion there were references to various governmental policies that have heavy impacts on pipelines, including the Jones Act. One might ask also about oil import policy, which has shifted substantially; about Canadian oil export policy, which has shifted significantly; about environmental policy, as it affects what oil can be burned in what markets. And finally, one might ask about what passes for a U.S. energy policy.

If a consortium for a new pipeline were trying to foresee the next thirty years, could it do so in the context of the high degree of variables in governmental policy? Is governmental policy something we should be worried about in this debate or can we just put it aside?

DONALD L. FLEXNER: The first critical issue raised by government regulatory policy would be whether it would significantly change the relevant market within which both the Antitrust Division and the pipeline builders were evaluating the effects of a venture.

We analyzed that problem in the Sohio Report, pointing out that the restrictions on exports from Alaska North Slope crude had essentially created a market that would probably not otherwise have existed, namely, PAD V. We therefore felt the need to analyze the so-called monopsony power possibility in the context of PAD V. We found that if a throughput restriction strategy were carried out, it would nonetheless work to the disadvantage of Sohio. In effect, its crude would have been backed up into PAD V, which would not have helped Sohio.

63

But the important point to remember is that the throughput restriction incentive does not require perfect foresight. Even after a diameter is established for a pipeline, there is plenty of leeway to increase capacity or not. In respect to TAPS, for example, with four pump stations the pipeline capacity is something like 600,000 barrels a day. Adding additional pumping capacity increases the throughput capacity to 1.2 million barrels a day. In addition, capacity can be increased by looping the pipeline to relieve pressure at the points where it builds up. The decision to do that kind of thing may be affected by the incentives that affect the price downstream or upstream.

I would like to raise a couple of additional questions about the points raised by Erickson and Wolbert. When I first read Erickson's paper, I felt a little bit like the trial lawyer who reads the reply brief the first time through. I was worried because what he set out to do and, in fact, has done looked pretty serious. But in fact the actual computations are, in my view, not terribly relevant to our theory.

Look at Table 17, for example. Our theory has nothing to do with whether the participants in the joint ventures are in the top eight, the top eighteen, or the top thirty, and demonstrating a high incidence of participation in joint venture pipelines by oil companies that are outside of a particular group does not really tell me anything. The issue is whether the membership of the joint venture is in some sense "correct" in the market in which it is relevant to analyze the competitive effects of the joint venture.

While it is generally true that the more members of a so-called conspiracy or cartel there are, the harder it is to maintain a common front and coherence, it is also true that if a cartel or conspiracy is going to work, it had better have in it the people it needs.

As I tried to point out in my opening statement, it is necessary to look at the collective market shares of the members of the joint venture to get some sense of whether, within a relevant market, there is a possibility of a significant downstream impact, such as we found in the Deepwater Port Report. (By the way, I do not think the purpose of the Deepwater Port Report was to apply the analysis to every pipeline; it merely provides an example of how to analyze the problem.)

One group of Erickson's tables supposedly deals with the incidence of access by nonowner-shippers. Frankly, I don't know how anything of significance can be drawn from the number of shippers on a pipeline. I don't know whether five is a lot, whether eighteen is a lot—or whether it's a lot compared with the independents' line. More meaningful figures would be barrel-miles and percentage of barrel-miles by nonowner-shippers. It would also be relevant to know whether nonowner participation on a line was a function of prorationing, which would be consistent with our theory. There is obviously some restraint on the ability of any pipeline to restrict access because of common carrier obligations. It is therefore necessary both to know percentage of barrel-miles by nonowner-shippers and to have some sense of whether that percentage is a function of prorationing.

It is true that if prorationing forced all the owners' shipments to a more expensive transportation mode they would not have any share of the so-called monopoly profits that might be gained through throughput restriction, but that is not usually the case. Usually prorationing will result in some sharing of the space on the pipeline—perhaps as much as 30 percent by nonowner-shippers. If we assume monopoly profits, that means the owners and the nonowners are sharing those profits. It does not mean that the owners are at a cost disadvantage because of having to use a more expensive transportation mode, but the incentive to undersize is certainly reduced.

I would be interested to know whether Wolbert felt that competition among pipelines was sufficient to remove the necessity for any rate regulation. This is one implication of the assertion that there is a great deal of competition among pipelines and other modes of transportation.

There has been some empirical work—Thomas Spavins has done some [2]—indicating that returns are excessive in pipelines. His study at Table 7 indicates that petroleum pipelines have average returns to total capital—of about 13 percent from 1975 to 1977.

GEORGE S. WOLBERT: By definition, does that mean to you that 13 percent is an unreasonable profit?

MR. FLEXNER: No, I don't think there is any way to pick any particular number. But I would point out, for example, that it is in excess of local and long-distance trucking, and my assumption would be that many people might expect excessive rates of return in long-distance trucking because of fixed prices.

EDWARD W. ERICKSON: I would like to return to the question of whether the number of people and diversity of the group are relevant to the total context within which I see the undersizing hypothesis operating. If we pursue to its logical extreme what I call the Spavins-Flexner argument on cumulative market share, the implication is that the whole industry is one big monopoly and that there is no transactions cost to collusion.

---

[2] See his paper in Part II.

65

I think we could generally agree that however else one regards the oil industry, the managements are rational; they seek to maximize profits, and they engage in activities and combinations of policies that lead to that goal. If that is the case, and if there is no transactions cost to collusion, then it is not necessary to fool around with indirect and fundamentally nonoptimal monopolization schemes, such as the pipeline undersizing strategy. Under those circumstances one would go directly to the real wellspring of monopoly profits, which is restriction in the end-use markets.

A second point, which is not in my paper, is also directed toward the general context within which the pipeline undersizing argument operates. It pertains to identifying the incremental source of supply. I am not sure that I would identify the incremental source of supply for refined products to the East Coast as shipments over Colonial or Plantation. It seems to me that the incremental source of additional oil consumption in the United States is imported crude oil that comes in and is refined either on the Gulf Coast or, for example, at Bayonne, New Jersey. Since tanker routes to the Gulf Coast and the East Coast are essentially equivalent, the landed cost of that oil ought to be approximately the same in both places.

If oil is landed on the Gulf Coast, refined into products, and shipped up Colonial—and if we are going to insist on having a deincremental source rather than all contributions pressing equally at the margin—then transport costs on Colonial are higher than the cost of increments to consumption in PAD I that come from imported oil refined in PAD I.

DAVID J. TEECE: If declining long-run average costs are the basis for suggesting that a natural monopoly exists in this industry, then many other industries must also be included as natural monopolies. In other words, the empirical evidence of plant-level economies indicates a great many declining long-run average cost situations. I think the relevant unit of analysis is the firm and perhaps even the system.

That raises a second point about competition from other pipelines and from other modes of transportation. Sometimes crude pipelines compete with product pipelines. For instance, if a product pipeline were undersized and generating economic rents in product markets, this could encourage refinery expansions in the product market; and if the refineries were supplied by crude pipelines, this would generate competition between crude and product pipelines. This second point relates to the difference between monopoly profits and quasi rents and I'm not sure that we are doing enough work to distinguish between them. If price is above the competitive level it may simply be a temporary situation, and market forces might eliminate it. It is easy to confuse monopoly rents with quasi-rents, but the policy implications of each are different.

MORRIS LIVINGSTON: In Shenefield's testimony last June, he said that Flexner's theory holds even if there is no lack of access to the pipeline; prorationing is perfect. He argued that, even under those conditions, the price of products at destination would include the cost of the higher-cost mode of transportation, and therefore both the shipper-owners and the nonowner-shippers would share in the monopoly profit dictated by this marginal cost.

This seems to me a misuse of the concept of marginal cost. In this instance, the marginal cost is not the cost of the higher-cost mode of transportation—assuming it does cost more. The marginal cost is a weighted average of the pipeline tariff and the higher cost of the alternative mode of shipment. Each shipper has the same marginal cost, and there is no monopoly profit. It can be argued that all shippers do not have equal access, but the pure theory as Shenefield stated it is a complete misuse of the concept of marginal cost.

According to Flexner's theory this works both upstream and downstream. The upstream situation can be conveniently tested using the Bureau of Mines data on deliveries of crude oil to refineries by mode of transportation. Analysis of those data demonstrate that, with very minor exceptions, the crude oil is delivered to refineries either by pipeline or by a mode of transportation which, presumably, is lower cost, not higher cost—for example, by tanker to coastal refineries. Empirical data thus demonstrates that what Flexner's theory assumes does not in fact happen.

If this theory holds anywhere, it would have to hold with Colonial. Others have questioned whether Colonial was deliberately undersized at the time that it was built, and I'm inclined to agree that, given what the owners knew at the time, their decision as to size was probably a rational one. The important point, however, is that Colonial has continually expanded that line ever since. I think it is now more than three times its original size, which does not sound like a deliberate attempt to hold down the capacity in order to get a monopoly profit.

Similarly, the record shows an increasing percentage of Colonial's total shipments by nonowner-shippers and an increasing number of nonowner-shippers. Perhaps the percentage ought to be even larger, but compared with nonowners' share of the Gulf Coast refining capacity, I think nonowners' share of shipments is pretty respectable.

Erickson has developed some data from ICC records as to who owns these lines. Developing data another way indicates that the ownership of pipelines is closely related to refining capacity. In other words, the large refiners' share of refining capacity pretty well matches their share of pipeline capacity. With groups of successively smaller firms the same thing is true.

Frequently, the small refiners own pipelines that are of no use to anybody else. The refiners built them to bring crude in or ship products out to a particular market, and the location of the lines is such that they are of very little use to anyone else. The whole concept seems to be completely foreign to an attempt to earn monopoly profits. Obviously these lines were built because nobody else would build them where they were needed. This, of course, raises the question, If refiners are prohibited from owning pipelines, who is going to build them and what assurance is there that they will get built?

SCOTT HARVEY: In regard to economies of scale, the real issue is, When there is a need for increased throughput into a given market, how many lines can add that capacity at a cost less than or equal to the market price?

The large low-cost lines can add capacity at a lower cost and influence the market price. This highlights the fact that what matters is not how many firms have an ownership share in the line but, as Flexner pointed out, who can block an expansion of the line. If it requires 80 percent approval to add capacity to the line, one firm that owns 25 percent can veto any expansion; it does not matter how many other firms own an interest in the line.

MR. BAKER: Would it make any difference if there were a statutory rule that any member of a pipeline consortium could opt for expansion, pay for the expansion, and provide the throughput guarantees as a unilateral thing?

MR. HARVEY: That's a very intriguing possibility, but there are perhaps problems with that, too.

MR. WOLBERT: May I ask Mr. Flexner to clarify a point? Are you talking about an across-the-board divestiture?

MR. FLEXNER: We think a case has been established for a so-called prospective divestiture—that is, a rule that would apply to pipelines built after the effective date of some statute. Certain caveats are attached to that: one very important condition is that there be rate reform, and another is that serious consideration be given to the kinds of incentives that would induce shippers who could not be owners to provide through-put agreements. We are participating in a proceeding before FERC on the Texas Deepwater Port Authority and have advocated that FERC give careful consideration to a two-tier rate system.

Before we support a proposal for a flat across-the-board divestiture of existing pipelines we will carefully weigh the benefits against the potential costs. We are sensitive to the potential costs that might attend if we make a mistake. For example, after a close look at Sohio we concluded that there would not be a competitive problem, and there may be other cases like that. We want to approach that part of the issue rather carefully.

Senator Kennedy has petitioned the Federal Trade Commission (FTC) to institute a rule-making proceeding, and we are interested in seeing that go forward if it is structured properly. We are working with the FTC now to see if we can work something out, which should give an idea of our current thinking on how to approach this problem. We are concerned about the future construction of pipelines, because it is vital to the nation's interest. One reason for our participation in FERC rate proceedings is that if rates are in fact excessive, as we believe, that would tend to discourage the building of independent pipelines.

MR. BAKER: Mr. Flexner, if the oil company managements believe that there is likely to be much less investment in oil pipelines in the future simply because no one will be willing to put up the money, shouldn't they favor your proposal? Wouldn't it make their monopolies over the existing lines more valuable if new pipelines were not built?

MR. FLEXNER: I would assume the major problem would be whether the existing pipelines were filled up. For a great deal of the oil, pipelines are about the only effective way to get the oil to market. It is inconceivable that there would be truckload shipments from the Gulf Coast to St. Louis, for example—if that is implied I would expect that if there were a rule against oil company ownership of pipelines, and if there were a market for new pipelines, the oil companies would support independent pipelines. I can't see that it is in their interest not to.

ULYESSE J. LEGRANGE: The economic consequences of whatever is done are important, and we need to consider carefully what those consequences might be. I have found three areas where more work needs to be done. The first is that we need to look at some more facts and more data. When we talk about exorbitant rates of return, we ought to look at the empirical evidence at hand and see what the rates were. Tom Spavins has been doing some of that, and so have we at Exxon. We do not agree, but at least we are getting to the point where we can talk to one another, compare methods, and see if we can determine from the evidence at hand whether the rates have been excessive. More studies of this kind should be done, both jointly and separately, and debated before a public policy decision is taken.

The undersizing question is another area that we need to spend more time discussing. I served on the boards of both LOOP and Seadock, and at the time that we were considering the size and the expansion steps I was not aware that we were going through an undersizing exercise. I thought we were worrying about future uncertainties, government regulation, changes in the world, and changes in supply pattern; trying to be careful not to commit too much of our shareholders' money too foolishly; trying to come up with what we considered to be an economic sizing with some significant expansion possibilities.

How does one go about deciding the initial size of a facility or a new pipeline? We have some ways to do it in the pipeline industry, and obviously Flexner has some ideas how this should be done. Some of his ideas surfaced, I think, in things that the Justice Department wanted to get imposed in the Seadock license and the LOOP license, but we did not discuss them in detail.

For example, how does one build a Seadock to certain size specifications in a world with Mexican oil just offshore of the facility? How does one take account of the unknown future so that one winds up neither overcommitted nor undersized?

The third point we need to spend more time on is pricing. Pricing comes up when we deal with such things as bringing in new parties or allowing easy entry into these ventures. How would we decide at what price new owners would come in? I'm not talking about those who come in initially and take all the risk of development; I'm talking about the shipper who six years from now decides he wants a piece of the action.

EDWARD J. MITCHELL: It's really a question of how you value the risk premium at that time.

MR. LEGRANGE: That's correct, but it's not an easy situation. And before we get too committed to the whole concept of letting anybody in as an owner at any time, we have to know the ground rules.

MR. BAKER: The Department of Justice has this problem in all kinds of essential facilities. One was a joint venture satellite to serve all the networks, and there was a question about CATV systems and other independents. Once the satellite is launched it is impossible to change its capacity. The department concluded about 1970 that the situation could be solved by something analogous to a probate procedure. The joint venturers would be required to announce that the satellite was going up and then anyone who wanted to get on and pay his share could do so.

GEORGE A. HAY: In talking about these pricing problems, such as how to price risk for latecomers, I think it is important to distinguish between behavioral problems, which can be solved by worrying about the specific rules, and what I call structural problems, which relate to incentives.

The Department of Justice recognizes that many things determine how large a pipeline should be, including whether the demand is going to be there and whether there might be a new source of supply. But there may be an additional consideration: an incentive to make the pipeline a little smaller out of fear of "spoiling" the market.

If the fundamental incentive exists to have a smaller quantity of oil over the pipeline, then it is useless to worry about details of expansion rules and ownership rules; the pipeline sizing problem may require a structural solution.

We recognize that in many situations such an incentive does not exist. If there is so much oil available that there is no way to raise the price and make more monopoly profits, spoiling the market has no meaning. Nor does it make any sense to hold back capacity when there is access to only a small amount of "cheap oil." If each of the members of the joint venture behaved totally independently, they could not possibly bring enough oil into market to spoil the market. Where the incentives do exist, however, they are fundamental to the decision-making process of the joint venture. And it is not clear to me that they can be solved by easy rules of open access.

MR. WOLBERT: It seems to me that if there is an incentive to undersize, the pattern would have to show up in constant prorationing of a sizable number of lines, but it just is not there.

DR. HAY: The nonowner-shippers' fair share under the prorationing rules may be very small, or they may be very uneasy about whether they will get access to the pipeline and unwilling to commit themselves to the supply which makes the pipeline a good idea in the first place. Under these conditions prorationing will not be evident because it is not in anyone's interest to make the investment if they are going to wind up with only 1 percent of access to the pipeline.

71

PROFESSOR MITCHELL: I don't see that point at all. If this theory works, prorating should operate continuously, all over the place. If one is setting out to undersize, one would surely undersize sufficiently to prevent that pipeline from completely serving a particular market. But for substantial periods of time pipelines are not full and are not prorating—a fact that seems seriously to undercut this theory. In the case of a line that was rarely prorated, the theory does not apply and would have to be set aside.

MR. WOLBERT: Virtually all the lines in the United States are well under capacity today, let alone prorationed. And virtually all have substantial percentages of nonowner shipments, incuding a substantial percentage of "independents," a very amorphous term. If we can assume for the sake of argument that this is so—investigation could verify it—then can those lines be undersized?

DR. HAY: If there are many nonowner-shippers, and if they have access to more cheap oil than they can bring in, and there is empty space on the pipeline which they can get access to, then, I would agree that the pipeline could not be undersized with respect to current demand facing it.

MR. WOLBERT: Let's not worry too much about this cheap oil because after we fiddle around with entitlements, a barrel is a barrel.

DR. HAY: Cheap oil is an important part of the argument. If only a limited number of people are in a position to spoil the market by bringing in oil at a price less than that of the alternative supply, those are the only people that one has to worry about keeping off the pipeline.

Mr. LEGRANGE: There might be a prorationing situation where even the Justice Department might conclude that undersizing was not deliberate but just came about over time. Perhaps we ought to be looking for evidence to prove the undersize theory in specific situations. I do not feel it can be proved for the whole pipeline industry, because most lines are running well below capacity.

DR. HAY: When people look to prospective bans they have to consider whether the situations in which the Justice Department's theory might work factually are uncommon. We don't adopt across-the-board remedies for uncommon situations.

MR. LEGRANGE: I was merely suggesting that rather than adopting some future ban against all ownership, maybe you ought to try banning the kind that worries you.

MR. FLEXNER: We would certainly be open to discussion between the industry and the Department of Justice, and I would also encourage you to think about ways to bring your expertise and assistance to bear in the FTC proceeding and to think about various approaches that might be useful in carrying on this kind of a discussion.

PART
TWO

EFFECTIVENESS OF
GOVERNMENT REGULATION

The Regulation of Oil Pipelines

Thomas C. Spavins

The system of oil pipeline regulation that evolved at the Interstate Commerce Commission (ICC) during the seventy-three years after the passage of the Hepburn Act of 1906 is undergoing a reexamination and reassessment. The reevaluation has been occasioned by the transfer of oil pipeline jurisdiction to the Federal Energy Regulatory Commission, the challenge to the tariffs filed by the Trans-Alaskan Pipeline System carriers, and the shock of the first appellate court review of an oil pipeline rate case. This paper seeks to survey, from an economic perspective, the most important aspects of the present oil pipeline regulatory scheme.

An introductory section considers the implications the structure of the oil pipeline industry has for its regulation. Following the introduction the paper will examine three features of the present system of oil pipeline regulation: (1) the use of the ICC method of valuation and the related Elkins Act consent decree as earnings limitations for oil pipelines; (2) the influence of ICC policy on the entry of new pipelines and the development of new pipeline projects; and (3) the question of proper pricing of pipeline services, as opposed to overall levels of return, under ICC regulation.

Two important related questions will be considered in the course of the discussion: the complex connection between regulation and the extent of oil pipeline vertical integration and some alternatives to the present system of regulation. The focus of this paper is, however, on the present system of oil pipeline regulation. The discussion of both of these interesting and important questions is necessarily truncated and incomplete.

The views expressed here are not necessarily those of the Antitrust Division, U.S. Department of Justice. The author's work on this paper was helped by many conversations on pipelines with R. Reynolds and L. M. Lewis. George A. Hay, K. Baseman, Donald A. Kaplan, David W. Brown, Richard A. Feinstein, Peter J. Tomao, and James P. Denvir provided helpful comments.

The Structure of the Oil Pipeline Industry and Its Regulation

The justification for oil pipeline regulation has been traditional concerns about the implications of monopoly power for pricing and output decisions, including concern about its impact on other aspects of the petroleum industry.[1] This discussion of the regulatory framework for oil pipelines will begin with a quick look at the industry structure that forms the basis of the concern about monopoly power.[2]

Much of the analysis supporting deregulation of many regulated sectors has been a convincing demonstration that such sectors would be competitive in the absence of regulation.[3] If one could conclude that the oil pipeline industry would be competitive without regulation, the analysis would be relatively easy to complete. The inevitable distortions created by the failure to rely on a competitive market could be ferreted out, the beneficiaries of the rents created by the regulations identified, and another deregulation proposal offered.

Unfortunately, the available evidence does not permit that course to be followed. The available evidence on the technology of pipelines, the data on pipeline profitability, and the studies of market power of individual pipelines to date all tend to indicate a less than competitive industry structure and behavior. Each of the elements will be reviewed briefly.

It is clear that for continuous movements of large volumes of petroleum, pipelines are the least costly means of transportation, with the possible exception of long-distance tankers larger than those that can be accommodated at most U.S. ports.[4] The evidence concerning the technological economies of scale for oil pipelines is unusually good, and clearly indicates that throughout the range of feasible pipeline sizes,

---

[1] This discussion of the motives of oil pipeline regulation accepts the public interest rationale stated for oil pipeline regulation. No other explanation for oil pipeline regulation has been offered to date. The starting point for much recent economic analysis of regulation has been on the issue not discussed here, the motivations of the participants in the regulatory process. See generally G. Stigler, *The Citizen and the State* (Chicago: University of Chicago Press, 1975), and S. Peltzman, "Toward a More General Theory of Regulation," *Journal of Law and Economics,* August 1976, pp. 211-40.

[2] The paper by Donald L. Flexner in this volume discusses pipeline market structure in more detail.

[3] See, for example, the studies contained in A. Phillips, ed., *Promoting Competition and Regulated Markets* (Washington, D.C.: Brookings, 1975).

[4] See *National Energy Transportation Report,* vol. 1, Senate Committee on Energy and Natural Resources, Publ. No. 95-15, 95th Cong., 1st sess., 1976, pp. 182-84 and 213-14.

pipelines experience declining long-run average costs.[5] This indicates that it is less costly, if one considers the problem from a situation in which no initial pipelines exist, for one pipeline to move all of the traffic between any two points. This range of demand includes all significant movements of petroleum in the United States. This also suggests that in a static world, with the pipeline network planned *ab initio,* every oil pipeline would be either a monopoly or a monopsony with respect to at least some of its origins or destinations.

The data on the technological basis for concern about monopoly power indicate that, as a result of their declining costs, pipelines exhibit a tendency for a lack of effective competition that is stronger than found in almost every other area of economic activity.[6] Furthermore, there are no economic studies suggesting that managerial or technological dis-economies are present to balance this tendency for one optimally sized pipeline to be the statically most efficient way to provide transportation services.[7] This technological basis for monopoly power by petroleum pipelines is offset, to a limited extent, by the pattern of growth of pipeline markets over time. This growth may result in more than one pipeline serving two points, as a subsequent pipeline may be built at a later date to accommodate demand growth that is beyond the capacity of the first pipeline or to serve a common origin or destination with differing intermediate points.[8]

---

[5] See L. Cookenboo, "Costs of Operating Crude Oil Pipelines," *Rice Institute Pamphlet* (April 1954); L. Cookenboo, *Crude Oil Pipelines* (Cambridge: Harvard University Press, 1955); D. J. Pearl and J. Enos, "Engineering Production Functions and Technological Progress," *Journal of Industrial Economics,* vol. 55, September 1975; and *National Energy Transportation Report,* vol. 1, pp. 209-11.

[6] For several comparative studies of minimum efficient firm size relative to the markets served, see F. M. Scherer et al., *The Economics of Multi-Plant Operation: An International Comparisons Study* (Cambridge: Harvard University Press, 1975); see also Phillips, *Promoting Competition.* Note, however, that some recent explorations of the relationship between firm size and efficiency have considered the issue of beneficial effects of firm size despite constant costs in the sense used here. See R. R. Nelson and S. G. Winter, "Dynamic Competition and Technological Progress," in *Economic Progress, Private Values, and Public Policy,* ed. B. Balassa and R. R. Nelson (New York and Amsterdam: North Holland, 1977); and S. Peltzman, "The Gains and Losses from Industrial Concentration," *Journal of Law and Economics,* October 1977, pp. 229-64.

[7] For an examination of these considerations in a number of case studies for industries other than the pipeline industry, see R. C. Levin, "Technical Change, Economies of Scale and Market Structure" (Ph.D. diss., Yale University, 1974).

[8] Once a pipeline is built, many of the possibilities for economies of scale are exhausted. A new pipeline's initial size should take into account the need for future growth, but as long as capital costs are positive this initial excess capacity will not necessarily be so great as to preclude the construction of an additional pipeline at a later date.

The potential for market power, which is indicated by the technical data on pipeline costs, is confirmed by two kinds of evidence. First, the data on pipeline profit rates before the onset of the present system of regulation shortly before World War II strongly suggested monopoly earnings.[9] In addition, the experience under the current system of regulation indicates a persistence of somewhat high profit rates on average, and very high profit rates on numerous pipelines.[10]

Two recent case studies have concerned the potential for monopoly power of three pipeline projects.[11] The in-depth examination of the two proposed deepwater ports indicated that both would, in their respective markets, have the potential to exercise market power. The examination of the Sohio project from Long Beach, California, to Midland, Texas, confirmed that that pipeline would possess monopsony power on the West Coast. However, that monopsony power was diminished by virtue of its ownership by Standard Oil of Ohio.

These considerations suggest that one cannot, on the basis of the existing evidence, support a policy of deregulation of oil pipelines on the grounds that the industry has a competitive structure. The merits of the regulatory scheme must therefore be addressed in detail.

Vertical Integration and the Regulation of Oil Pipelines

Most petroleum pipelines are owned by firms engaged in other phases of the petroleum business. Pipelines affiliated with oil companies accounted for 98.6 percent of all crude oil barrel-miles and 86.9 percent of all product barrel-miles in 1975.[12] Other data indicate that much of the petroleum moved in pipelines is for use of the shipper-owners.[13] The

[9] See P. Locklin, *The Economics of Transportation* (Homewood, Ill.: Irwin, 1972); also R. C. Cook, *Control of the Oil Industry by Major Oil Companies,* Temporary National Economic Committee Monograph No. 39 (Washington, D.C., 1941); and R. A. Prewitt, "The Operation and Regulation of Crude Oil Pipelines," *Quarterly Journal of Economics,* vol. 106 (February 1942), pp. 177-211.

[10] See below. Note that one cannot draw any useful conclusion about the competitiveness of the industry from the accounting profit rate earned on some pipelines.

[11] "Report of the Attorney General Pursuant to Section 7 of the Deepwater Port Act of 1974 on the Application of LOOP, Inc., and Seadock, Inc., for Deepwater Port Licenses," November 5, 1976, and "Report of the Antitrust Division, Department of Justice on the Ownership and Operation by Standard Oil of Ohio of a Long Beach, Calif., to Midland, Texas, Crude Oil Pipeline," June 1978.

[12] See statement of David L. Jones, on behalf of the Bureau of Investigations and Enforcement, Interstate Commerce Commission, in *Ex Parte No. 308,* April 28, 1977.

[13] See Howrey and Simon, "Pipelines Owned by Oil Companies Provide a Pro-Competitive and Low Cost Means of Energy Transportation to the Nation's Industries and Consumers," March 1977.

other activities of the shipper-owners have not historically been subject to price regulation, although the industry has recently been subjected to an extensive regulatory framework not directly related to oil pipeline regulation.[14]

The existence of common ownership of regulated and nonregulated activities poses problems in general for the coherence of any regulatory scheme. If the regulated activities of the firm are performed by either customers or suppliers of the nonregulated aspects of the firm, the potential exists for subverting the regulatory scheme by charging a transfer price that permits monopoly profits to be transferred from the regulated sector to the unregulated sector. This creates an incentive for a regulator to extend the boundaries of the regulatory scheme, at the cost of further distortions and complexity.[15]

A classic example of the problem of regulatory boundaries has been discussed in a recent Department of Justice report.[16] Many electric utilities have partially integrated themselves into the coal mining sector to provide the needs of their coal-fired generating stations. A number of difficulties in securing long-term supplies of coal provide the efficiency rationale for this decision. This vertical integration, however, offers the utilities the chance to evade rate-of-return regulation by charging themselves an excessive price for this "captive" coal. The utilities' regulators must then either try to monitor the transfer price of captive coal or extend rate-of-return regulation to the captive coal market. The only alternative to this increasing regulatory complexity is to permit subversion of the regulatory scheme or to bar the vertical integration.[17]

In the case of petroleum pipelines the economic benefit earned by a shipper-owner from the movement of petroleum through a pipeline is not measured by the tariff, but by the difference in the value of the petroleum between origin and destination. This difference can be estimated by the market prices of the petroleum at those points, if that infor-

---

[14] See "Comments of the U.S. Department of Justice on the Proposed Amendments to the Entitlements Program to Reduce the Level of Benefits Received under the Small Refiner Bias," January 26, 1979; and Paul W. MacAvoy, ed., *Federal Energy Administration Regulation: Report of the Presidential Task Force* (Washington, D.C.: American Enterprise Institute, 1977).

[15] See J. MacKie, "Regulation and the Free Market: The Problem of Boundaries," *Bell Journal of Economics*, vol. 1, Spring 1970.

[16] See U.S. Department of Justice, *Competition in the Coal Industry* (Washington, D.C., 1977), pp. 109-18.

[17] B. Owen and R. Braeutigam, *The Regulation Game* (Cambridge: Ballinger, 1978), ch. 3, "Regulation of a Depletable Resource: The National Gas Industry," provides another example of a possible boundary problem if natural gas wellhead prices are ever deregulated. The problem will lie with the natural gas produced by affiliates of the vertically integrated transmission companies.

mation is available. If the owner can influence the throughput of the pipeline, and thereby the price differential between origin and destination points, then the possibility exists for the subversion of any pipeline regulatory scheme that is concerned solely with pipeline tariffs. This presents still another version of the problem of regulatory boundaries.

Thus, if the tariff does not equal the difference in value of the petroleum, it will not accurately reflect the gains to owners from the use of the pipeline.[18] This suggests that any analysis that accepts the reported financial statements of oil pipelines is subject to error and must be interpreted with a degree of caution.

The possibility of evasion of regulation by a vertically integrated firm is generally a less effective way to exploit monopoly power than by simply charging higher tariffs. In addition to any difficulties in manipulating throughput, a firm engaging in such behavior receives the benefits of higher prices only on its share of the pipeline's throughput, and on whatever other sales it is able to make in the downstream market from sources of supply less costly to the firm than the market price. If the higher price is the result of an excessive tariff, the firm gains on all the throughput of the line. This indicates that firms will engage in subversion of the regulatory scheme only if the rate-of-return regulation provides a limitation on pipeline earnings. It also shows that firms will have a greater incentive to engage in vertical restrictions the more strict the limitation on earnings.[19]

Oil Pipeline Profits under the Present System of Regulation

Basic Concepts. Since 1941, the profits of oil pipeline companies have been nominally limited by two rate-of-return standards and a negotiated earnings constraint. The two rate-of-return standards applied to the overall return, including payments to both debt and equity holders, as a percentage of a base called the ICC valuation. A return standard of 8 percent was established for crude oil pipelines;[20] the standard for

---

[18] A number of factors could cause the benefits of transportation to an owner to differ from the tariff. For example, if the owner has differential access to the pipeline, or receives the benefits of a prorationing rule, the differential could be greater than the tariff. Alternatively, a sole owner could move petroleum so long as the benefits are greater than the incremental costs of the movement, which may be less than the tariff if the line is not at capacity. An owner would adopt this strategy rather than reduce the tariff only if nonowner-shippers had reason to value movement through the pipeline more than the owner.

[19] This discussion ignores the role played by various forms of common carrier regulation in limiting the ability of firms to engage in these restrictive practices. ICC activity in this area has been nonexistent with the exception of some ancient regulations concerning minimum tender size.

[20] *Reduced Pipeline Rates and Gathering Charges,* 243 ICC 115 (1940).

of pipeline projects reflected the potential returns available under the guidelines if the project were successful. The process to a limited extent is one in which the standard regulatory model of risks dictating returns was reversed, and the pipeline returns determined the riskiness of the projects that were undertaken.[34] But one must not leap to the conclusion that the standards in use were desirable simply because such an adjustment took place.

There are a number of ways the pipeline industry can respond to any guideline return. One must remember that the average observed profitability must be below the guideline maximum. The obvious first response is that the industry will not undertake projects whose risks are such that the possibility of earning the guideline maximum is not a sufficient reward.[35] The industry will respond in other ways to a guideline, however. The projects that would have been undertaken with a lower guideline will tend to be undertaken sooner, as alternative owners of a pipeline seek to be the owner to construct the pipeline. Partially alternative projects that would ideally have been completed sequentially may now be undertaken simultaneously. This process will tend to reduce pipeline profitability.

The concept is generally applicable that the industry will, in part, respond to the opportunity for profits by bidding away the additional profits in competing for the right to be the firm earning the profits.[36] Two factors may be at work in the oil pipeline industry that prevent all the gains from any higher rate-of-return standard from being exhausted by the activities of the pipeline companies. First, for all the returns to be competed away, the process of building new pipeline projects must be competitive, and there must be an active threat that a pipeline project will be built by another firm or coalition of firms. This competition will cause the starting date for a pipeline project to be pushed forward until any prospective monopoly returns are exhausted by the costs and additional risks of being first in. Large joint-venture pipeline projects are often undertaken by coalitions of pipeline companies, and even if there is more than one competing group, there are not many competing groups.

---

[34] It does not necessarily follow that the pipeline return framework must always be based on some sort of guidelines. The extent to which many pipeline risks are one-shot risks associated with the initial investment tends to make difficult a process of *ex post* risk assessment and adjustment of the return. This characterization of the pipeline investment process may be incomplete, however. For developed pipelines systems such as Colonial, Plantation, or Williams, much of the investment had been added in a process of growth long after and separate from the initial risks.

[35] Again assuming that all the benefits of the project are reflected in the tariff.

[36] See Richard A. Posner, "The Social Costs of Monopoly and Regulation," *Journal of Political Economy,* volume 83, no. 4 (August 1975), pp. 807-27.

This fact, taken together with the possibility of overlapping membership among the alternative coalitions, suggests that there may be limits to the extent of competition to be first.

More fundamentally, competition for the right to be a firm with excess returns will continue as long as the marginal gains from that competition expressed in terms of additional prospective profits exceed the marginal costs of achieving those gains. If firms can receive benefits from some initial advantage in the race for position, then this process will stop short of the point where all the additional profits are expended.[37]

One should take care to make the distinction between two types of competition for a market. The type described above is one in which the additional profits are wasted in the scramble for the right to earn them. In general this process produces a social loss greater than one in which the profits are kept by the monopolist. An alternative concept is one in which prospective suppliers bid for the right to serve a market, with the bid taking the form of a promise of a lower *ex post* price. The ICC return standards do not produce this type of franchise bidding, as the price paid by shippers is substantially independent of the efforts of prospective pipeline builders. In any event there is substantial reason to doubt the efficacy of franchise bidding as a solution to the question of *ex post* natural monopoly.[38]

This analysis suggests the need to exercise great care in the assessment of any guideline profitability standard. Industry behavior is affected by a return standard. It is not a simple matter of assuming that the only implication of a standard is one of different levels of profits and prices. The impact of any standard depends on both the standard and industry response to it.

The Valuation System. The ICC valuation concept has been the basis for each of the existing rate-of-return standards for petroleum pipeline profitability. The origins of this system are in the abandoned legal precepts that returns of regulated industries must be based on a measure

---

[37] One important source of advantage for possible coalitions of pipeline builders is the rents that may accrue to a coalition of firms most likely to benefit from its construction. These firms may have an advantage of information and position that precludes the formation of alternative coalitions of pipeline builders. If the upstream and downstream markets for the pipeline are highly competitive with a very large number of users of equal importance, this phenomenon may not be important. For individual pipeline projects, however, the downstream or upstream market may be workably competitive and still not have sufficient firms to provide for numerous alternative groups of pipeline sponsors.

[38] See O. E. Williamson, "Franchise Bidding for Natural Monopolies: In General and With Respect to CATV," *Bell Journal of Economics,* vol. 7, no. 1, Spring 1976.

TABLE 1

SINGLE SUM FORMULA

$$V = 1.06 \left(OC \left(\frac{OC}{OC + CRN} \right) + CRN \left(\frac{CRN}{OC + CRN} \right) \right) CP + L_1 + L_2 + W$$

V = Valuation

OC = Original cost new

CRN = Reproduction cost new

CP = Condition percent, ratio of cost of reproduction new less depreciation to CRN

L_1 = Land

L_2 = Rights of way

W = Working capital

of the current fair value of the property, and that one could estimate the current fair value of the property by engineering concepts.

The valuation system is a way of calculating returns. The use of valuation as a method of calculation will be considered here. Table 1 describes the single-sum formula that is used to calculate the annual valuation of a pipeline.[39] The formula computes the ICC valuation as a weighted average of the reproduction cost of the pipeline and the original cost. The weights are the proportions of each to the sum of the two. The change in the valuation of a pipeline over time is governed by the rate of growth in the cost of reproduction new, and by the condition percent, which is an estimate of the physical depreciation of the pipeline.[40] Table 2 presents the reproduction cost new, condition percents, and valuation for a pipeline with an estimated service life of twenty-five years and an assumed initial original cost of $1,000.

Table 3 shows the returns produced over a twenty-five year period by the traditional ICC 8 percent and 10 percent return on valuation standards. In order to estimate the benefit received under these stan-

[39] This formula first appeared in the testimony of Jesse C. Oak, valuation engineer, in *Valuation of Common Carrier Pipelines,* Docket No. RM 78-2 (March 25, 1977).

[40] The details of the valuation scheme are discussed in the Oak testimony. See also *Joint Pre-Hearing Brief of the United States Department of Justice, the State of Alaska, and the Arctic Slope Regional Corporation,* in *Trans-Alaskan Pipeline System,* Docket No. OR 78-1 (December 15, 1977), pp. 49-54; and *Memorandum Defending and Stating the Position of the United States Department of Justice . . .,* in *Valuation of Common Carrier Pipelines,* Docket No. RM 78-2 (April 3, 1978), pp. 19-38.

TABLE 2

VALUATION CALCULATION

Year	Reproduction Cost New[a]	Condition Percent[b]	Valuation[c]
1	1000	100	1060
2	1060	94	1027
3	1124	88	994
4	1191	83	971
5	1262	78	948
6	1338	75	948
7	1419	70	926
8	1504	67	925
9	1594	63	912
10	1689	60	911
11	1791	57	911
12	1898	55	926
13	2012	53	942
14	2133	50	939
15	2261	48	955
16	2397	45	947
17	2540	43	960
18	2693	41	971
19	2854	39	981
20	3026	37	990
21	3207	36	1023
22	3400	34	1029
23	3609	32	1030
24	3820	30	1029
25 [d]	4049	28	1022

[a] The reproduction cost new is assumed to grow at six percent per year. The arithmetic average of the geometric rate of increase of the twenty-eight ICC pipeline cost categories over the period 1966–1976 was 6.15.

[b] The valuation system has two different condition percent tables, one for pipe and pipe fittings, the other for machinery and buildings. In the table the condition percent factor used is a weighted average of these two condition percentages. The weight of the pipe and pipe fittings category is .75, which approximates the weight this category reflects in a typical pipeline.

[c] This valuation assumes that reproduction cost new is equal to original cost in the first year of the pipeline's life, which will happen if the pipeline is constructed in the year before the pipeline goes into service and if the interest cost during construction is equal to the six percent allowed in the valuation system. This latter assumption is clearly unrealistic, but the difference is small. The calculations assume that cost of land, rights of way, and working capital are zero.

[d] The valuation for year 26 and following is not zero, but continues in about the same fashion as before in its movement over time.

TABLE 3
ICC Allowed Returns

Year	Crude Oil Pipelines		Products Pipelines	
	Return[a]	Cash Flow[b]	Return[c]	Cash Flow[d]
1	84.8	259	106	280.2
2	82.2	150.8	102.7	173.3
3	79.5	147	99.4	166.9
4	77.7	142.2	97.1	161.6
5	75.8	137.2	94.8	156.2
6	75.8	134.1	94.8	153.1
7	74.1	129.3	92.6	147.8
8	74.0	126.2	92.5	144.7
9	73.0	122.1	91.2	140.3
10	73.0	119	91.1	137.1
11	73.0	116	91.1	134.1
12	74.1	114	92.6	132.5
13	75.4	112.2	94.2	131
14	75.1	109	93.9	127.8
15	76.4	107.1	95.5	126.2
16	75.8	103.4	94.7	122.3
17	76.8	103.4	96.0	122.6
18	77.7	98.5	97.1	118.6
19	78.5	98.5	98.1	118.1
20	79.2	99.2	99	119
21	81.8	101.8	102.3	122.3
22	82.3	102.3	102.9	122.9
23	82.4	102.4	103	123
24	82.3	102.3	102.9	122.9
25	81.8	101.8	102.2	122.2

Rate of return 14.7 percent[e]   Rate of return 17.3 percent[e]

[a] 8 percent of ICC valuation from Table 2.

[b] Return, plus $40 straight-line depreciation, plus estimated investment tax credit, plus tax savings from accelerated depreciation. The tax savings from accelerated depreciation are estimated to be one-half the difference between straight-line depreciation, and the depreciation calculated as if the pipeline had a 17.5 year service life, switching from double-declining-balance depreciation to sum-of-the-years-digits in year 3.

[c] 10 percent of ICC valuation.

[d] Return plus items mentioned in note [b].

[e] Rate of return is the return that equalizes the present value of the cash flows with the cost of the pipeline.

dards, it is necessary to look beyond the returns on valuation to the total cash flow received under the valuation system. The column labeled cash flow includes the return, rate-making depreciation, and tax benefits captured by the carrier under traditional ICC methodology. Table 4 provides the same information as Table 3 for the consent decree methodology. These tables assume the pipeline investment is able to earn the maximum regulated return for every year.

Careful interpretation is necessary to understand properly the implications of the cash flows presented in Table 3 and Table 4. Table 5 presents estimates of the rate of return earned over the twenty-five year life of the hypothetical investment described in those tables. The rate of return presented is the rate that equalizes the present value of the cash flows generated by the return standards with the initial cost of the investment.[41]

The numbers presented in Table 5 are in principle comparable to the returns measured in Tables 7 and 8 in that they are returns to total capital, including both interest payments and profits and including the impact of tax timing differences. The weighting process implicit in the Table 5 returns is different, however, from that used in Tables 7 and 8. The returns in Table 5 are weighted by the time they are received and by the undepreciated investment in the firm when they are received. In contrast, the returns presented in Tables 7 and 8 are weighted only by firm size.

In estimating the return of a long-lived asset such as a pipeline, it is important to evaluate the return by looking at the stream of earnings over the life of the project and not in any one year. This can be seen in Table 6, which presents the ratio of return plus current period tax savings from accelerated depreciation to original cost less straight-line depreciation. Note that while the consent decree produces a much higher overall economic return than the 8 percent crude oil pipeline standard, for a twenty-year-old pipeline the yearly returns are about the same. Also note that while the traditional standard for petroleum products pipelines results in sharply lower returns over the life of the facility than the consent decree standard, the products standard results in sharply

---

[41] See M. J. Peck and J. R. Meyer, "The Determination of a Fair Return on Investment for Regulated Industries," in *Transportation Economics* (New York: Columbia University Press, 1965), pp. 199-239. They provide a discussion of various economic return measures and use the measure adopted here as the best long-run estimate of economic returns. The return used here is a form of discounted cash flow rate of return. But the reader should be aware that d.c.f. is a generic term, which may include many different ways of calculating returns. The cash outlay on the construction of a pipeline the year before it goes into service will be above $970, if the marginal cost of the pipeline is $1,000 and the interest during construction is 6 percent.

TABLE 4
CONSENT DECREE RETURNS

Year	Return[a]	Cash Flow[b]
1	155	329.2
2	150	220.6
3	145	212.5
4	139	203.5
5	134	195.4
6	131	189.3
7	127	182.2
8	123	175.2
9	119	168.1
10	116	162.0
11	113	156.0
12	110	149.9
13	108	144.8
14	105	138.9
15	104	133.7
16	98	125.6
17	96	120.6
18	94	115.5
19	92	112.0
20	88	108.0
21	88	108.0
22	85	105.0
23	82	102.0
24	78	98.0
25	75	95.0

Rate of return = 21.9 percent

[a] 7 percent of ICC valuation plus interest payments. Interest payments are equal to the interest on 90 percent of the cost of the pipeline at 9 percent and are assumed to decline in equal annual amounts for the life of the pipeline.
[b] Return plus depreciation and tax benefits as per Table 3.

TABLE 5
RATES OF RETURNS
(percent)

Crude oil pipeline standard	14.7
Petroleum products pipeline standard	17.3
Consent decree standard	21.8

TABLE 6

ACCOUNTING RETURNS IN SELECTED YEARS
(percent)

Year	Crude Oil	Products	Consent Decree
5	11.6	13.8	18.5
10	12.3	15.2	19.1
15	15.3	19.6	21.3
20	24.7	32.9	28.3
25	154.5	205.5	137.5
Overall economic rate of return	14.7	17.3	21.8

higher returns for old pipelines. One should also note that there is no simple relationship between the returns produced by the 8 percent and 10 percent standards. These features of the mechanics of the valuation system are elementary, but they should not be forgotten in assessing returns produced under that system.[42] A high or low accounting return for any individual pipeline in any one year may be consistent with a reasonable return over the pipeline's life.

Some Evidence of Profitability. In order to provide some insight into the profits earned by oil pipelines and to compare these returns to other industries, Tables 7 and 8 present measures of profitability of the oil pipeline industry and of other industries.

The period 1975–1977 was selected to provide an average of several recent years. A longer period was not chosen, as that would include 1973–1974, years in which the earnings of energy related companies were distorted by the impact of the four-fold increase in the price of imported crude oil.

The rate-of-return measure used is the standard economic measure of the return to total capital earned by the suppliers of capital to the firm. The numerator is the sum of interest payments and after-tax profits.[43] The denominator of the return measure is the sum of stock-

---

[42] There is a more important, less elementary reason to look at pipeline returns over time. If, as noted earlier, pipeline investments have been responsive to the returns standards offered, then one might expect systematic life cycle effects to be reflected in pipeline returns in the early years of a pipeline's life, with returns at times falling below the maximum allowed by the standard.

[43] This is the standard economic measure of the return to capital earned. It is widely used throughout the economic literature. In addition to the Peck-Meyer

holder's equity, long-term debt, notes payable, and the currently due portion of long-term debt.

Weighted average returns were calculated for each year, 1975, 1976, and 1977. The weight given to each company was that company's share of the capital of the industry. The simple arithmetic average of each industry group's returns for the three years is presented. Included are all firms in the industry groups selected on which data was available from Compustat. Only oil pipeline companies with revenues in excess of $2.5 million are included. Four oil pipeline companies, Arco, British Petroleum, Exxon, and Sohio had to be excluded, as a great proportion of their assets for the period in question constituted construction work in progress on the Trans-Alaskan Pipeline System (TAPS). The publicly available data for these companies were insufficient either to exclude the TAPS construction in progress or to include the interest on that work in the return.

Table 7 presents the returns to total capital for petroleum pipelines and various energy, transportation, and public utility sectors. These are the sectors that one might want to compare with petroleum pipelines. Table 8 presents for comparative purposes the return to total capital experienced by petroleum pipelines and various other selected industries. Table 9 presents the underlying data on pipeline company returns.[44]

Inspection of these tables indicates that petroleum pipelines on the average tend to experience relatively high returns. The manufacturing industries with systematically higher returns tend to be industries associated with large amounts of research and development or advertising expenditures. These "capital" expenditures that are expensed may tend to cause conventional accounting returns to appear high.[45]

---

study noted above (note 41), see W. D. Nordhaus, "The Share of Profits," *Brookings Papers on Economic Activity* (1974 #1), pp. 169-200; G. Stigler, *Capital and Rates of Return in the Manufacturing Industries* (Princeton: Princeton University Press, 1963); and Chase Manhattan Bank—Energy Economics, *Financial Analysis of a Group of Petroleum Companies—1977* (December 1977).

[44] For the reasons noted above one cannot conclude from the observation that in the industry the overall level of profits of individual petroleum pipelines is excessive.

[45] See R. Ayanian, "Advertising and the Rate of Return," *Journal of Law and Economics,* vol. 18, no. 2 (October 1975), pp. 479-506; and H. G. Grabowski and D. C. Mueller, "Industrial Research and Development, Intangible Capital Stocks, and Firm Profit Rates," *Bell Journal of Economics,* vol. 9, no. 2 (Autumn 1978), pp. 328-43. One should not accept the premise that the impact of research and development and advertising activities on firm profitability is a simple one of capitalizing those expenditures as if they were expenditures on capital goods. The mechanism by which profitability is affected by these activities is likely much more complex. See R. R. Nelson and S. G. Winter, "Forces Generating and Limiting Competition under Schumpeterian Competition," *Bell Journal of Economics,* vol. 9, no. 2 (Autumn 1978), pp. 524-48.

TABLE 7

AVERAGE RETURNS TO TOTAL CAPITAL:
ENERGY, TRANSPORTATION, AND PUBLIC UTILITY SECTORS, 1975–1977

	Percent[a]
Bituminous coal mining	15.9
Petroleum pipelines	13.0
Crude petroleum and natural gas	12.4
Local and long-distance trucking	11.7
Water transportation	11.6
Petroleum refining	11.5
Natural gas transportation	11.0
Natural gas distribution	9.9
Telephone communication	8.7
Electric utilities—normalized accounting	8.3
Railroads	7.5
Air transportation	6.5

[a] Percent per year. Each entry is the simple arithmetic average of the weighted average return to total capital for each industry for the years 1975, 1976, and 1977. Within each industry each firm is weighted by its share of the total capital of that industry. The return to total capital is measured as the sum of net income and interest expense, divided by the sum of long-term debt, notes payable, current portion of long-term debt, and stockholders equity. All data except for oil pipelines are from Compustat. For oil pipelines, source is the Form P's, filed with the ICC, Non-carrier income has been excluded from the pipeline returns.

The Present System of Petroleum Pipeline
Regulation and the Entry of New Pipelines

The level of profits earned by petroleum pipelines is not the only aspect of the pipeline industry affected by the system of oil pipeline regulation. The way new pipeline projects are built and sponsored has been affected by the system of regulation developed at the ICC. In order to understand these effects, it is necessary to look at the economics of the formation of coalitions to build new pipeline projects.

At present, with only a few exceptions, new pipeline projects are sponsored by integrated oil companies or coalitions of integrated oil companies. These groups provide the necessary entrepreneurial skills to develop and assemble the project, arrange for its financing, bear the risks of the project pending its completion, and provide financial guarantees that underlie its continuing financial security.

In return for this sponsorship the owners receive returns based on the rate-of-return standards that have been perceived to be operating.

TABLE 8

AVERAGE RETURNS TO TOTAL CAPITAL:
VARIOUS MINING AND MANUFACTURING INDUSTRIES, 1975–1977

	Percent[a]
Perfumes and cosmetics	16.4
Drugs	16.0
Office computers	15.8
Photographic equipment	14.9
Food and kindred products	13.7
Motor vehicles	13.5
Oil pipelines [b]	13.0
Motor vehicle parts	12.0
Footwear	11.6
Chemicals and allied products	11.6
Aircraft manufacturing	11.5
General industrial machinery	11.4
Shipbuilding	11.3
Apparel	10.8
Lumber	10.8
Paper	10.2
Plastic products	10.2
Conglomerates	10.2
Retail grocery stores	10.0
Glass containers	9.8
Concrete gypsum	9.3
Meat products	9.1
Railroad equipment	9.0
Rubber products	8.6
General building contractors	8.6
Metal mining	8.3
Textile mill products	8.2
Household furniture	7.7
Primary smelting refining-nonferrous metals	7.0
Blast furnaces and steel works	6.0

[a] Calculated as in Table 5.

[b] For comparison.

TABLE 9

PETROLEUM PIPELINES AVERAGE
RETURNS TO TOTAL CAPITAL, 1975–1977

	Percent
Butte	20.50
Wyco	20.44
Texas–New Mexico	19.86
Shell	19.73
Chevron	19.35
Amoco	19.26
Powder River	18.53
West Texas Gulf	17.37
Marathon	16.75
Southcap	16.39
Colonial	16.38
Osage	16.28
Platte	16.26
Badger	16.25
Plantation	16.20
Continental	15.86
Southern Pacific	15.77
Sun	15.70
Dixie	15.67
Chicap	15.61
Yellowstone	15.58
Texas Eastern	15.23
Mobil	14.97
Buckeye	14.54
Kiantone	14.52
Western	14.51
Wolverine	14.33
Cities Service	14.32
Texas	14.04
Phillips	13.98
Texoma	13.90
Mid Valley	13.86
Santa Fe	13.73
Olympic	13.69
Diamond Shamrock	13.46
Products	13.44

TABLE 9 (continued)

	Percent
Pioneer	13.30
Ashland	13.22
Hydrocarbon	13.17
West Shore	13.16
Minnesota	13.07
Calnev	12.75
American Petrofina	12.59
Black Lake	12.06
Owensash	11.90
Texaco Cities Service	11.76
Dome	11.62
Kaneb	11.35
Collins	11.29
Jet Lines	11.25
Lakehead	11.23
Tecumseh	11.00
Portal	10.93
Laurel	10.13
Cook Inlet	9.84
Chase	9.75
Belle Fourche	9.69
Michigan-Ohio	8.86
Williams	8.74
Kaw	8.73
Gulf Refining	8.58
Portland	8.14
Mapco	7.87
Shamrock	7.84
Allegheny	7.79
Wesco	6.74
Pure	6.64
Explorer	6.27
Jay Hawk	4.13
Gulf Central[a]	4.03
Amdel	2.88
Hess	2.32
Pasco	1.83
Eureka	−10.75

[a] Gulf Central is not a petroleum pipeline. It carries exclusively anhydrous ammonia.

They also receive any additional advantages and opportunities that may be conferred and available by virtue of their ownership.[46] One could attempt to separate the returns earned under current rate-of-return standards into two components. First, there is a return that acknowledges only the profits the project requires, given all the burdens assumed by the sponsors. Second, there is a lump sum payment or bonus for the risks of sponsorship.[47]

Some compensation must be available for the burdens of sponsorship, and it ought to be commensurate with these burdens, or no one would sponsor a new pipeline project. A system of pipeline regulation that does not provide this compensation will create a free rider problem and discourage construction of new pipeline projects.[48] No nonshipper would have any reason to construct a pipeline. Any potential shipper would find it in its best interest to let another shipper bear these costs. If the system of regulation permits only some groups to serve as sponsors, then the costs of these limitations on the potential sponsors should be understood and possible alternatives should be considered.[49] The

---

[46] The *Report of the Attorney General . . . For Deepwater Port Licenses . . .*, pp. 56-59, provides a discussion of these other factors as they were perceived by the sponsors of the Deepwater Ports.

[47] For example, if one discounts the cash flows received by the owners under the existing rate-of-return standards, at an arbitrary 11 percent, which reflects a very approximate return on a pipeline given the support of its sponsors, then the resulting number is the sum of the cost of the project and the sponsorship bonus. In the example presented in the tables, the bonus as a percentage of the cost is 22 percent for the crude pipeline standard, 34.4 percent for the products pipeline standard, and 64.5 percent under the consent decree standard.

[48] It has been said that the "competitive rules" put forth by the Antitrust Division, U.S. Department of Justice, would create this free rider problem. It is asserted that the rules would do this as they permit shippers to become owners and to increase their ownership shares after construction, thus denying the sponsors compensation for their efforts. Rules proposed by the Antitrust Division, however, do provide compensation for the risks and burdens of construction consistent with rate-of-return regulation. If the valuation system were continued and used as the buy-in price, the payment of a new owner who entered after the third year of operation would, together with the cash flows received in the first three years of operation, provide the original owners with payment that would not exactly equal the payment received if ownership were closed, but one that constituted a substantial portion of the premium they could earn if no other shippers were permitted to become owners. If a new rate base is selected, the rules clearly contemplate adequate allowance for capital costs during construction as part of the rate base, which will provide the necessary compensation to sponsors.

[49] The question of limitations on sponsorship and alternatives to vertically integrated owner/shipper sponsorship is obviously of interest in a world in which retrospective or prospective pipeline divestiture is contemplated. The issue of alternative sponsorship of pipeline projects is of interest even in a world with continued oil company ownership of petroleum pipelines because, as noted above, there are not apt to be many possible competing coalitions of oil pipeline companies in the development of any one pipeline project.

present system of petroleum pipeline construction can be described as a combination of the valuation system returns and the sponsorship of new pipelines by shipper-owners.

Four entrepreneurial and managerial services are provided by the sponsors of a project: developing and assembling the project, arranging the financing, bearing the risks during construction, and providing financial guarantees once it is complete. Regulation does not appear to affect the first two items; [50] it has a direct bearing on the latter two.

Risk bearing can be affected by regulation to the extent that the present system produces returns, over time, to compensate the sponsors for all the risks of the project. The current system of petroleum pipeline rate regulation does not distinguish between the initial risks and the subsequent risks to the project. In part there is good reason for this, as most risks or unfortunate occurrences during construction translate into higher costs during the period of operation. The lack of explicit awareness of the differences between the risks borne during construction and those borne during operation poses difficulties if alternative ownership arrangements for petroleum pipelines are to be considered. [51]

The fourth responsibility now assumed by oil company sponsors is to provide either sufficient throughput for the project or the necessary funds to support it. If such commitments are necessary to obtain financing, other groups of sponsors will be excluded unless they can obtain these commitments. There is no reason, in general, for anyone to provide these commitments unless some compensation is forthcoming.

It is well established that the successful finance of large transportation or similar projects requires commitments to the project by potential users. The natural gas pipeline transmission industry developed with support from long-term commitments for the gas. [52] The great historical common carrier transportation projects, the railroads and canals, developed with extensive government support. [53] A recent study by the De-

---

[50] A complete discussion of the question of vertical integration and new pipeline construction would have to address the relationship between vertical integration and the transmission of information and skills necessary to the development and completion of the project. But the present discussion relates only to the impact of the present system of regulation on new pipeline construction.

[51] For example, if ownership is fixed prior to construction, it is a matter of no consequence whether compensation for risks during construction is in the form of higher returns after construction or an explicit allowance for this cost. If ownership were available to shippers subsequent to construction, however, at a price equal to the rate base, this arrangement could deny the owners proper compensation if the risk compensation were only in the form of higher return.

[52] See P. MacAvoy, *Price Formation in Natural Gas Fields* (New Haven: Yale University Press, 1962).

[53] See L. E. Davis, R. A. Easterlin, and W. N. Parker, *American Economic Growth*

partment of Justice provided a detailed review of the role of long-term commitments in the development of new coal mines and found that the financing of these projects was significantly aided by long-term contracts.[54]

The desire for long-term commitments to support the financing of an important project does not reflect economic irrationality on the part of prospective sources of finance. The contracts in part shift the risks of the project to the providers of the long-term commitments. They also provide real information about the potential economics of the project. It is difficult and costly for financial markets to assess a complex project. The willingness of potential users to give a commitment provides useful information as to the project's economic potential.[55] This latter element suggests that the simple expedient of offering a higher *ex post* return may not be a workable or efficient solution to the problem of compensating sponsors.

The system of regulation of petroleum pipelines developed to date has included a historic ICC hostility to long-term contracts for transportation services by ICC regulated common carriers. This hostility is in no way mandated or required by the Interstate Commerce Act.[56] The ICC has, in fact, recently reversed its historic opposition to contract rates for railroad haulage of coal.[57] The origins of the general ICC hostility toward these contract rates appear to have been in concerns about their effect on the extent of competition and the overall rate structure in regulated transportation.[58] This inability to make meaningful long-term contracts with prospective shippers has had the effect of seriously limiting the ability of nonintegrated oil company sponsors to develop a petroleum pipeline.[59]

(New York: Harper and Row, 1972), chapter 13, pp. 468-547; and L. E. Davis and D. North, *Institutional Change and American Economic Growth* (Cambridge: Cambridge University Press, 1971), chapter 7, pp. 135-66.

[54] See *Competition in the Coal Industry* (May 1978). The focus of the discussion of long-term contracts in the report on the coal industry included many factors in addition to the financing of new coal mines.

[55] S. A. Ross, "The Determination of Financial Structure: The Incentive Signaling Approach," *Bell Journal of Economics,* vol. 8, no. 1 (Spring 1978), pp. 23-40, provides a discussion of the use of signals by the financial market to provide useful information.

[56] See *Initial Comments of the United States Department of Justice on Proposed Inducement Tariff Structure,* in Texas Deepwater Port Authority, Docket No. OR 79-2, January 22, 1979.

[57] Interstate Commerce Commission, Ex Parte No. 358-F, Change of Policy: Railroad Contract Rates, General Policy Statement, November 9, 1978.

[58] See Robert F. Lundy, *The Economics of Loyalty-Incentive Rates in the Railroad Industry of the United States* (Pullman: Washington State University Press, 1963).

[59] This discussion has not tried to address the full range of issues necessary to

100

The Present System of Regulating Petroleum Pipeline Pricing

The ICC system of oil pipeline rate regulation has not developed any explicit mechanism to assess the proper rate structure of petroleum pipelines—that is, the division of expenses across the classes of pipeline shippers and across shippers using different origins and destinations on the same pipeline system.

A reason to be concerned about the problem of petroleum pipeline rate structure can be seen by an examination of one typical large pipeline company. For 1975–1977, the three-year average rate of return to total capital of the company was about 19 percent. The company operates almost 12,000 miles of pipeline. Its total revenues are about $100 million. Eighty-two percent of the barrels moved by the company were owned by other components of its parent. This indicates a possibility that whatever overall limitation on pipeline company profitability exists may not be meaningful. For example, if there are parts of the company that effectively carry only petroleum for its parent and that are charging tariffs that are low in relation to costs, then the effective earnings of the system that is carrying traffic for others may be much higher than the systemwide average. This possibility provides the chance to restrict the impact of rate-of-return regulation.[60]

The problem in any attempt to regulate pipeline rate structure is that it is difficult to develop an economically meaningful way to allocate

---

assess the potential benefits of integrated petroleum pipeline firms. That would be beyond the scope of this paper. At least one author has argued that the problem of contractual incompleteness may be a cause of the vertically integrated structure of the oil pipeline industry. See David Teece, *Vertical Integration and Vertical Divestiture in the U.S. Oil Industry* (Stanford: Stanford University Institute for Energy Studies, 1976). Contractual incompleteness is a term that describes a broader class of phenomena than just a regulatory ban on a type of long-term contract. The difficulties that stem from fully specifying the terms of the contracts and small numbers bargaining are also important. See O. E. Williamson, *Markets and Hierarchies* (Glencoe: Free Press, 1976). Directly related to these questions is the issue of legal recourse in the event of a breach of contract. See R. Posner and A. M. Rosenfield, "Impossibility and Related Doctrines in Contract Law," *The Journal of Legal Studies,* vol. 6, no. 3 (January 1977). Note, however, that another author has suggested that complex long-term contracts often require enforcement mechanisms that are effective regulatory schemes. V. Goldberg, "Regulation and Administered Contracts," *Bell Journal of Economics,* vol. 7, no. 2 (Autumn 1976). In the specific context of petroleum pipelines, see A. Alchian, B. Klein, and R. G. Crawford, "Vertical Integration, Appropriable Rents, and the Competitive Contracting Process," *Journal of Law and Economics,* vol. 21, no. 2 (October 1978), pp. 297-326. At note 31 of their article they suggest that vertical integration in the petroleum pipeline industry in general may be a response to the present system of regulation.

[60] I do not mean to suggest any conclusion as to the existence of such behavior under current regulation.

costs across segments of a petroleum pipeline system. Consider a simple pipeline system with one origin and two destinations, one simply farther along the pipeline than the other. One might wish to estimate the costs of the additional resources that must be expended to move an additional barrel to the second point over the cost of shipments to the first point. One might wish also to try to estimate the cost of the additional resources that must be consumed to move an additional unit to the first destination, given the resources expended to move the shipments to the second destination. In general, there is no reason to expect that these two numbers, if they are the tariffs, will provide sufficient revenues to cover the total costs of the pipeline. It is difficult to develop useful standards to allocate the total revenue required between the traffic to the two destinations.

In general there is reason to be concerned about the efficiency and competitive implications of alternative division of costs by regulated enterprises. For example, it has been said that there are significant competitive implications of alternative ways of pricing telecommunications equipment and services.[61] There may be less need for extensive rate structure regulation in oil pipelines as the competitive margin narrows between regulated firms interacting with competitive firms providing the *same* service. But the possibility of the exercise of discretion by integrated oil firms to the disadvantage of their shipper competitors has been an important historical part of petroleum pipeline regulation. A lack of rate structure regulation could be an important gap in the regulatory scheme.

There is much reason to be cautious in the approach to rate structure regulation for petroleum pipelines. First, once it is announced that the regulator cares about rate structure, a process will be set in motion by which formal proceedings will develop to try to determine the "correct" structure. Those who will be helped by a change will seek relief, and those who are hurt by the change will put forth alternative rules. They will do so, possibly at considerable expense, as long as there are private gains from alternative rate structure rules. Yet this expense should be incurred only if there is good reason to believe that there will be public benefit from aggressive regulation of rate structure.[62]

There is a second problem with extensive rate structure regulation that can be demonstrated by an example. Assume the simple pipeline system discussed earlier, with one point of origin and two destinations.

---

[61] See Owen and Braeutigam, *The Regulation Game,* chapter 1 and chapter 7 for a discussion.

[62] This problem is part of a general problem of using the reliance on the private gains of potential enforcers to develop general regulatory frameworks. See R. Posner and W. Landes, "The Private Enforcement of Law," *The Journal of Legal Studies,* vol. 4, no. 1 (January 1975), pp. 1-46.

Assume further that when the pipeline was first built, it faced alternative sources of supply at the further destination, but not the nearer. That might have required a tariff structure that was relatively insensitive to distance. Assume, for example, that the tariff to the further destination was one-third more than the tariff to the nearer destination, although the distance was twice as far. Assume that this type of tariff was based on an equitable and efficient cost allocation rule.

If, however, circumstances change over time this initial rate structure may become outmoded. Table 10 presents the hypothetical data of this example. The column labeled "initially" describes the circumstances when the pipeline was constructed. The column labeled "later" describes the changed circumstances. The cost of the alternative supply to the further destination has increased. The potential traffic on the pipeline has increased.

Assume that the pipeline expansion will be undertaken only if the tariff to the further destination is $.75. Assume that this is the cost of providing the capacity to carry the incremental traffic. This neglects the additional revenues the carrier receives on the first 100 units delivered to the second destination. This expansion is desirable. It will cause the additional supply to come by means of the cheaper pipeline. The pipeline should be able to change its rate structure in response to this change in conditions to permit the expansion to take place.

The difficulty that this process causes for rate structure regulation is two-fold. First, rate structure regulation is apt to seek to evolve relatively simple rules for allocation of cost. A simple rule that was sufficient to find reasonable the rate structure in the initial case, and that excluded other plausible rate structures, might rule out the determination that the second rate structure was reasonable.

Furthermore, the change to the second rate structure may provoke vigorous opposition from the shippers of the pipeline and others. This can be seen simply. In the conditions in the example, the cost of the alternative supply is $1.25, the "old" tariff is $.55. In the example, 200 units is the constant demand in the second period. This implies that at $.55 tariff demand will exceed the capacity of the pipeline. One hundred units will be supplied by the alternative source; the market price will reflect the $1.25 cost of the alternative. The shippers who are fortunate enough to have access to the pipeline reap a gain of $.70 from that access.

Now, under the proposed expansion and tariff increase, several groups will be affected adversely. The suppliers of the alternative source of supply will lose business, and the existing shippers will be hurt. The present shippers will suffer a loss from the increase in tariff and will lose

TABLE 10
HYPOTHETICAL RATE STRUCTURE PROBLEM

	Initially	Later
Tariff to closer destination	.40	.40
Tariff to further destination	.55	.75
Cost of alternative supply	.60	1.25
Volume to closer destination on pipeline	100	100
Volume to further destination on pipeline	100	200 (100)[a]
Volume of alternative supply source	0	0 (100)[a]

[a] If the pipeline does not expand.

the benefits that accrue from the impact on market price of the use of the higher-cost alternative. The beneficiaries of the scheme will be the ultimate purchasers and perhaps the pipeline company. Regulation has often shown itself to be very sensitive to the type of losses experienced by the first groups.[63] Thus, an aggressive system of rate structure regulation may not be without significant costs, but it may be important to the integrity of the regulatory scheme.

Conclusion

It has been said that the prevailing view among economists of economic regulation is that it is either pernicious or inefficacious.[64] The present system of oil pipeline regulation has been a little of both. The standard for the overall level of profit is unreasoned and does not appear to present a careful check on pipeline profits. The ICC hostility to long-term contracts has hindered the entry into ownership of pipelines by companies other than oil companies. While the neglect of rate structure regulation may not be entirely unsatisfactory, it is a significant gap in the regulatory scheme. In any event, pipeline companies have no indication of the standards within which they can vary their rates.

[63] See studies cited in note 1.

[64] D. J. Dewey, "The New Learning: One Man's View," in H. J. Goldschmid, M. G. Mann, and J. F. Weston, *Industrial Concentration: The New Learning* (Boston: Little, Brown & Co., 1974).

Across a wide range of industries, including airlines, trucking, natural gas production, insurance, and perhaps railroads, a clear consensus among economists has emerged that deregulation is preferable to continued attempts to change or patch up regulatory schemes. In the areas of classic natural monopoly regulation, the prevailing view is mixed. Electric utility regulation may have evolved sensible or at least workable rules for estimating overall rates of return.[65] But electric utility regulation has had significant difficulty adapting to recent cost increases.[66] One author has indicated that the regulation of natural gas transmission may have resulted in reasonably efficient pricing and production.[67] Others, however, have cast serious doubt on the efficacy of natural gas transmission regulation.[68]

Victor Goldberg has indicated that in a situation in which private long-term contracts endure changing circumstances, mechanisms often evolve that are equivalent to overt regulation or require the development of a government regulatory system.[69] In the petroleum pipeline industry, where the technology dictates the commitment of capital for long periods of time to provide service to many users, just such mechanisms may be necessary. It is clear, however, that the present system, which has been neither flexible nor adaptive and which has little rational basis, will not provide such a mechanism.[70]

---

[65] See footnote 27.

[66] See P. Joskow, "Inflation and Environmental Concern. . .," *Journal of Law and Economics,* vol. 18, no. 2 (October 1974), and P. W. MacAvoy and P. L. Joskow, "Regulation and the Financial Condition of the Electric Power Companies in the 1970's," *American Economic Review,* May 1975, pp. 298-301.

[67] See J. L. Callen, "Production, Efficiency, and Welfare in the Natural Gas Transmission Industry," *American Economic Review,* June 1978.

[68] See P. MacAvoy, and R. Noll, "Relative Prices on Regulated Transactions of Natural Gas Pipelines," *Bell Journal of Economics,* vol. 4, no. 1 (Spring 1973), pp. 213-34; and MacAvoy and Breyer, *Energy Regulation by the Federal Power Commission.*

[69] See V. P. Goldberg, "Regulation and Administered Contracts," *Bell Journal of Economics,* vol. 7, no. 2 (Autumn 1976).

[70] For a discussion of an interesting alternative approach to the control of natural monopoly, see the paper by Lucinda M. Lewis and Robert J. Reynolds in this volume.

Effectiveness of Government Regulation

Ulyesse J. LeGrange

In addressing the question of the effectiveness of government regulation I would like to break the subject down into three aspects and discuss each one briefly. The three areas are

- the state of the industry as a result of government regulation
- the nature of government regulations
- the returns to the industry allowed by regulation.

The State of the Industry

One significant aspect of regulation is the end result—that is, in what shape is the industry after a number of years of regulation? I believe it is fair to state that the pipeline industry is in excellent shape. A sizable and efficient network of liquid pipelines is in existence and in operation. As the need for new capacity has developed, the need has been met by investors who were willing to risk their capital under the existing regulatory environment. As a result, liquid pipelines have increased over the years and are now second only to railroads in total ton miles of freight moved in the United States. Incidentally, they do this with no government subsidy. In the petroleum sector, pipelines account for about half of all the crude and petroleum liquid movements.

It has not all been rosy along the way. Individual pipeline companies have had rocky going as their requirements either diminished unexpectedy or did not develop as the owners had anticipated. Some of them have lost money, but this is the risk they took since the regulatory environment in which the capital commitment was made gave them no special privilege of territory, no guarantee of throughput, and no guarantee of profit—only an opportunity to earn a profit. This opportunity has been deemed attractive enough to bear the risk of the investment.

It is unfortunate that over the past several years this regulatory environment has come under extreme criticism. This has caused considerable uncertainty in the industry, and the new high level of uncertainty of regulations has actually killed one important project with which I have been directly associated—namely, Seadock.

Even though the state of the industry is sound, it will not necessarily remain that way if any of the what I call "radical" ideas of changes in the pipeline regulatory environment take over.

The Nature of Pipeline Regulation

A proper characterization of oil pipeline regulation prior to 1978 is that the regulatory agency, the ICC, took a low profile role in overseeing the industry. Of course, I am sure that the pipeline consent decree, signed in the early 1940s, has been a key factor in this approach. The consent decree itself has been an effective regulator from the standpoint of controlling the level of tariffs. Speaking from my own personal experience, the consent decree was the principal consideration in our internal rate setting discussions, and it put an effective lid on our rates. Indeed, looking across the industry in total we see that pipeline revenue per hundred barrel miles actually decreased from the mid-1960s to the mid-1970s, despite the high level of inflation taking place during that period. This is a reflection of the regulatory lid as well as the efficiencies generated by the industry during that time. With the consent decree in operation, the ICC could take a muted role and still meet its regulatory obligation.

Looking back over the last twenty or thirty years we see another interesting phenomenon—the almost complete lack of challenges either to tariffs or to services. Historically, there have been only a small number of these disputes, and, surprisingly to me, the critics of the industry have used the lack of disputes to question the effectiveness of the regulators and to imply that shippers—en masse—have such a fear of not being heard in the disputes that they have simply taken their injustices in silence. This explanation is unacceptable to me. It completely ignores what I believe are the facts of the situation—the industry has posted just and reasonable tariffs and service has been granted under reasonable conditions. The president of the Independent Petroleum Association of America, representing some 4,000 independent producers, told the United States Congress that the independents are not aware of any difficulty moving crude and that he did not believe discrimination exists. I have a hard time accepting the argument that he does this only out of fear of the big boys.

It is an interesting paradox that in this age of cries for deregulation, or lessening of regulation, or lessening of the government's role in business, we have just the opposite happening in the pipeline regulatory area. The low profile regulation of the past has been sharply criticized, and there are cries for more and tougher regulation of pipelines, and indeed for their divestiture from oil companies. Another interesting paradox is the challenge today to the methodology of pipeline regulation—particularly the methodology that has at least attempted to give some recognition to the impact of inflation on the rate base by building in factors for reproduction cost. The entire thrust of modern economics and modern financial reporting is towards recognizing the impact of inflation in the destruction of capital. Yet the ICC valuation system has been at least partially doing this for years and has been severely criticized as inappropriate, instead of being recognized for what it was—very much ahead of its time.

The Return Allowed under Regulation

Probably the strongest criticism of pipeline regulation has been that it has allowed pipeline rates of return which are far too lucrative. I believe this criticism is also misfounded and flies in the face of the facts.

In order to address this question, I will describe what happens in business on investment decisions and relate this to the pipeline industry. Then we can see directly from the facts whether pipeline returns have been excessive.

In looking at any investment, a corporation has two decisions to make—whether to make the investment and, if it is made, how to finance it. These are separable decisions, and they are treated accordingly. The investment decision obviously comes first. It is based on the expected return on the total capital required, and it is generally based on a discounted cash flow calculation using total cash inflows and cash outflows. Only after the investment decision is favorable does the question of the financing of the investment arise.

The financing decision is based on tax considerations, availability of funds, government regulation, exchange controls, and other similar factors. Except in the most unusual circumstances, or for very limited periods of time, a corporation's financial resources are subject to limited availability, whether from internal cash generation, depreciation plus earnings retained in the business, or equity and debt capital provided from external sources. A corporation must use these limited resources in the optimum manner, balancing the cost of new equity capital, the cost of new debt, and the availability of other funds, with the investment opportunities at hand.

Pipeline investments are both capital intensive and relatively high-risk investments because of their specialized nature and use. Despite

109

this, pipeline companies have been able to obtain large amounts of debt and maintain high debt-to-equity ratios. This is a direct result of the consent decree, and, if necessary, we can discuss later why this happened. Wherever this occurs, however, there is also present some sort of backing of the debt instruments of the pipeline company by its parent company, since the pipeline company alone could not borrow at such a high level. It alone does not have that type of credit worthiness and, thus, requires either a straight debt guarantee or a throughput agreement of the parent to provide, in effect, the assurance of an adequate flow of funds for the repayment of the debt obligation. This means the parent lends its credit to the pipeline company to substitute for lack of credit by the pipeline company. This comes at a cost, since even the parent does not have unlimited credit, and, once its credit or effective borrowing capacity is given to the pipeline company, it cannot be used again. This question of the giving up of the parent credit to the pipeline affiliate and the fact that this results in a cost to the parent was little understood, or even ignored, by critics for a long time when these critics insisted on looking at pipeline equity returns. I have recently seen, however, that even the Department of Justice has finally come to recognize that a cost does indeed exist for the guarantor. In recent publicity on the Seadock project, which the state of Texas is attempting to revive, the Department of Justice seems to be going out of its way to argue the legality of a two-tier tariff structure, which would give a lower tariff to anyone willing to guarantee a portion of the debt of the facility. I find it strange, if not ironic, to see the Department of Justice strongly backing a higher tariff for nonguarantors in this new Seadock project after having had a direct hand in killing the original Seadock venture where these nonguarantors would have had an equal tariff.

The net result of parent guarantees then is, in effect, to finance pipeline investment by 100 percent equity. Because of this, it is meaningless to look at returns on the pipeline company stated on nominal book equity. At this point we are faced with several choices—we can attempt some correction of the nominal debt/equity ratios; we can make some allowance for a higher return on nominal equity recognizing the true nature of the pipeline financing; or we can simply ignore the pipeline company nominal debt/equity ratio. The easiest way to solve the problem is the last way. We do this by looking at returns on total capital employed. This was the basis of the original investment decision anyway, and we would be looking at the basic economics of the investment decision, which was made without regard to the financing decision. We can then look at all corporations before interest charges are deducted, so that the level of debt and the cost of debt are eliminated as factors. Then, we can compare pipeline returns to any other company without regard to capital structure or worrying about why and how the structure was derived.

Now we can turn to the question, Have pipeline rates been generating excessive returns? There is also a subsidiary question, Do they far exceed the most profitable public utilities? To answer the questions, we have looked at returns for the period 1973–1977 for some seventy-nine pipeline companies with annual revenues of $2.5 million or more. The data came from ICC annual reports. In addition, we have used data from a *Forbes* magazine study of 1,005 public companies, published in 1978. We added data from the S&P Compustat, Inc., services of nearly 2,500 public companies to get an even broader data base. We calculated, on a comparable basis for all companies, returns on total capital. The results are interesting and informative.

The median return for pipeline companies was 10 percent. The median for *all* industry was 9.6 percent or about the same level. The average return for pipeline companies was 9.5 percent. The average return for *all* industry was 9.9 percent or about the same level. The mean weighted average for pipeline companies was 7.6 percent compared with about 10 percent for general industry. Thus, we see pipeline returns at about the level of industry averages and not excessive.

Next we compared pipeline returns to 171 gas, electric, and telephone utility companies. Here we find the mean, median, and mean weighted average return of pipeline companies of 9.5 percent, 10 percent, and 7.6 percent, somewhat higher than the comparable returns of 6.9 percent, 6.6 percent, and 6.7 percent for the utilities.

Since pipeline companies do earn higher returns than utilities, we considered relative risks to see if we can account for differences, and I believe we can. First, pipeline companies do not have a market monopoly. No one is forced to ship over a pipeline. Other services are available and can be used. This should lead to more volatile pipeline returns. Second, pipeline companies cannot sign long-term agreements for capacity. They must take on all comers under their common carrier obligations. This means that shippers are not locked in and can quit shipping at their own choice. Again, this should directionally cause more volatility of pipeline results. Third, pipeline companies have a ceiling on earnings but no floor; for example, there is no guarantee they will make any return. On the other hand, a utility under any rational regulatory environment will effectively be provided with a basis for making some minimum level return. This was demonstrated in the study results. A total of ten of the seventy-nine pipeline companies had returns of less than 4 percent whereas of 171 utilities, none had returns of less than

4 percent. The dispersion of utility returns was not great—mainly clustered around 6-8 percent return with the lowest about 4 percent and highest at 16 percent. The dispersion of pipeline companies is wide, from a negative 8 percent to a positive 18 percent. From this I conclude what I intuitively believed in the first place—that pipelines are indeed riskier than utilities and this accounts for the differential in overall industry results.

In conclusion, it is time to quit speculating on theoretical situations and to get back to the facts of the real world if we are to have any hope of arriving at real world solutions. If shippers have been harmed, let's find them and find out how and why they have been harmed. If pipelines have monopolies, let's find the specific pipeline situation and see why such a situation exists and what, if anything, needs to be done about it. If pipelines have been undersized, let's find the pipelines that are pro-rating space and see that the prorating is being done equitably. But let's quit debating hypothetical situations, which can lead only to confusion and misunderstanding rather than clarification.

Commentaries

Shyam Sunder

Spavins and LeGrange agree that there is a need for better and more comparable empirical work to determine who has made how much money and whether or not it is too much. If we can assume that the income and the cash flows reported by various firms in their financial statements can be taken at their face value, the only remaining question is whether the money made by pipeline companies is too much in terms of some fair return standard.

How is that standard to be determined? Spavins and LeGrange agree that a return standard should be specified in terms of the total capital invested—that is, whether it is the equity capital or borrowed capital, the amount of money that a firm has made should be determined as a proportion of the total investment of the firm.

In both papers, the return on total capital has been calculated, and Spavins' results are somewhat higher. The rate of return for the pipeline companies, according to Spavins, is 13 percent, compared with a median of 10.8 percent for other industries. According to Mr. LeGrange, the average rate of return for pipeline companies is 9.5 percent, compared with an overall industry average of 9.9 percent—figures that are much closer together.

On the face of it, one might conclude that there is some discrepancy in the data; fortunately, both papers have been based on the same data, and it is possible to reconcile the two numbers. I do not see any basic difference in these numbers.

The key, of course, is the adjustment for leverage. LeGrange computed his rates of return after taking out the effect of leverage; he added to the net income available to shareholders the interest paid on borrowed capital. That interest payment was adjusted for tax savings, so he has used net income plus after-tax interest, divided by total capital invested as the appropriate rate. This rate has taken out the effect of leverage and reduced all the data of all firms within the pipeline industry and across the industries to a comparable basis after adjusting it for differences in leverage.

Spavins has not carried out that adjustment in his paper. He has chosen to compute the return on total capital without carrying out the leverage adjustment. The effect of the leverage adjustment is not really very important because a comparison of a fair return standard across firms and across industries will need an appropriate adjustment for risk anyway.

There are two main components of risk we have to to consider. One part of the risk is the business risk of the company—the nature of the business the firm is in. The second part of the risk is the leverage. If we compare LeGrange's numbers, we have to worry only about the business risk of the firm because the leverage risk has already been taken out of those numbers. When we compare the numbers prepared by Spavins, we have to consider the total risk of these firms, which combines both the business risk and the leverage risk.

The basic question, then, is, What is to be regarded as a fair return on an industry whose risk is different from the risk of others?

LeGrange has provided some evidence on comparison of one measure—one empirical measure of risk—between the oil pipeline industry and the utility industry. He has shown that that particular measure of risk is much higher for the pipeline industry. The measure is a variance for returns on total investment. It is much higher for the pipeline industry than it is for the utility industry.

Now, one might quibble over this measure of risk versus another measure of risk, but that is really no problem. We can work with three or four or five different measures of risk and see whether, after risk adjustments have been made, pipelines have made more money or less money than other industries.

The question is, How should the risk adjustment be made? I suppose the best way of doing this would be to compute the rate of return for various industries, then compute the risk of various industries, and plot them on a chart. In the relationships across industries between risk and return, we could see where the pipeline industry stands. If the oil pipeline industries stand out either on the high side or on the low side, then we will have some ground for saying that after adjustment for risk— that is, after risk differentials have been considered—this industry is making too much or too little money.

Unfortunately, that part of the analysis has not been carried out in either paper. LeGrange's paper does show that this particular measure of risk is higher for the oil pipeline industry, and the rates of return for the oil pipeline industry are higher than for the utilities; but how much higher should the return be for this kind of differential and risk? That question can be answered by either graphical or regression analysis of risk and return across industries. I don't really see much of a problem in doing that analysis.

On the other hand, if we use the measure of returns used by Spavins, we have to use the risk measure appropriate to that rate of return, and, of course, that rate of return is much more volatile for the pipeline industry than is LeGrange's rate of return. The fact that they have used somewhat different measures of rate of return will not cause any problems once the appropriate adjustment for risk has been made.

A second problem in comparing the rate of return across firms and across industries is that all these data and analyses are based on accounting rates of return. Now, of course, we know that businessmen do not make their investment decisions on the basis of accounting rates of return. They would probably carry out discounted present value analysis, compute the present value of the investment or the economic (internal) rate of return.

Table 9 of Spavins' paper gives the accounting rates of return for more than seventy firms. The firms at the top of this list with very high rates of return tend to be the firms that are older and relatively well established. The firms at the bottom of the list with very low rates of return are the ones that are relatively new. They are all small.

Why is that so? Of course, this is largely a result of the way accounting rates of return behave, because once the firm has matured, its properties have depreciated by a considerable amount so the denominator in calculating the rate of return is smaller and the rate of return is higher.

This arbitrary effect of the maturity of the firm on the rate of return can be eliminated if the comparisons are made on the basis of the internal rate of return of these firms. Internal rate of return requires additional data about the current value of the fixed assets of the firm which usually are not available. Fortunately for the oil pipelines, we do have those data available. And for the past two years, since the SEC imposed additional disclosure requirements on replacement value of fixed assets and inventories, we do have those data available for at least the larger U.S. corporations also.

Those data could form the basis of comparing the internal rate of return of the pipeline firms with the other industry firms. If such a comparison were made, we could probably settle the argument about who has made how much money in a more objective manner.

David J. Teece

As I see it, there are basically three issues in these two papers where there is some level of either explicit or implicit disagreement. One is the regulated rate of return—Is it too high, or is it too low, or, basically, are economic rents being captured through the existing tariff structure? That is one central issue. Another issue is the pipeline undersizing issue which has already been discussed. And there is a third issue that I want to address briefly, also implicit in Spavins's paper, namely the role of vertical integration. In particular, to what extent is vertical integration a device to circumvent regulation and therefore does it have pernicious effects.

Let's briefly turn to the rate of return discussion. Flexner drew a monopoly power explanation from those numbers. I did not find Spavins making that claim in his paper, nor do I think he would want to make one without first seeing if risk could explain the difference that exists.

It is also fair to point out that there is greater competition in pipelines than in other utilities, and that entry is not regulated, and, of course, that affects the level of business risk. There appears to be greater business risk in the pipeline industry than in other utilities.

Another issue relates to vertical integration. Mr. Spavins is questioning whether the regulated return from pipelines is too high, but one might also question whether it is too low. Consider the fact that many pipelines are owned by refiners. The explanation I heard this morning from some participants was that the reason for vertical integration into pipelines is that the major companies do not want to share the monopoly rents associated with pipeline ownership, and so will deny business to independents. An alternative hypothesis is that at existing regulated rates of return, there is insufficient incentive for entry by non-vertically-integrated firms. If the regulated rate of return was below the competitive level, then one way a pipeline could be made viable is by providing some kind of throughput guarantee which reduces the risk for the pipeline company. In this regard, examining the earnings variability of pipelines will understate risk if throughput guarantees are in place, since some of the risk is being carried by the throughput provider. Another way in which pipelines could be made viable in this circumstance is by directly leaning the pipeline project against the financial structure of the parent through loan guarantees and the like. However,

116

even if the regulated rate of return is at or above the competitive level, vertical integration has compelling efficiency properties which may explain the relative absence of independent lines. Clearly, vertical integration facilitates the coordination of complementary investments throughout the industry. Very often, for instance, pipeline and refinery investments must be made contemporaneously. There may also be important scheduling and other operating advantages associated with a vertically integrated system. In short, efficiency and not monopoly power explanations may explain vertical integration into pipelines. While there is the potential that vertical integration may circumvent regulation, we need evidence that this is actually the case before delivering an indictment against vertical integration. A case-by-case analysis would be the appropriate approach to these potential problems. It is very hard to fashion a comprehensive indictment against vertically integrated pipelines. Efficiency considerations must also be examined along with monopoly power concerns.

Thomas C. Spavins

I have two comments. First, profit rates do not mean anything by themselves, but they are very useful when taken together with a number of other factors. They can be examined together with evidence on the technology and with evidence on the structure of the industry. For example, my department has done a lengthy monograph on the coal industry, mostly the bituminous coal industry, and we strongly concluded it was workably competitive. During the three years studied, bituminous coal was the one industry in the energy, transportation, and utility section which had significantly higher rates of return than the oil pipeline industry. As we all know, the airline industry was not competitive in some ways before deregulation, and it had the lowest rate of return.

Second, across different industries there are different amounts of leverage, and, because of federal corporate income tax laws, the income received from capital is taxed differently as "equity income" and as "debt income." In any interindustry study, a question to be asked is, What is the effect across industries with different leveraging of the fact that the taxation on income from corporate capital differs depending upon that leveraging?

As Sunder pointed out, the comparison of these numbers is not just a matter of the right way to do it and the facts of the risk, but requires an analysis of the behavior of firms in response to many different things, including taxes.

Ulyesse J. LeGrange

The point on accounting rates of return is well taken. There is additional information available now, and there will be even more available in the future. By the end of 1979, we will actually be computing as individual companies what these new profit numbers are on an adjusted basis for fixed assets if the FASB goes through with its project this year, as it is planning to do.

The important point, I think, is to be sure we assemble the data and make the comparisons. And we should not come to the conclusions before we make the comparisons.

Discussion

QUESTION: Wouldn't it be possible to do an indirect test on the level of profitability on average of oil pipelines by assessing whether or not there is, let's say, excess demand to come in on the ownership at the time of formation of a new pipeline project? One would think that, if there are excessive returns, there would be great demand to be part of the project.

I would ask, one, whether that is a way of testing the proposition, and, two, what the experience has been on lining up owners. It raises the question of how pipeline projects are actually generated.

MR. LEGRANGE: I have been involved in a few of the more recent efforts, and we went out of our way, we thought, to invite anybody and everybody into these ventures. When we look at our cut point of investments, we find that investments in pipelines, even under the so-called lucrative returns allowed by the Consent Decree, fall out at the low range of what is acceptable to us. And we have to tie them in to the need of the corporation to move ahead and invest our capital in them.

Since they are of that nature, and since we catch so much hell for being in them from people worrying that we are doing something in an anti-competitive manner, we would be delighted for other people to step up and do it. My concern is that some significant investments need to be made to get the job done in the future, and I don't see people stumbling all over themselves to step up.

EDMUND W. KITCH: Of course, I would think that if you were doing your job, you would price the opportunity to invest to achieve precisely the amount of investment needed and no more. If there were a surplus, the price of the opportunity should be raised.

MR. LeGrange: And that would be a wonderful world if we could find it.

James Shamas (Getty Oil): I am chairman of the board of Texoma pipeline, and it would be nice to trace Texoma's history so the people here who are good with statistics and everyone else could understand what we go through to promote a major project. We started in 1972 and surveyed twenty-five companies trying to promote an $89 million pipeline from the Gulf Coast to Cushing, Oklahoma. By 1974, we had eighteen companies that split into two groups. One went off and formed its own project, which was called Seaway.

We completed ours at $135 million, which was 40 percent over our estimate. One year later, the other group completed its project at $180 million. We both ran in the red the first quarter and had to consider asking the owners for a deficiency claim. Seaway may still be running in the red. How can we get people to go out and put up this much money? People were depending on Seadock, but did not come through. We've never denied access to anyone on the systems.

On an ICC rate of return, we earned about 5.8 percent. We sold it to our Getty Oil management on a 14 percent DCF rate of return. We did a three-year follow-up, and we have achieved 13.8 percent DCF rate of return. Today, my management would not accept Texoma pipeline because it only earned 13.8 percent DCF rate of return. Our new hurdle rate for the corporation is now 14 percent.

The pipelines do not yield the kinds of returns that are normally jumped at unless they are a necessity. Texoma was a necessity because of the oil embargo, and all the inland refiners needed a source of crude other than Texas, Oklahoma, Kansas and the other traditional places.

Our two groups couldn't agree on sizing. We started out to build a twenty-four-inch line; then we said we would build a twenty-six-inch; then we said if we built a thirty-inch, we would serve all the needs of everybody in that area. We built a thirty-inch line, and then Seaway came along and built a thirty-inch right next to it. So, now we have two thirty-inch lines; so I don't think we're undersized. It bothered me when people said we were undersizing things intentionally or earning excessive rates of return.

Professor Kitch: Well, it may be the only "you" that is doing it is the statistical "you."

MR. Shamas: Well, that's the thing. A lot of things can be proved with statistics.

MR. LEGRANGE: A couple of the problems with the DOT license provisions have come out in the discussion. After having gone through a fairly exhaustive list several years before, trying to encourage people to get into Seadock, we had settled to the final group. Then, as one of the license requirements, we reopened ownership. We went back out and beat the bushes again—with little success, I might add. The problem was how should we measure, at that point in time, the value picked up by someone coming in who had not been in during the prior two, three, or four years of work, and effort and expenditures? And, remember, by that time, Seadock had invested something like $10 or $12 million.

That was the problem: how do we calculate the price a new owner should pay? I think it is an area that needs some work because we will be faced with this question again if we are going to have this kind of open ownership in the future.

The other problem is, When should we expand, and who should make the decision to expand? Recognizing that the decision to expand is a commitment of somebody's hard-earned dollars, we felt like the one putting those dollars on the line ought to have something to say about what was being done in spending his money.

The original version of the license was that the U.S. government would decide when that expansion step would be taken and when we would spend our money.

We felt that that was an unreasonable risk to have to bear. We refused to bear it and negotiated pretty close to what probably could have been an adequate solution, but we never finally got there.

One of the other major uncertainties overhanging the whole project was the question of the operation of the facilities. There was the clear indication that the U.S. government would have a strong hand in writing the rules for operating the facility. Those rules would include what kind of ships would be used, when they would come, how they would approach, what kind of tankage we would have, how long we would hold goods free of charge, what charge would be made, the size of the hoses, the pumping rates, and on and on and on. Another critical one was, if we went into proration, what would our proration policy have to be, or what policy might be forced on us at that point in time. Again, there was another major risk, a major uncertainty.

There was a lot of concern along the way about the change that we sensed in looking through the project into the entire downstream in the United States. In essence, then, what happened is that the risk became too great for what was a minimum-return investment to start with. So, the project was shelved.

121

MICHAEL J. PIETTE: The critical question, as I see it, is, How competitive is the oil pipeline industry?

PROFESSOR KITCH: Spavins suggests the adjustment of some of the rates of return that are higher than those for oil pipelines, because of missing items from the balance sheet, in the form of goodwill. It occurs to me that an asset may be missing from the pipeline balance sheets, which is the know-how and the expertise of the organizing corporations.

To play that game with rates of return, it seems to me, requires equal inventiveness for all of the industries. And the guarantee of the debt, of course, is really a form of that, as is the parent companies' access to financial markets, and their know-how in the issuance of securities, and so on and so forth.

An issue that runs through the whole discussion is whether these organizer-owners have, in the beginning, a special advantage in organizing pipelines which could come from know-how, or from the fact that they have an important position in the field that is being exploited.

I am implicitly suggesting that some of the returns could be characterized as rents. And to make a general point about price theory, there are times at which the Department of Justice's position seems to me to fall into the fallacy—or to teeter on the brink of the fallacy—that the existence of a low-cost firm in a market makes that firm a monopoly. In fact, low-cost and high-cost firms can be in the same market. The competitive price will be the marginal cost of the high-cost firm. The low-cost firm will be very profitable, and it would not be correct to characterize that as a monopoly return.

Now, the other part of this is, there is also implicit, it seems to me, in the Justice Department's position an assumption that entry by non-petroleum entities, nonpresent participants in the oil industry, is an equally attractive substitute, which would solve some of the complexities of vertical integration. This assumes that the oil companies do not possess specialized know-how, planning capability, technology, or position that is of value to the pipeline company. If they do have such a special position, then the insistence on the use of outside organizers will raise the cost of the organized pipelines to consumers.

At times, it sounds as if you believe the special position exists and is bad; at other times, you sound as if it does not exist.

THOMAS C. SPAVINS: With respect to price theory, we have firms that are of different efficiencies. A distinction has to be made between the average efficiency of a firm and the marginal efficiency. If a firm is more efficient at the margin than the market price, it will keep expanding. Now, it may be argued that, in economics, it takes a long time to reach equilibrium, or it may be never reached. It is a question of whether the firms are high-cost or low-cost at the margin, including all the constraints. And that is different from the natural monopoly assertion, which is based on the question of how to describe, ex ante, the technologies.

To analyze a specific fact situation, if two separate pipelines were running next to each other, then there could be "competitive pricing" to get rents one way or the other.

In regulatory design, we have to worry that, if we let people keep rents today, that will affect their strategic decision on the size of the pipeline in the first place. There is a potential trade-off between concerns about today's minor inefficiency due to rents and what that does as feedback on the sizing decision.

MICHAEL E. CANES: I will put a question to George Hay: What would you accept as a proper empirical test of the undersizing hypothesis?

DR. HAY: I found very interesting the comment by LeGrange that, in the planning of Seadock, he felt the consent decree was an effective constraint. I would ask whether that suggests the incentive to undersize might exist in a situation like Seadock. The whole point of this discussion is that there is an incentive to undersize when an unregulated firm would charge a higher rate than it does under regulation. The evidence on Seadock, in fact, indicates that is what they would have done. Unregulated by the Consent Decree, they would have charged a higher rate. I would argue that is exactly the kind of situation which gives an incentive to undersize a pipeline to capture the rents downstream.

DR. CANES: Perhaps, I don't fully comprehend the answer, but, it seems to me there is always incentive to monopolize in any situation, any industry. However, one can design empirical tests to see whether or not monopoly is in fact occurring in a particular instance. It might be a profit test, it might be some other test. The spirit of my question is towards the latter. Even if an incentive may exist in the pipeline industry, what constitutes an empirical test of whether, in fact, undersizing has occurred?

DR. HAY: The ideal empirical test would be a comparison of the value of the oil at one end of the pipeline—less the market value of the oil at the other end of the pipeline compared with the actual economic cost of transportation. Where that exceeds the actual cost of transportation, then one can infer that the size of the pipeline—

PROFESSOR KITCH: But that will exist in every case where some of the oil is moving to the destination market by an inferior technology.

DAVID J. TEECE: It is a quasi-rent.

PROFESSOR KITCH: That doesn't show the pipeline is undersized since the sizing decision was made when it was built. It may simply be proof of an error.

DR. HAY: Okay. I accept that.

PROFESSOR KITCH: And if you found a dispersion in a set of markets, where some pipelines seem to be oversized and some seem under, you would just be showing dispersion of errors around some mean.

DR. HAY: Suppose I refine my testimony to say, at the time the pipeline is built, we observe a difference between the market value of the oil at either end compared with the cost of building the pipeline?

PROFESSOR TEECE: Maybe I am missing something. Where would lie the incentive for anyone to ever build a pipeline if there was not such a discrepancy? If the value of the oil that comes out of the other end is exactly the same as the value of what goes in, plus transportation costs—

DR. HAY: Well, transportation costs including a competitive rate of return.

PROFESSOR TEECE: Yes, but where lies an incentive for anyone to ever build a transportation facility? It has to be at least epsilon greater than that, right?

DR. HAY: Okay. I will allow you epsilon. Then the question is how big to build it.

PROFESSOR TEECE: Yes, that is right.

MR. SHAMAS: I have had twenty-three years with four different pipeline companies, some of them majors and some of them minors. I am an engineer by training. I have never been associated with any project that undersized a facility. I had a chance to be with Colonial, originally, and we were all convinced that we could never fill the thirty-six-inch line. It was the largest line we had ever thought about, and we thought we could never fill it. It was going to replace tankers that hauled oil around at fifty cents a barrel by doing it for thirty cents.

A pipeline is oversized, not undersized, because of economics. It is built as large as the economics will support.

When a pipeliner hears that facilities have been undersized, it doesn't make any sense. No one does that. It would be like building a railroad and stopping halfway there, or something. We build what our economics will support.

What usually happens is that we fail to foresee all the volumes that come forward. On the Texoma system, for example, we didn't foresee the Arab embargo. If we had been smart, I guess we could have foreseen that, but we didn't.

So, we don't understand accusations of undersizing. We just finished looping an eight-inch line that we built in 1967 with a twenty-inch line. We didn't foresee that Texas crude, and Oklahoma crude, and Kansas crude would run out. Now, we are a lot smarter.

DONALD KAPLAN: A wise businessman puts in a piece of pipe at the beginning that plans for the future. That is what was done on TAPS. Prudhoe Bay cannot sustain 2-million-barrel-a-day production, but it has a 2-million-barrel-a-day piece of pipe in TAPS.

MR. SHAMAS: I made a junior engineer's evaluation of the first Alaskan pipeline in about 1959. It was going to be a twenty-four-inch line at $750 million, and now we have a forty-eight-inch line at $8 billion. Nobody wanted North Slope oil, and we could only justify a twenty-four-inch line at $750 million. Everyone thought we were crazy because that sounded like such a huge sum of money. Today, it doesn't sound like a lot. We were not intentionally undersizing it. We based it on the economics that would support the cost of the system. We do not size for the first year; we size for the tenth-year volumes, if we can forecast them accurately. And we figure it has at least a twenty-year life.

PROFESSOR KITCH: Now that the economists have become so important in the formulation of antitrust policy, I am afraid the inability to find memos recording an intent to undersize will not be the end of the case. Being a natural monopoly is like some diseases that cause sickness only a doctor can detect. Being a natural monopoly is a condition that can be diagnosed only by an economist.

DR. CANES: I would like to follow-up one more time on the test George Hay prescribes. If a line were undersized in the sense described,

125

it must have excess demand—that is to say, it must be filled up; there must be prorationing of that line. Hence, at least as I understood the remark, it must be the case that the existence of systematic prorationing would be an empirical test of whether or not there has been systematic undersizing.

PROFESSOR MITCHELL: I think you would want to qualify that some. It is a necessary condition, is it not? I mean, if it is prorationed, it doesn't prove that undersizing has occurred?

DR. CANES: Yes, I accept that.

PROFESSOR KITCH: And the other critical problem, of course, is to define "capacity": that is, it might be that the economic capacity of the line would leave vacancies—just as airlines have empty seats, with no feasible way to fill them.

There would have to be some definition of capacity level in order to know whether it has been reached. It has to be something other than, there is oil in the line at maximum pressure, every hour of every day, because there will be scheduling problems, and other difficulties.

PROFESSOR KITCH: And that greatly complicates the empirical inquiry because it may indeed, have been possible to slip some oil through at a particular moment, but that doesn't tell us whether it would have made economic sense to do so.

MR. LIVINGSTON: There is a further qualification in regard to capacity— capacity between which two points? There may be inadequate capacity at one segment of the line and excess capacity everywhere else.

MR. SPAVINS: Assume there is a pipeline that has unused capacity in the sense that it is continuously willing to offer transportation, and can pump more petroleum through the line. This fact is a valid test of throughput restrictions and vertical restrictions and is only valid if access is equal in the sense that there are no differentials in the access cost of getting into the pipeline.

MR. LIVINGSTON: Let's understand that a pipeline deliberately operating at less than capacity, turns business away knowing that the incremental cost of putting through that additional volume is very low, and the profits that the owners could earn on that volume are very high. Even if the owners had ulterior motives, don't forget that the pipeline is operated by a management that is responsible for making a profit. Somehow or other, these owners would have to instruct the managers of the pipeline to turn away business which could be very profitable.

MR. SPAVINS: I would just like to note that Dr. Canes framed the question which I interpreted as an attempt to develop a set of necessary and/or sufficient conditions to indicate pipeline undersizing, and my comment was limited just to that question.

PROFESSOR KITCH: One of the ambiguities here has to do with the assumed pricing policy of the "natural monopoly" pipeline. If we assume that tariff rate regulation is missing—and to some extent, it seems to me, the findings of Spavins' paper suggest that it is not really constraining because of the high realized profit rates and also the inactivity of the Commission—then it would seem that the direct way for the natural monopolist to optimize the value of the position would be to set prices which reflect the natural monopoly price. That done, we will not observe any queues for the right to get on the pipeline, since the price has been set at a level which eliminates the queues at the under-sized level.

There is a certain tension between some of the findings in the Spavins paper and the other aspects of the Justice Department position, which, as I understand it, assumes a constraint on the natural monopo-list's ability to price the transportation tariff, so that he is held back. But the interest in divestiture seems to become important only if we assume the pricing constraint.

DR. HAY: You are perfectly right, but there was never any attempt to disguise that tension. It was fully laid out in Flexner's paper this morning.

In fact, the chronology of our involvement in rate regulation is exactly that. In thinking about divestiture, we became concerned that if, in fact, the profits are captured in the rates, then a pipeline divested at a price which capitalizes all the profits already in it wouldn't solve any problem.

PROFESSOR KITCH: That leads to the subject of the next session, How can we improve the situation? I certainly don't know any way of improving the pricing behavior of a true natural monopoly.

MR. SHAMAS: The fact is, today we are reducing tariffs on brand new systems because of the limitations of the Consent Decree. In all these systems, there are more outside shippers than there are owner-shippers. Therefore, those people are benefiting by the lower rates directly, so I don't agree that the existing rules do not put a constraint upon earnings.

PROFESSOR KITCH: I understand, but then the Justice Department position might suggest that one way to improve the situation is to eliminate that constraint, which then eliminates the incentives to pass these disturbances up and down stream.

MR. SHAMAS: I don't know of a shipper ever being refused in any of the sixteen systems I have been associated with. We want their business. We have never turned away a customer. Even against the 7 percent limitation there is nothing we can do except to increase our investment or lower our tariff.

PROFESSOR KITCH: From a social welfare point of view, allowing collection of monopoly rents might be preferable.

MR. SPAVINS: That depends upon what happens to the collected monopoly rents?

PROFESSOR KITCH: No, this is not a case where we have to worry about waste due to competition in the way of monopoly rents because we defined this as a natural monopoly case. We are not going to get rid of the natural monopoly. It exists. It is an unfortunate state of affairs.

MR. SPAVINS: First, if we accept the premise that competition for the right to be first—competition to have the right to have the ex post natural monopoly—is very limited, you might be right.

PROFESSOR KITCH: No, there will be competition for the natural monopoly. How are we going to get rid of it? There is a natural monopoly there.

MR. SPAVINS: There is a general intellectual question with respect to any monopoly power anywhere whether we aren't better off just letting it go on, rather than trying to do anything about it.

PROFESSOR KITCH: Yes.

MR. SPAVINS: I would grant that, as an honorable, debatable, intellectual proposition.

MR. KAPLAN: What we want to do is achieve a method which addresses reality, which addresses some of the same problems LeGrange and some others have raised, which does not involve any unnecessary monopoly rents, except those generated through the normal competitive process.

We are always interested in learning what reality is. I know there are a lot of representatives of pipeline companies here, and of their parent oil companies. And if LeGrange thinks that we don't know what reality is, then we would like him to tell us. That is the best way for us to learn, and to ground our proposals in a workable solution to the problems we have identified.

MR. LEGRANGE: That is a good offer. We will take you up on it.

MR. SPAVINS: The failure of the ICC to allow fair but different treatment of people willing to ship under different circumstances may very well have been a problem. This issue can be merged with the issue of discriminatory access.

We might go further. If the nominal tariff is too high and does not reflect the cost of capital for a pipeline, then there is the issue of discrimination between those who own and those who don't. This is the ancient legal basis for the Elkin's Act Consent Decree. Charging the same tariff to owners and nonowners where the owners are getting an excessive return is the equivalent of a rebate.

PROFESSOR KITCH: But I take it the access question is generated by the assumption that the regulation requires a uniform price. One way for the owners to favor themselves is to give less preferred-access positions to nonowners. Is that the mechanism which generates this concern?

MR. KAPLAN: A distinction should also be made between the reasonableness of general rate levels and a fair compensation for being the initial risk-bearer. At the same time, it is not inconsistent to propose lowering of general rate levels, while, at the same time, indicating that a fair compensation for bearing those starting-up risks, those organization risks, would be appropriate.

MR. LEGRANGE: As I recall, one of the basic premises in that Deep Water Report was that this was a natural monopoly and would have high profits. Therefore, it was necessary to give open access to those lucrative profits, so that anyone who wanted to ship could get the benefit of all those good profits he was contributing to. That still seems to be an inconsistency, but maybe it will be clarified in the next session of this seminar.

MR. SPAVINS: Fair enough.

MR. LEGRANGE: I am not against the new, two-tier tariff. I think it is fine that we are finally starting to recognize that some have different obligations from others and ought to be rewarded for them. If we can move to a new world where those distinctions can be made, fine. We just have to recognize that the distinction has to be measurable, and it has to be of a magnitude to justify the additional risks being borne by the investor taking on that obligation.

MR. SHAMAS: No one is willing to come in and take a share of our losses when the thing fails. Our competitors may have been running in the red for three years. I would think someone ought to step up and take part of the losses. That is open too, but no one has been volunteering to do that. Everyone wants to come in only when an operation is successful. That is my answer to open ownership.

MR. KAPLAN: The Department of Justice made a recent filing in the Texas Deep Water Port Application for permission to charge a two-tier tariff. We found that it can be lawful if it is a reasonable differentiation.

HOWARD O'LEARY: I would like to ask a question in the nature of clarification, either of Tom Spavins or Don Kaplan. What facts will satisfy the Department of Justice that undersizing does not exist? I inferred from the colloquy earlier that perhaps it was excess capacity and the existence of nonowner-shippers using the line. Am I correct in that? If I am not correct, what proof will satisfy the Department?

MR. KAPLAN: I cannot say without the kind of study that we made in the Deep Water Port context or the Sohio Pipeline context. In the Sohio Pipeline context, we undertook an analysis which resulted in our conclusion that there was not a problem with the Sohio Pipeline.

All I can say here is, if those facts can somehow be replicated, fact for fact, number for number, that would probably be safe. As the law and the economic analysis in this area develop, other situations may be identified so that there will be guidelines under which businessmen can plan.

PROFESSOR KITCH: Is the essence that the burden of persuasion is on the industry?

MR. KAPLAN: I don't want to say that. It depends on the forum we are in. If we bring a pipeline case, we will have the burden of proof, of course. If something results from the FTC proceeding, there might be a world in which, given certain initial facts, the burden of proof will be on the pipeline company to show it does not come under those facts, as the Department has proposed in its conglomerate merger statute proposal.

MR. LeGRANGE: What we may have the opportunity to do is to sit down and compare notes on that pipeline and some other ventures you were apparently concerned about. Perhaps we can convince you or you can convince us that the circumstances are not entirely different between the Sohio Pipeline dumping a million barrels a day into West Texas and a Deep Water Port dumping a million to a million and a half barrels a day into Southeast Texas. We can see if we can come to the same conclusion. I don't know what position you would have taken on the Sohio situation two years ago, and you probably couldn't tell me, but you did take a position on something else that seems somewhat similar to it.

Since that time, we have had this new thought about two-tier tariffs. It was not around for us to discuss a couple of years ago. We ought to throw all that into the hopper now and see if we are getting closer to a meeting of minds about how you would look at these facilities.

PROFESSOR KITCH: It certainly would help the people who have to plan pipelines to know what the rules are.

PART THREE

ALTERNATIVES TO THE PRESENT SYSTEM

Appraising Alternatives to Regulation for Natural Monopolies

Lucinda M. Lewis and Robert J. Reynolds

Introduction

This paper considers the following problem: given a "natural monopoly" that is not prevented either by the existence of close substitutes or by the threat of entry from charging a price that is excessive for efficient resource allocation, is there an alternative to traditional rate regulation that can maintain the advantages of single-firm production while minimizing monopolistic exploitation? The authors are interested in considering alternatives to traditional rate regulation because of theoretical and empirical evidence of the substantial costs and limited effectiveness, in part due to strategic manipulation by the regulated companies, of traditional regulation.

The need for examination of alternatives seems to be particularly relevant to oil pipelines, which are natural monopolies and possess market power, where the concern is either that rate regulation will be impotent in controlling monopoly profits or that vertical integration of oil companies into pipelines will succeed in evading even "perfect regulation." Further, there are a wide variety of other sources of regulatory failure (for example, Averch-Johnson effects; distortions of rate structure) for which the analysis contained herein may be relevant.

The alternative that we wish to consider in this paper involves joint-venture operation of the natural monopoly facility; the joint-venture owners are the customers of the facility[1]; the operation of the joint

Respectively University of Pennsylvania and U.S. Department of Justice. The views expressed herein do not necessarily represent those of the Department of Justice. This paper has significantly benefited from the authors' discussions with Scott Harvey, George Hay, and Tom Spavins and from the comments of Wally Hendricks and Andy Postlewaite on a related paper. The responsibility for remaining errors is the authors'.

[1] Hence the alternative approaches, if successful in reducing monopolistic exploitation, are directly applicable to situations in which vertical integration allows evasion of regulation, as has been suggested in oil pipelines; given the costs of regulation, they may be an attractive alternative to traditional regulation more generally.

venture is subject to rules (referred to as "joint-venture rules") whose basic aim is to induce competition among the owner-users in a manner that reduces the natural monopolist's exercise of market power. We wish to explore whether there is a set of rules that will eliminate or at least diminish the degree of monopolistic exploitation and to discuss the characteristics of any such rules.

The idea behind the joint-venture rules approach can be thought of as follows. Suppose we start with the natural monopolist charging the (profit-maximizing) monopoly tariff and producing the monopoly output. If the natural monopolist facility is owned by multiple owners, and if the owners individually have both the incentive and the ability to cause the facility to expand output, there will be a welfare improvement.

In this situation, the owners as a group have no such incentive. Individually, any owner's incentive is predicated on the following type of calculation: if an owner firm elects to take an additional unit of output (causing total output to expand by one unit), it will receive as additional revenue the downstream product price, less its share of the fall in revenues resulting from the reduction in downstream price (in other words, a standard marginal revenue calculation). The cost of the additional output is the tariff less the owner's share of the increase in joint facility profits. If there is only a single owner firm, of course, there is no incentive to expand. If there are several owners, while each owner will be less affected by the fall in downstream price, they will each perceive the effective cost of the additional output to be approximately the monopoly tariff, since only a small portion of the increase in joint facility profits accrues to an individual owner. The result is that there is only a small incentive to increase output beyond the monopoly level; even this small incentive can be negated if the tariff slightly exceeds the monopoly level.

A greater incentive to increase output must be provided if the joint-venture rules approach is to improve welfare significantly. Such an incentive can be created by lowering the individual owner's effective cost of additional output by providing the expanding firm with an increased share of joint-venture profits. Thus, if joint ventures are to enhance social welfare significantly, two necessary ingredients are: (1) several owners and (2) readjustment of the share in ownership as output shares vary. In addition since an increase in individual firm output without a change in group output does not improve welfare, an additional necessary ingredient is some mechanism to translate increased firm demand into larger facility output.

The joint-venture rule approach is to provide rules concerning the operation of the joint venture incorporating these general ingredients.

To determine the specific form the rules must take, we need to examine some additional factors. At various times the joint venture must make decisions on certain variables such as price, output, and distribution of output. Over a longer run, decisions must be made with respect to the initial size and location of the facility and with respect to the size and timing of subsequent expansion. The joint venture must also decide on the initial distribution of any profit from the joint venture. As much as possible such decisions ought to be made by the firms, without public intervention. Yet such choices must also be made strategically in the sense of serving the group interest of maximizing joint profit. Thus, joint-venture rules that would apparently induce competition among the owners but are sensitive to strategic manipulation will be unacceptable.

Thus the aim is to develop joint-venture rules that leave individual decisions to the group of firms but are sufficiently insensitive to strategic manipulation so that the desired welfare improvement will ensue. In effect, we are looking for "constitutional" rules that will permit the individual profit-seeking tendencies of the members to undermine the joint maximal solution and bring about a competitive result.

The requirement that rules be insensitive to strategic choice sharply delimits the form that such rules must take if they are to be effective in dissipating monopoly power. By way of illustration, this paper will discuss the efficacy of a particular set of rules, those put forth in the Department of Justice's Deepwater Ports Report.[2]

The Competitive Rules

A particular form of joint-venture rules is that proposed by the Department of Justice in its Deepwater Ports Report. These competitive rules were as follows:

1. Deepwater ports must provide open and nondiscriminatory access to all shippers, owner and nonowner alike.
2. Any deepwater port owner or shipper providing adequate throughput guarantees at the standard tariff can unilaterally request and obtain expansion of capacity.
3. Deepwater ports must provide open ownership to all shippers at a price equivalent to replacement cost less economic depreciation.

---

[2] Report of the Attorney General Pursuant to Section 7 of the Deepwater Port Act of 1974 on the application of LOOP, Inc. and Seadock, Inc. for Deepwater Port Licenses, November 5, 1976 (Deepwater Port Report).

4. The ownership shares of the deepwater ports' owners must be revised frequently (annually) so that each owner's share equals his share of average throughput.[3]

Why do the rules have this form? Could alternative forms of the rules also be effective in dissipating market power? What is the effect of the rules, given the strategic choices of the group of firms? These are the questions this section will address partially.

Consider first rules 3 and 4. The preceding analysis indicated that the incentive for an individual firm to expand output depended on its effective marginal cost of output expansion. It also showed that this incentive would be small if the tariff were set at the monopoly level, since each firm's effective marginal cost would approximate that of the monopoly tariff.

This cost is lowered, however, insofar as output expansion increases the expanding firm's share of joint-venture profits. Thus, in situations in which the tariff substantially exceeds cost, a share readjustment rule is required to provide the correct incentive. In itself this does not lead to rule 4; what does lead there is the consideration of strategic choices by the joint venturers. What rule 4 implies is that the effective cost of firm output expansion is independent of the tariff chosen by the joint venture. Since each firm's share of the joint venture's profits (approximately) equals its share of throughput, then the tariff is, in effect, paid to the firm itself; what matters, therefore, in its output decision making is not the tariff, but pipeline marginal cost. Thus the effectiveness of rule 4 cannot be strategically weakened by the joint venture's choice of tariff; if there were less frequent share readjustments, or if share readjustment were not proportional to firm output, then the effectiveness of share readjustment in dissipating monopoly power could be vitiated by the joint venture's strategic choice of a tariff level.[4]

Similarly, the effectiveness of share readjustment in inducing output expansion could be negated if the price set for the new shares capitalized the venture's monopoly profits; in that case, while share readjustment would increase the ownership of an expanding firm, it would not lower the effective cost of the expansion below that of the monopoly tariff. The role of rule 3 is, in part, to ensure that the effectiveness of share readjustment is not thwarted by the joint venture strategically manipulating the share price.[5] It is conceivable that there may be other ways of

---

[3] Ibid., p. 106.

[4] It can also be shown that the effectiveness of rule 4 is not subject to vitiation by the joint venture's assignment of original ownership shares; this is not true of other share readjustment rules.

[5] Rule 3 has other implications, which are discussed below.

achieving this protection; a provision that might work would be one that required that shares be freely transferable.

Notice that in situations in which tariff setting is not a discretionary variable for the joint venture, then strategic considerations do not require that share readjustment rules have the form of rule 4. Further, where the tariff is constrained to equal cost, then share readjustment and the associated task of defining a share transfer price are unnecessary.

The analysis in the first portion of this paper suggested two other necessary ingredients for joint-venture rules: (1) a significant number of venturers; and (2) a means for translating increases in individual firm output demands into a larger joint-venture output. These requirements are addressed by rules 1, 2, and 3.

Consider the expansion of the facility; the group of firms has an incentive to prevent expansion of the facility beyond the monopoly level. They may be able to achieve this by reducing the number of firms that are members of the venture, by denying ownership, by making physical access more difficult and less desirable, or by ensuring that other firms will expand at the same time as the individual firm wishes to expand. They may also make it difficult for an expansion in an individual firm's desired output to translate into a larger joint facility output (by, for example, requiring a large majority of owners to approve an expansion). Rules 1-3 are clearly aimed at preventing these strategic choices.

Conclusions

This paper has suggested an alternative approach to the problem of controlling natural monopolies; this alternative approach is based on the development of "joint-venture" rules in which the incentive of the joint-venture owners is to compete among themselves and thus dissipate monopoly power. This approach may be thought of as an alternative to traditional regulation or as complementary to regulation in situations in which vertical integration leads to regulatory evasion.[6]

In order for this approach to be a successful alternative to regulation, three ingredients are necessary: (1) several owners of the natural monopoly facility; (2) the readjustment of firms' shares in ownership as output share varies; and (3) a mechanism to translate increases in firm "demands" into larger facility output. The particular form that joint-venture rules must take to incorporate these ingredients and achieve significant reduction in market power is determined by the fact that a number of variables are left under the control of the owners. Such con-

---

[6] Note that this analysis has implications for the analysis of joint ventures more generally.

trol will presumably be exercised strategically—that is, to maintain insofar as possible monopoly power and profits. Hence, the form of the rules is affected by *which* variables are available to the firm.

We have examined these considerations in terms of a particular set of rules, the "competitive rules" introduced by the Department of Justice in its Deepwater Ports Report. That report advocated joint-venture rules in a "natural monopoly" situation; in a modified form they became part of the rules under which one Deepwater Port operates.[7] We have seen how the "Deepwater Port" rules attempt to encourage "competitive" use of the facility and deal with the problem of strategic manipulation by the ventures.

[7] In another modified version these became part of the statutory requirements for outer continental shelf pipelines.

Pipelines and Public Policy

Michael E. Canes and Donald A. Norman

The operations of oil pipeline companies are subject to numerous constraints. In addition to those imposed by competition, there are regulatory constraints that, among other things, are designed to ensure the "reasonableness" of tariff rates, limit dividends paid out by many pipeline companies, and ensure open access to pipelines for all shippers on a nondiscriminatory basis. Finally, for many years the oil pipeline market has been subject to antitrust scrutiny.[1]

Recently the U.S. Department of Justice has proposed alternative policies to replace or repair existing constraints on pipeline operations. One proposal would divest pipelines from other segments of the oil industry. An alternative proposal would incorporate a set of "competitive rules" within the regulatory framework. Essentially, the second proposal represents a "reform" in existing constraints, whereas divestiture would impose a structural constraint. A major rationale for both proposals is that pipelines built by integrated oil companies are inefficiently sized and that this sizing allows oil companies to circumvent regulatory constraints on pipeline profits.

In this paper, we examine the merits of these two proposed alternatives to the present system of pipeline regulation. Because pipeline sizing by integrated oil companies is a linchpin of both policy alternatives, we describe the Justice Department's reasoning concerning such sizing and then critique the assumptions, logic, and empirical implications of this reasoning. Next we discuss reasons why pipelines are vertically integrated. We contend that, to understand the consequences of structural changes, one must understand why integration has occurred in the first place. We then criticize the notion that either prospective or retrospective divestiture of oil pipelines would benefit U.S. oil consumers. The criticisms are based on normative precepts that antitrust policy should be based on economic analysis and that

---

[1] For a history of such scrutiny, see Arthur M. Johnson, *Petroleum Pipelines and Public Policy, 1906–1959* (Cambridge, Mass.: Harvard University Press, 1967).

consumer welfare is the ultimate aim of antitrust.[2] Finally, we examine problems associated with the competitive rules for pipelines that have been proposed by the Justice Department. We conclude that the peculiar institutional constraints implied by such rules are either inconsistent with one another or would require more complex pipeline regulation than has been attempted heretofore.

Background

Because oil pipelines are subject to large economies of scale, the oil pipeline market is said to be characterized by "natural monopoly"—that is, the most economical market alternative over the entire range of oil transportation demand is a single pipeline facility. Since these conditions are inherent in the technology of pipelines,[3] and since substantial savings in transportation costs are associated with pipelines of larger diameter,[4] public policy has allowed firms to take advantage of the economies of scale but has imposed regulations on both rates and common carriers to avoid monopoly pricing and to provide equal access to all prospective shippers. Thus, although firms have achieved large economies of scale with pipelines, these regulatory constraints are intended to force pipeline firms to act about as they would have if the pipeline market were characterized by a competitive structure.[5]

---

[2] These precepts are also advanced in Robert H. Bork, *The Antitrust Paradox* (New York: Basic Books, 1978), p. 7.

[3] Essentially, the cross-sectional area of a pipeline, which is related to its output, increases faster than its diameter, which is related to its cost.

[4] See, for example, Leslie Cookenboo, Jr., *Crude Oil Pipelines* (Cambridge, Mass.: Harvard University Press, 1955); or John G. McLean and Robert W. Haigh, *The Growth of Integrated Oil Companies* (Norwood, Mass.: Plimpton Press, 1954), pp. 185–86. Also, Charles Spahr has provided indexes that relate pipeline cost and throughput capacity to diameter, as shown:

Diameter (inches)	Pipe Cost Index	Capacity Index
4	1.0	1
8	1.9	6
12	2.4	15
16	3.5	28
20	4.3	48

Testimony of Charles E. Spahr, chairman, Standard Oil of Ohio, before the U.S. Congress, House of Representatives, Subcommittee on Monopolies and Commercial Law of the Judiciary Committee, September 10, 1975.

[5] The extent to which regulation, as opposed to competition and vertical integration, has constrained the behavior of individual pipelines is open to question, however. Competitive constraints on pipeline behavior and the effects of vertical integration are discussed below.

Recently, however, the Department of Justice has advanced a different analysis of the oil pipeline market. Noting that the great majority of U.S. pipelines are owned by vertically integrated oil companies, the department has reasoned that, even with perfect rate regulation for pipelines, integrated companies can earn monopoly profits in other, adjacent stages of the oil industry by sizing pipelines so as to achieve the monopoly throughput rate. These profits are said to arise because "undersizing" a pipeline forces some oil to travel by a more expensive transportation mode, and because the marginal cost of this other transportation mode must be reflected in the price of oil at an adjacent industry stage.[6] The reasoning is summarized in recent testimony released by the Justice Department.

As we see it, the ability to avoid regulation is rooted in the ability of the shipper-owner to limit pipeline throughput below the level an independent would size or operate the line. If the vertically integrated pipeline owner is a significant seller in the downstream market and the pipeline has market power downstream, the pipeline's throughput may be such as to ensure that at least some not insubstantial amount of supply arrives by the more expensive transport modes. Where this occurs, the downstream market price would likely reflect the cost of delivery by the next least expensive transport mode, allowing the pipeline's shippers to pick up the difference in transportation costs.[7]

In the recent past, the Department of Justice has applied this reasoning not only to oil pipelines but also to natural gas pipelines, deepwater ports, and railroad transportation of coal.[8] Further, it has used the reasoning to advance strong policy prescriptions concerning all of these markets.[9] In addition, the department's reasoning and some of

---

[6] Henceforth, the term "undersizing" will be taken to mean sizing a pipeline at the monopoly output rate.

[7] Proposed testimony of John H. Shenefield, assistant attorney general, Antitrust Division, before the U.S. Congress, Senate, Antitrust and Monopoly Subcommittee of the Judiciary Committee, June 28, 1978.

[8] See "Report of the Attorney General Pursuant to Section 19 of the Alaska Natural Gas Transportation Act," July 1977 (hereafter, the Alaska Gas Report); "Report of the Attorney General on the Applications of LOOP and Seadock for Deepwater Port Licenses," November 5, 1976 (hereafter, the Deepwater Port Report); and "Competition in the Coal Industry," May 1978 (hereafter, the Coal Industry Report).

[9] As discussed below, the department has endorsed oil pipeline divestiture; proposed that no gas producers be allowed to own shares of the Alaskan gas pipeline; proposed that oil companies owning deepwater ports follow competitive rules; and proposed that no railroad be allowed to secure new federal coal leases. The

its policy prescriptions recently have been adopted by the staffs of both the Federal Trade Commission and the Antitrust and Monopoly Subcommittee of the Senate Judiciary Committee.[10]

The Justice Department's Theory of Pipeline Undersizing

In analyzing petroleum pipelines, the Justice Department treats pipelines as an industry stage separate from crude oil production, refining, and marketing. The pipeline stage is assumed to be subject to significant economies of scale. According to the department,

> This phenomenon of declining unit costs has been observed to take place over the entire existing range of technologically feasible pipeline sizes. . . . As a result, in almost all circumstances, one properly sized pipeline will be more efficient than two or more in satisfying the available transportation demand between any two given points.[11]

Other industry stages are assumed to be workably competitive.[12] Indeed, according to the department, if one of these other stages were noncompetitive, integrated firms would have no motivation to undersize pipelines. For example,

> If present Alaskan producers have market power in that field not derived from the pipeline . . . [t]here is no need for them to tinker with the pipeline stage in order to exploit their market power (except, ironically perhaps, to encourage

---

department also used the reasoning to evaluate whether the Standard Oil Company of Ohio should be allowed to construct and operate an oil pipeline from the West to the Southwest and decided that it should. See "Report of the Antitrust Division on the Competitive Implications of the Ownership and Operation by Standard Oil Company of Ohio of a Long Beach, California–Midland, Texas, Crude Oil Pipeline," June 1978 (hereafter, the Sohio Report).

[10] See proposed testimony of Alfred F. Dougherty, Jr., director, Bureau of Competition of the Federal Trade Commission, before the U.S. Congress, Senate, Antitrust and Monopoly Subcommittee of the Judiciary Committee, July 10, 1978; and "Oil Company Ownership of Pipelines," Staff Report of the Senate Judiciary Antitrust and Monopoly Subcommittee, 1978 (hereafter, the Senate Staff Report).

[11] Proposed testimony of John H. Shenefield, p. 6.

[12] In its Alaska Gas Report, the department concentrates its attention on pipelining and the upstream production market. In carrying out its analysis of integrated firm behavior, the department states: "to focus our analysis on the pipeline, we will abstract from any anticompetitive problems in production by assuming that entry can occur so as to render workably competitive that stage of the industry" (pp. 30–31). Similarly, in its Coal Industry Report, the department finds coal production workably competitive. From these and from its analysis of integrated firm behavior if other industry stages are not competitive, it is clear that the department assumes that competition exists in all stages but transportation.

the lowest-cost line available to insure the delivered cost advantage margin of their gas).[13]

The output of a single pipeline is assumed to affect price in an adjacent stage market. For a downstream market, such market power arises if both the market demand and the supply of oil from alternative sources are less than perfectly elastic; for an upstream market, it arises if both the supply of oil and the supply of alternative transportation services are less than perfectly elastic.

Where pipelines are owned jointly by more than one integrated company, it is assumed that the companies collude to achieve monopoly pipeline size. The cost of this collusion is assumed to be independent of the number of integrated companies that have ownership in the pipeline or their share distribution, and also of the extent to which the market positions of the partners in the venture upstream and downstream of the pipeline may differ.[14]

The department assumes that regulation constrains price to equal average cost, so that economic profits are held to zero in the pipeline stage: "we will assume that [pipeline] tariff policy is ideal, i.e., the tariff is designed to permit the pipeline to earn no more than a fair and reasonable return to its owners' and creditors' investment."[15]

A graphic representation of the Justice Department's characterization of the pipeline stage is portrayed in Figure 1. In a market unconstrained by rate regulation, a profit-maximizing pipeline would simply charge the monopoly rate. Such an equilibrium is shown in Figure 1 as Q_m, P_m. Monopoly profits are equal to the area P_mABC. The size (scale) of the pipeline that minimizes the average cost of throughput Q_m has a short-run average cost curve, represented by AC_1.

If the demand for service were just met under a constraint requiring that the long-run average cost be equal to price, then the regulated throughput would be Q_r and the short-run average cost AC_r. The department, however, proposes that integrated firms directly restrict supply by constructing pipelines sized at the monopoly rate Q_m:

[13] Alaska Gas Report, pp. 41–42.

[14] This assumption is not explicit in any of the Justice Department reports. Nevertheless, since the department acknowledges the widespread use of joint ventures in pipelines but does not relate its undersizing allegations to the number of companies in such ventures or to their relative market positions, the department must be assuming that total collusion costs are invariant to the number of the colluders or their make-up.

[15] Alaska Gas Report, pp. 32–33. The department disputes whether oil pipeline rate regulation has in fact been ideal but, for purposes of the undersizing theory, assumes that it has.

FIGURE 1

PIPELINE MARKET

Pipeline Tariff

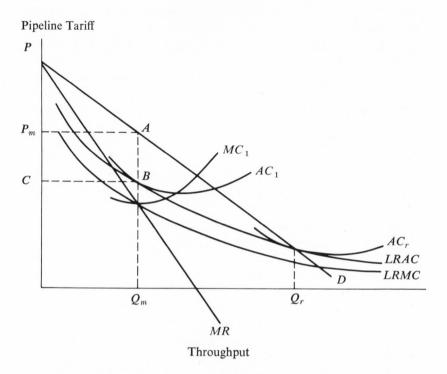

Throughput

Having been denied the opportunity to raise the tariff to a monopoly level (with a consequent reduction in the amount shipped over the pipeline), the profit-maximizing strategy would be to restrict quantity directly by limiting the throughput of the pipeline.[16]

By this description, because of the regulatory constraint that price must equal average cost, price is set at C, which is below that necessary to clear the market. Thus, there is a permanent, long-run shortage of capacity in the pipeline market.

A shortage of capacity at the rate charged implies nonprice rationing of pipeline capacity. Under common carrier regulation, oil pipelines are legally required to grant nondiscriminatory access to all shippers.[17]

---

[16] Ibid., p. 36.

[17] Interstate Commerce Act, sections 1(4), 2, and 3(1); 49 U.S.C., sections 1(4), 2, and 3(1).

The Justice Department is skeptical that nondiscriminatory access actually occurs but argues that, even under perfect nondiscrimination, the undersizing problem exists:

> "If prorationing is then fair and equitable, has the upstream profit vertical integration problem been solved? The answer is emphatically in the negative." [18]

> ". . . at the most fundamental level, effective prorationing would not alter basic throughput restriction incentives. Prorationing would merely require the monopolist to dilute its monopoly returns by sharing it with others." [19]

According to this, although equal access to all pipeline users may reduce the presumed rewards to pipeline owners from undersizing, positive rewards still remain. Thus, nondiscriminatory access is consistent with pipeline undersizing.

Even though pipeline capacity is restricted to a monopoly rate, the Justice Department assumes that no second, larger pipeline will be built. Thus, by assumption, no direct competitive constraints are placed on pipeline sizing.

Finally, integrated company profits in adjacent industry stages are alleged to rise under the pipeline undersizing strategy. According to the Justice Department,

> By restricting the throughput or capacity of a pipeline, its owners can force some oil to flow to the final market by more costly, less efficient means of transportation. As the price of crude or product downstream is generally set by the cost of the incremental barrel, throughput restriction has the effect of raising the downstream profits. Vertically integrated firms can thus capture at other unregulated stages some of the potential monopoly profits on pipeline transportation denied by regulation.[20]

Assumptions Underlying the Justice Department's Undersizing Theory

A number of the Justice Department's assumptions concerning the petroleum industry are necessary to its reasoning on pipeline under-

---

[18] Alaska Gas Report, p. 40.

[19] Proposed testimony of John H. Shenefield, p. 27.

[20] Ex Parte No. 308, Valuation of Common Carrier Pipelines, Statement of Views and Arguments by the U.S. Department of Justice, footnote 44, p. 22 (hearings now under FERC, RM 78-2). The department's testimony, however, offers no evidence of excessive downstream refining or marketing profits among integrated firms.

sizing. In this section, brief comments are made on several of these assumptions.

Economies of scale undoubtedly exist in the pipeline market. But those very economies raise the question whether an undersized line would be viable in a competitive market for pipelines. Conceptually, at least, any line would be constrained by the possible entry of a larger, lower cost line. Unlike other modes of transportation, petroleum pipelines do not require certificates of public convenience and necessity from a regulatory agency, and so competitive entry is not legally barred.[21] A possible counter is that existing integrated pipeline owners would refuse to ship over prospective competitive pipelines, perhaps because of throughput and deficiency agreements, thus effectively denying them access to the market.[22] But if adjacent industry stages are competitive, as the Justice Department assumes, then such refusal to ship would penalize those refusing and would reward those not doing so, and hence would not block pipeline entry. Similarly, competition among prospective pipelines during planning stages would encourage the achieving of economies of scale at competitive levels. Again, the absence of such competition would suggest that adjacent stages were not competitively organized, which is counter to the assumptions of the department. As an empirical matter, it is clear that new pipelines are built along routes of existing lines.[23]

Many large petroleum pipelines are owned jointly by several companies. For example, the Colonial Pipeline, a large products line from the Gulf Coast to the East Coast, is owned by a consortium including Gulf Oil, Standard Oil of Indiana, Texaco, Cities Service, Mobil, Standard Oil of Ohio, Atlantic Richfield, Continental, Phillips, and Union. The assumption that consortiums with so many owners collusively undersize pipelines raises several questions.

If the object were to size a pipeline monopolistically, it is unclear why joint ownership rather than ownership by a single company would

---

[21] 49 U.S.C., section 1(18) and the following.

[22] Throughput and deficiency agreements essentially represent commitments by shipper-owners to provide sufficient cash to cover the debt obligations of a pipeline, regardless of the actual quantities shipped by the shipper-owner.

[23] Examples of multiple lines along a given route include the Colonial and Plantation product lines (Gulf Coast to East Coast), the Exxon and American Petrofina crude lines (West Texas to Gulf Coast), and the Basin, Shell, and Amoco crude lines (West Texas to Oklahoma). The case of Colonial and Plantation is instructive. Colonial was built after Plantation and began operation in 1964. The effect on Plantation was immediate; from a pre-Colonial peak of 352,000 barrels a day in 1963, deliveries fell to 242,000 barrels a day in 1964, a drop of 31 percent. See John P. O'Donnell, "Pipelines Humping to Meet Southeast Products Demand," *Oil and Gas Journal*, February 13, 1967, pp. 74–75.

be adopted. Joint ownership would involve more individuals in an undersizing strategy and hence would raise the probability of detection by antitrust authorities.[24] Also, it would raise the costs of determining just what size pipeline should be constructed, since it would provide more companies with rights to determine that size. The problem faced by a group of firms attempting to undersize a pipeline is that, unless their relative positions in the stages adjacent to the pipeline are similar, the pipeline size that would maximize each firm's profits will differ.[25] For example, in its Sohio Report, the Justice Department itself pointed out that a significant seller of crude oil in the market from which a pipeline originates would have an incentive to increase shipments rather than restrict them if this would raise the price of crude at the origin point without depressing it significantly in the market served by the pipeline's terminus.[26] Since the relative market positions of companies in fact usually differ, joint ownership would raise the costs of undersizing. Also, the greater the number of companies participating in the assumed noncompetitive behavior, the less likely that the collusion could be disciplined to meet the objective of the group.

These considerations suggest that integrated companies would seek to minimize the number of parties with knowledge of pipeline sizing relative to market demand and with rights to make decisions about pipeline sizing. Yet, invitations to ownership of new oil pipelines are often widely tendered, and larger lines, which tend to be those with potentially larger effects on upstream or downstream market price, are exactly those with multiple ownership.

Whether individual pipelines can affect downstream market prices much is also subject to question. Crude oil lines flow to refining centers, located either on a coast or inland. About 90 percent of the total U.S. refining capacity is located on a coast, and such capacity is supplied not only by domestic pipelines but also by imported crude oil brought in by tanker. Since the worldwide market for crude is extensive—with U.S.

---

[24] According to industry testimony, the planning of a new joint venture pipeline normally involves the setting up of a management committee, a feasibility study committee, a right-of-way committee, an engineering committee, a finance committee, and a legal committee. See testimony of Charles J. Waidelich, president of Cities Service Co., before the U.S. Congress, House of Representatives, Subcommittee on Monopolies and Commercial Law of the Judiciary Committee, September 10, 1975. It is presumed that most, if not all, of these committees would have knowledge of the proposed size and market demand for a new pipeline.

[25] See Michael J. Piette, "Crude Oil and Refined Product Pipelines in the United States: An Examination of the Major Issues for Public Policy," unpublished paper presented to the U.S. Department of Energy Pipeline Conference, January 30, 1979, p. 13.

[26] Standard Oil of Ohio Report, pp. 39–52, especially p. 50.

firms alone purchasing upwards of 6 million barrels a day—it appears unlikely that a single crude oil pipeline supplying a coastal refinery center would have much effect on price there.

Inland refinery centers are often supplied by more than one crude pipeline system.[27] In such instances, any one pipeline has only limited ability to affect price. Certain smaller and isolated refineries may be serviced by a single pipeline, but in such cases the structure of the refining market would render pipeline undersizing superfluous.

Petroleum product pipelines service wholesale markets. Again, such markets are located either on a coast or inland. Although world trade in petroleum products is less extensive than that in crude, U.S. coastal markets are supplied in part by imports from refineries in the Caribbean and elsewhere, and such supplies constrain the ability of a domestic product pipeline to affect market price.[28] The larger inland markets are sometimes serviced by more than one product pipeline; more commonly, a given product line often competes with crude pipelines that service refineries located near the product market.[29] These considerations suggest that few if any product pipelines are in position to affect the downstream market price much.

Why the Justice Department's Assumptions Imply Pipelines Will Not Be Undersized

In the preceding sections, the Justice Department's reasoning concerning the sizing of oil pipelines has been described and several of its underlying assumptions noted. Some of these assumptions rest on questionable empirical foundation or appear inconsistent with other assumptions. Nevertheless, for the moment we shall accept the department's assumptions. Our aim is to show that, counter to the department's own analysis, its assumptions imply that integrated companies will not undersize pipelines.[30]

---

[27] For example, the Chicago area, which contains about half of all U.S. inland refining capacity, is supplied crude oil by the Capline-Chicap, Amoco, Arco, Lakehead, and Texaco–Cities Service pipeline systems. The St. Louis area, another inland refining center, is supplied crude oil by the Platte, Amoco, Ozark, Shell, Mobil, and Capline systems.

[28] In addition, product pipelines serving coastal markets face possible entry by refineries that can utilize imported crude oil. If such entry is open, it is unclear that coastal market product pipelines can obtain significant market power.

[29] For example, Salt Lake City is supplied product by the Pioneer pipeline and crude oil by the Amoco and Chevron lines, and St. Louis is supplied product by the Phillips, Cherokee, and Williams Brothers lines and crude by the Shell, Amoco, Capline, Ozark, Mobile, and Platte lines.

[30] We are indebted to Edward J. Mitchell for suggesting the approach taken in this section.

That something is amiss in the Justice analysis is easily seen by recalling that the department assumes that workable competition exists in industry stages other than the pipeline stage and at the same time asserts that monopoly profits are earned in these adjacent stages. The simultaneous existence of competition and monopoly profits, however, violates an elementary principle of economic analysis. The rest of this section identifies exactly why monopoly profits will not be earned under the assumptions employed by Justice and why integrated companies would not be motivated to adopt the strategy suggested by the department.

Suppose for the sake of argument that a group of integrated firms indeed sought to restrict pipeline throughput so as to earn monopoly profits at an adjacent industry stage. By assumption, the adjacent stage is competitive and there is common carrier, nondiscriminatory access to the pipeline. Because the monopoly profits are gained by means of access to the pipeline with restricted throughput, firms will be motivated to compete for this access. Some set of rules will govern the obtaining of access but, regardless of the form of these rules, existing firms and entrants will devote resources to obtain greater access to the pipeline, so long as higher than normal profits can be earned thereby. But the competitive devoting of resources to acquire access will affect both costs and returns from such access, and this competitive process will result in an equilibrium in which no higher than normal profits will be earned. Hence, contrary to the department's assertion, its own assumptions concerning competition at the adjacent stage will ensure that no monopoly profits are earned.

Exactly how does competition ensure that no higher than normal profits can be earned from pipeline use? Suppose for example that pipeline access is conditional upon total shipments tendered, and that oil not shipped via pipeline must be shipped instead via an alternative, more expensive transportation mode. Then the marginal cost of access includes not only the cost of shipping via pipeline but also the cost of shipping via the alternative mode, and this cost will equal a weighted average cost of shipping on both modes.[31] As shippers compete for access by tendering additional oil, the cost of access must rise with the proportion of oil shipped via the alternative mode, while the price received at the destination must fall. Assuming competition in the adjacent stage, this process will continue so long as higher than normal profits

---

[31] It is wrong to assert, as the Justice Department does, that the marginal cost of shipment is only the cost of shipping via the alternative mode. The change in total shipment cost to a shipper from an incremental barrel shipped must take account of both the cost of the alternative and the enhanced access to the lower cost pipeline. In a competitive market, it is this change in total shipment cost that the shipper will equate to the price he receives at the destination point.

can be anticipated from tendering an additional barrel, and this competition therefore will drive such profits to zero.[32]

As an empirical matter, for higher than normal profits to be completely eroded, at least one of the two conditions must hold. The profits will be completely dissipated if firms can compete for access without incurring significantly rising marginal costs or if firms can enter the market without significant cost disadvantage. Since pipeline shippers can be either producers or buyers of crude oil or products, relatively flat marginal costs over some range or open entry need characterize only one of the markets in an adjacent stage. In fact, petroleum firms have production, refining, and marketing operations of widely varying size, and firms in each of these sectors have grown over time, suggesting that over wide ranges the marginal costs of firms do not rise at rapid rates.[33] Also, historically, there has been considerable entry into all stages of the industry.[34] This evidence suggests that, as an empirical matter, potential monopoly profits from undersizing would be completely dissipated.

Even in the situation of relatively flat marginal costs and open entry, one might argue that it takes time to carry out investments to secure enhanced pipeline access; thus, even though profits from undersizing ultimately approach zero, they could persist for a time. If so, the incentive to undersize could still remain. But the persistence of even short-run abnormal profits from undersizing is implausible for at least two reasons. First, profits dissipate not only because of actual gaining of access but also because of the investment outlays required to obtain such access, and the undertaking of such outlays need not be a time-consuming process. Second, if some firm or group of firms contemplated an undersizing strategy, pipeline users would be in a position to anticipate it. For example, even if undersizing were foreseen only by the

---

[32] This argument is presented rigorously in Paul Kobrin, "A Formal Critique of the Justice Department's Undersizing Assertion," American Petroleum Institute, Policy Analysis Department Critique no. 004, August 22, 1978.

[33] More direct evidence is available for the petroleum refining industry. According to Scherer and others, slight economies are available over the range of one to three refineries of minimally optimum scale. See F. M. Scherer, Alan Beckenstein, Erich Kaufer, and R. Dennis Murphy, *The Economics of Multi-Plant Operations* (Cambridge, Mass.: Harvard University Press, 1975).

[34] See, for example, Edward W. Erickson and Robert M. Spann, "Vertical Integration and Cross-Subsidization in the U.S. Petroleum Industry: A Critical Review," North Carolina State University, Department of Economics, Economics Special Report, December 1977. Erickson and Spann present quantitative data concerning entry into crude production, refining, and marketing as well as pipelining itself. They conclude: "a wide range of firms has participated in [the growth of the U.S. petroleum market] and there has been entry and expansion of new firms and small firms in addition to growth by the established, large firms" (p. 33).

builder-owners of the pipeline, competition for access could begin well before the pipeline was physically completed. And if nonowners foresaw undersizing, they too could compete for access before the line was built. Since, as a practical matter, new oil pipelines generally require several years from conception to completion, anticipation of undersizing and competitive expenditures to secure enhanced access would render even temporary abnormal profits implausible.

Since it is plausible that potential monopoly profits from undersizing pipelines would dissipate and that integrated oil firms would perceive this, there appears to be no incentive for them to engage in such behavior. On the contrary, under the assumptions employed by the Justice Department, the companies would appear to be motivated to build competitively sized lines. This is so because by undersizing the firms would find themselves with excessive investment in adjacent stages, underinvestment in pipelines, and hence an inefficient structure relative to competitors not engaging in pipeline undersizing. In addition, firms involved in joint venture pipelines would bear the cost of reaching agreement on optimal undersizing and the risk of possible antitrust prosecution. Finally, any firm engaging in undersizing would run the risk that another firm would build a larger sized pipeline and capture the market. On the other hand, competitive pipeline sizing would provide firms the opportunity to ship via lower cost modes rather than higher cost ones and hence to enhance their competitive viability. Thus, under the conditions specified by the Justice Department, pipeline undersizing would appear to be unprofitable, whereas competitive sizing would appear to enhance firm viability.

Empirical Evidence on Undersizing

Thus far, we have criticized several of the assumptions in the Justice Department's analysis and the logic of the analysis itself. In this section, we examine the empirical implications the department draws from its analysis.

Prorationing. Justice's characterization of pipeline undersizing implies systematic prorationing of pipeline space, and the department has argued that prorationing is symptomatic of undersizing.[35] Systematic prorationing can, however, occur for reasons unrelated to monopoly behavior, and nonsystematic prorationing can also occur. Hence, prorationing is consistent with a number of behavioral hypotheses and is not, in and of itself, evidence that undersizing has occurred.

---

[35] See, for example, the Alaska Gas Report, p. 40.

Even in a world of perfect certainty, a pipeline likely will be subject to fluctuating demand over its lifetime. For example, oil fields typically have a life cycle characterized by increasing production during a period of development and declining production as the oil is depleted. Under purely competitive conditions, a pipeline is built according to a scale that balances the present value of expected costs of excess capacity against those of too little capacity; because costs of excess capacity are positive, this scale generally will be smaller than that necessary to service peak demand. Thus, even in a certain world, systematic prorationing over part of every pipeline's lifetime would be expected.

In a world of uncertainty, unanticipated increases in demand can lead to nonsystematic prorationing. For example, when Canada reduced crude oil exports to the United States, refineries in the Midwest that formerly received oil from Canada turned to shipments from the Gulf Coast. Consequently, the demand for pipeline capacity from the Gulf to these refiners increased and some prorationing resulted.

Scattered evidence is available to test the implication that oil pipelines are systematically prorationed. The Justice Department has testified that the Colonial Pipeline is possibly undersized, since a part of it has had to proration capacity since 1967.[36] But closer examination reveals that from the time Colonial opened in late 1963 to July 1967, a period of about three and one-half years, there was no prorationing. From July 1967 to December 1978, the Colonial line 1 running from Houston to Greensboro, North Carolina, was prorationed, but the more recently built line 2 running from Houston to Atlanta has never been prorationed. Starting in December 1978, line 1 no longer was prorationed, but the segment from Greensboro to Linden, New Jersey, which had not previously been prorationed, became so. Thus, the Colonial record is mixed, with some prorationing over some parts of the system at certain times and no prorationing in other parts.

One possible explanation for the extended prorationing of line 1 of Colonial is that expansion was deterred by the prospect that new refineries would be built in the Middle Atlantic states to refine imported crude oil for those markets. In practice, however, new refinery projects in the area have been stopped on environmental grounds, thereby increasing demand for space on Colonial over what had been expected.[37] In response, Colonial has several times expanded capacity more rapidly

---

[36] Proposed testimony of John H. Shenefield, pp. 28–29.

[37] For example, the proposed Hampton Roads Energy Company oil refinery planned for Portsmouth, Virginia, would be the largest East Coast refinery. It has been delayed, however, because of environmental opposition. If built, it would be the first new East Coast refinery in twenty years.

than was originally planned.[38] With environmental opposition to new refineries in the Middle Atlantic area continuing, the Colonial line is still expanding.[39]

Several other large lines—including Plantation, Wolverine, Yellowstone, and Platte—have been mentioned as possibly being undersized.[40] According to data collected by the Association of Oil Pipelines (AOPL), no prorationing occurred on Plantation in twelve of the twenty-eight years from 1950 to 1977. Some prorationing occurred in other years, but mostly on a seasonal basis or in some lateral lines. Again according to data collected by AOPL, almost no prorationing has occurred on Wolverine and Yellowstone, with only seasonal prorationing in one year for each. Platte has had no prorationing in its twenty-seven years of operation. Another large line, the Explorer Line, also has had no prorationing. Although these prorationing data are limited in scope, they do not appear consistent with systematic undersizing of oil pipelines.

Tanker Rates and Downstream Prices. A second implication derived from the undersizing analysis is that downstream prices are determined by costs of incremental shipments arriving by the next least expensive mode of transportation.[41] As a result, the difference between the shipping costs is said to represent additional profit to the pipeline's shippers. For example, for Colonial, the Justice Department compared tanker rates from the Gulf Coast to the East Coast with pipeline rates and estimated that pipeline shippers realized as much as $93 million in increased profits downstream in 1976.[42]

A problem with the Justice Department's estimate is that it presumes what is subject to test—namely, that the next least expensive transport mode sets the price in a downstream market. In this case, the test is whether costs of refined products traveling by tanker from the Gulf Coast to the East Coast determine East Coast product prices. In a preliminary industry study, New York spot cargo prices for four products in 1978, as reported in Platt's Oilgram Price Service Report, were plotted and compared with Gulf Coast prices plus transportation costs by tanker and by pipeline.[43] The plots show that the New York spot

---

[38] See "Colonial Races to Up System Capacity," *Oil and Gas Journal,* January 17, 1966, p. 40.

[39] See "Colonial Slates $214 Million Expansion," *Oil and Gas Journal,* September 25, 1978, p. 69.

[40] Senate Staff Report, p. 67, n. 34.

[41] See proposed testimony of John H. Shenefield, p. 22.

[42] Ibid., p. 29.

[43] Exxon Company, U.S.A., "Comments on John H. Shenefield's Pipeline Divestiture Testimony," preliminary draft, January 15, 1979.

prices were generally lower than the laid-in U.S. Gulf Coast spot prices plus tanker costs and that they more closely tracked the prices of products out of the U.S. Gulf Coast by pipeline.[44] This evidence is inconsistent with the notion that the prorationing of Colonial has resulted in extraordinary profits to its shippers.

The Vertical Integration of Pipelines

Crucial to an assessment of alternative policies toward pipelines is an understanding of the motivations for vertical integration. One explanation, which was examined above and rejected, is that vertical integration enables firms to restrict pipeline throughput and thereby capture rents otherwise denied by regulation. An alternative theory of vertical integration has recently been suggested by Klein, Crawford, and Alchian.[45] This theory implies that, rather than creating monopoly power, vertical integration is a means to prevent the exercise of such power.

Briefly, the Klein, Crawford, and Alchian analysis is based on the proposition that, once a specific and immobile investment is made that is contingent on investment by others, these others have an inducement not to honor fully a contract linking the investments in an attempt to appropriate a share of the value of the specific investment. This is so because, once a specific investment is made, with a value that is strongly dependent upon other investments, quasi rents are created. These quasi rents are equal to the difference between the value of the investment in that use and its value in its next most valuable use. The more specific and the less mobile the investment, the greater this difference. Because it is difficult to write and enforce perfectly contracts that assign the entire difference to the party making the specific investment, there is incentive for those on whom the investment is dependent to behave opportunistically; that is, to seek a share of the difference. This incentive, and the foreknowledge that such behavior is likely to occur, provide incentive for the vertical integration of assets. Under integration, the fixed investment and the investment on which it is dependent are owned by the same party, and hence the inducement to behave opportunistically disappears.

[44] Such a result, if confirmed by more thorough analysis of the data, would support the argument of this paper that, where a pipeline is prorationed and some oil moves via alternative means, downstream prices will be equal to a weighted average cost of shipping on both modes.

[45] Benjamin Klein, Robert G. Crawford, and Armen A. Alchian, "Vertical Integration, Appropriable Rents, and the Competitive Contracting Process," *Journal of Law and Economics,* vol. 21 (October 1978), pp. 297–326. The paper presents a general explanation of vertical integration and then analyzes oil pipelines as one of several applications.

The Klein, Crawford, and Alchian paper applies this theory of vertical integration to oil pipelines. They argue that appropriable quasi rents exist in the specialized assets of oil fields, pipelines, and oil refineries. For example, the independent owner of a pipeline that serves a field and a refinery could purchase oil from the owners of the field at a price little higher than the marginal production cost, while seeking to capture "specialized-to-the-pipeline" quasi rents of existing refineries by pricing crude oil at the price of alternative sources of supply. Under such circumstances, the refinery and oil field would be "hostage" to the independent pipeline owner. But the pipeline owner too holds specific and immobile investment and may be subject to opportunistic behavior by owners of crude oil properties and refineries. The vertical integration of crude oil pipelines with production and refining avoids costs associated with these problems.

Another important aspect of pipeline vertical integration is that, under plausible assumptions, such integration will achieve marginal cost pricing of pipeline services.[46] Because pipelines exhibit increasing returns, marginal cost pricing is difficult to achieve by alternative means, including rate regulation. This suggests that vertical integration facilitates rather than prevents economic efficiency in the operation of oil pipelines.

Alternatives to the Present System

In order to achieve what it considers more competitive behavior by oil pipelines, the Justice Department has proposed two policy alternatives: the divestiture of pipelines from other industry stages, or the inclusion of a set of "competitive rules" within the regulatory framework. This analysis takes issue with the logic of the undersizing thesis and therefore challenges this rationale for either policy. In this section, we consider possible consequences of the policies.

Divestiture. The Justice Department has proposed to the Congress that companies operating in other stages of the oil industry be legally barred from owning new pipelines (prospective divestiture) and that, where social gains outweigh social costs, integrated companies be divested of existing pipelines (retrospective divestiture). Divestiture also has been supported in a recent Senate Staff Report and by Senator Edward F. Kennedy (Democrat, Massachusetts), who recently asked the

---

[46] See Paul Kobrin, "Integration, Natural Monopoly, and Efficiency in the Petroleum Industry," American Petroleum Institute, Policy Analysis Department Research Paper no. 008, April 11, 1978.

Federal Trade Commission to issue a rule preventing vertical ownership of pipelines.[47]

Divestiture will compel the formation of nonintegrated pipeline companies and will create a conflict of interest among industry stages. The Justice Department believes that this is desirable because such conflict will create incentives to provide government regulators of pipelines with more information than they now receive and hence will assist the regulators to regulate effectively. According to the department:

> In reality, where shippers and carriers are truly different entities, interests cannot fail to frequently clash, and the adversary fallout provides the regulator with insight into the workings of, and the dynamic tensions within, the regulated system, as well as the basis for enlightened compromise. The absence of true adverseness then, is a substantial disadvantage; if the regulator does not take affirmative steps to regulate effectively it is unlikely to have the opportunities to develop the expertise ever to do so. Indeed, as I have noted, it was not until there were complaints regarding the rates of *independent* pipeline companies that the ICC, after nearly 70 years, began to awaken to its affirmative responsibilities respecting rate regulation.[48]

This line of argument, however, and particularly its concluding observation, reflects confusion over the consequences of independent ownership of pipelines. It is true that such independence will create a conflict of interest between shippers and carriers; in particular, carriers will be motivated to exploit their monopoly position, whereas shippers will seek to avoid such exploitation. But contrary to the thinking of the Justice Department, such conflict of interest is unlikely to enhance consumer welfare.

For one thing, as the Justice Department points out, more resources would be devoted to regulating the behavior of the independent pipelines. To date, independent pipeline companies have been the source of most complaints concerning tariffs filed with the Interstate Commerce Commission (ICC). A survey of these complaints during the period from January 1, 1969, to February 15, 1975, reveals that, of the eighteen formal protests filed, fifteen were directed against

---

[47] See Senate Staff Report, and Senator Edward F. Kennedy, "Petition for the Initiation of a Rulemaking Proceeding Prohibiting Ownership of Petroleum Pipelines by Petroleum Companies," before the Federal Trade Commission, January 4, 1979.

[48] Proposed testimony of John H. Shenefield, p. 15; emphasis in original.

independent pipelines.[49] This evidence has led one investigator to estimate that additional costs of regulating divested interstate oil pipelines would be in excess of $49 million annually.[50]

Second, it is not at all clear how nonintegrated companies will size their pipelines. Their decisions will likely be influenced by the particular type of rate base regulation used to regulate pipelines. For example, if rate-of-return regulation is used then plausibly an inefficiently large amount of capital would be devoted to oil pipelining.[51] Recent empirical research on this matter is inconclusive.[52] Nevertheless, a recent study that examines the consequences for pipeline firm behavior of divesting pipelines from other stages of the industry suggests that the problem of overcapitalization would be significant.[53]

A quite different possibility is that rate regulation would continue about as at present but that pipeline firms would incorporate the costs and risks of transacting solely with independent shippers into their investment calculus and would invest at inefficiently low rates. If, for example, costs of contractual arrangements among pipeline firms and independent shippers exceed those associated with vertical integration, less investment in the pipeline stage would be forthcoming. It is possible that regulatory authorities could offset such behavior by granting pipeline firms higher rates of return or by otherwise reducing investment risk (for example, by strictly limiting the construction of competing

---

[49] Hearings before the U.S. Congress, Senate, Subcommittee on Antitrust and Monopoly of the Judiciary Committee, Part 9, *The Energy Industry: Vertical Integration in the Petroleum Industry,* 94th Congress, 1st session, 1975, p. 653.

[50] Earl A. Thompson, with Robert F. Rooney and Roger L. Faith, "Competition and Vertical Integration in the U.S. Oil Industry," unpublished study, February 1976. The Thompson estimate refers to complete retrospective divestiture of interstate oil pipelines. If actual divestiture were more limited than this, the estimate would have to be adjusted accordingly.

[51] For the seminal analysis, see H. Averch and L.L. Johnson, "Behavior of the Firm under Regulatory Constraint," *American Economic Review,* vol. 52, no. 5 (December 1962). For a detailed theoretical treatment, see Elizabeth E. Bailey, *Economic Theory of Regulatory Constraint* (Lexington, Mass.: D.C. Heath, 1973). For a geometric exposition, see E.E. Zajac, "A Geometric Treatment of Averch-Johnson's Behavior of the Firm Model," *American Economic Review,* vol. 20, no. 1 (March 1970), pp. 117–25.

[52] See Paul L. Joskow and Roger G. Noll, "Regulation in Theory and Practice," California Institute of Technology Social Science Working Paper no. 213, May 1978, pp. 19–20.

[53] See Thompson and others, "Competition and Vertical Integration." According to that source, the social costs of inefficiency would be several billions of dollars a year. That estimate, however, refers to complete retrospective divestiture of interstate oil pipelines. Since the Justice Department supports complete prospective but only some retrospective divestiture, the estimate would have to be adjusted accordingly.

pipelines), but it is unclear that such offsets would be socially advantageous. Again, the net result is an inefficient utilization of resources.

Another important factor in the sizing decision is the ability of an independent pipeline company to obtain sufficient financing to build an efficiently sized pipeline.[54] Typically, the financing of pipeline projects requires prospective owners to make throughput and deficiency commitments so that, in case actual shipments fall short of the level needed to pay off the debt, the loans will be paid off by the pipeline's owners according to their initial commitments.[55] Shippers have argued that they would be hesitant to use their credit to guarantee the debt of an independent pipeline company; this suggests that independents would find it difficult to raise sufficient capital. To overcome this problem, the Justice Department has suggested that tariff-related incentives to obtain commitments might be legally employed.[56] Though such incentives conceptually would induce financial commitments from shippers, they would also raise potential competitive problems because they could result in some shippers paying more than others, relative to costs. Thus, it seems plausible that such schemes could lead to more detailed regulation than now occurs.

Under present regulatory constraints, the reluctance of shippers to offer throughput and deficiency commitments is not limited to independent pipelines. Those who propose significant pipeline systems typically invite all interested shippers to join in the planning and ownership of the project. Yet, the number who actually become owners is often relatively small.[57]

We contend that the difficulty of obtaining throughput agreements—that is, the willingness to bear risk—is attributable to the property rights attached to pipeline ownership. Common carrier status has the effect of making an interstate pipeline a nonexclusive resource; once the pipeline is built, any shipper can gain access, whether or not he is an owner. This can give rise to a problem of free riders, whereby each shipper hopes that others will bear the costs

---

[54] See, for example, John H. Shenefield, "Pipeline Policy—A Light at the End of the Tunnel," remarks before the Tulsa County Bar Association, November 8, 1978, p. 15.

[55] A shortfall in shipments is more than a hypothetical possibility. For example, the owners of Explorer Pipeline, a large products pipeline, have advanced $41.7 million to date to cover operating deficiencies, as required by the throughput and deficiency agreements.

[56] Shenefield, "Pipeline Policy," p. 15.

[57] For example, fifty-six companies were invited to participate in the Eugene Island pipeline system. Ultimately, seven became owners.

and risks of building and owning the pipeline and provide him with shipping capacity.[58]

The free-rider problem will influence the type of shipment information a prospective pipeline owner will obtain, and hence the basis he has for sizing the line. For example, suppose completely nonintegrated carriers polled potential shippers regarding their shipping plans. They likely would find that each shipper would be liberal in estimating his shipping needs because, in the absence of ownership risk, liberal estimates are inexpensive and may help to ensure that capacity will not be constrained. The carrier will likely discount these estimates in sizing the line, but the way he does so will probably influence future shipper estimates.

This suggests that an efficient pipeline ownership structure is one that minimizes such information problems. Complete shipper vertical integration into a line from the beginning of a project would appear to be a means to achieve such efficiency. On the other hand, completely nonintegrated ownership would maximize the information problem. Hence, the ability of nonintegrated companies to build efficiently sized pipelines again is subject to question.

The Competitive Rules. In its Deepwater Port Report, the Justice Department proposed the adoption of a set of "competitive rules" to regulate operation of oil pipelines.[59] Ironically, whereas the goal of divestiture is to break up and prevent vertical integration, the goal of the competitive rules appears to be to compel nearly perfect vertical integration of existing lines. Although we have argued that the vertical integration of oil pipelines generally is economically efficient, compelling nearly perfect vertical integration of existing lines when they are required by law to be a nonexclusive resource is a different matter. The four proposed competitive rules are either inconsistent with each other or they would demand a more complex set of regulations than currently exists.

---

[58] What do free riders obtain from pipelines? The "free" element may be created by regulation that places a ceiling but no floor on allowed rates of return. Also, competitive constraints may impose mandated losses on some pipelines. This raises questions about why anyone would agree to become an owner. The reason probably is that, overall, shipper assets increase in value from reduced transport costs relative to those of alternative transport modes. Thus, even if pipeline investment is itself unattractive in some instances, a consortium of shippers may be formed to capture the overall benefits of lower shipping costs but, because of regulation, it must offer these benefits to nonowner shippers as well (that is, to free riders).

[59] Deepwater Port Report, p. 106.

The first rule states that pipelines must provide open and non-discriminatory access to all shippers, owners and nonowners alike. Since current regulations already call for such access, it is not clear that this rule is a change in policy. It is clear, however, that such a rule maintains oil pipelines as a nonexclusive resource. Under one interpretation of the rule, a change in policy would occur—namely, the rule would modify or forbid the use of historical prorationing. Space on prorationed lines is allocated to previous shippers on the basis of their historical use; thus, the means whereby these shippers can compete for existing capacity is reduced.[60] But historical prorationing is a way of resolving the free-rider problem. Under historical prorationing, each owner is assured that, should a line prove sufficiently attractive that demand outstrips capacity, other owners will not be able to gain incremental access at his expense. By reducing the free-rider problem, historical prorationing induces more shippers to become owners and, further, to reveal their true demand for shipping capacity.[61] As a result, there is possibly more inducement to combine the financial commitments and the information needed to build an efficiently sized pipeline than would be the case were shippers faced with the prospect that all other shippers (including other owners) could compete for capacity independently of their ownership interests.

The second rule states that any pipeline owner or shipper providing adequate throughput guarantees at the standard tariff can unilaterally request and obtain expansion of capacity. Such a rule is intended to prevent line capacity from constraining shipments, but in fact it will induce firms not to join pipeline consortiums when they are first formed. The reason is simply that, if the planned pipeline proves larger than what is demanded and a low return is earned, a nonowner can always gain access while having avoided the low return, whereas if the line proves smaller than what is demanded and a high return is earned, the nonowner can then call for expansion and join in the ownership. Of course, the rule could be modified to include a "risk premium" for late ownership, but that would create problems of ensuring that the premium is exactly adjusted for the risk previously avoided. It is unclear that creation of such problems is an improvement over the existing regulatory structure.

---

[60] New shippers, on the other hand are granted space as a prorata share of their total tendered shipments. New shippers generally become previous shippers after twelve consecutive months of use.

[61] The problem is reduced rather than solved. For a pipeline project to be economically viable, it probably is necessary to gain participation by major shippers. Given the cost of attracting all shippers to the project, however, some firms, typically the smaller shippers, will be in a position to free ride.

The third rule would compel open ownership, under which a non-owner could obtain ownership to reflect his portion of shipments on the line; the fourth rule would allow for annual revisions of ownership shares so that each owner's share could equal his average throughput share. These rules are similar in that each seeks to bring ownership shares in line with shipping throughput. As such, they appear to share the same problem as the second rule; that is, they offer risk-free ownership. Again, if a shipping firm is guaranteed ownership of a line once it is built, its incentive to offer initial throughput guarantees is diminished. And systems to induce such guarantees—such as prospective tariff reductions or payments for ownership that reflect risk borne by the original owners—will inevitably complicate the regulatory task.

The third and fourth rules might also create a source of instability for lenders. A pipeline lender will likely take account of the fact that his borrower could be compelled through time to yield ownership in the pipeline structure and will readjust his lending behavior accordingly. Insofar as such lending is specific to both the pipeline project and the particular owner, changes in ownership are cause for changes in the terms and conditions of loans; if there are costs to such changes, the costs of lending to pipeline projects will be increased.

In summary, the first competitive rule, which would guarantee nondiscriminatory access to existing lines, perpetuates free-rider problems that now exist. Given the first rule, the second, third, and fourth rules, which call for adjustments to ownership of existing lines and for mechanisms for expansion, seem to provide low-risk means for non-owners to secure ownership of existing or new capacity. Under these circumstances, free-rider problems are exacerbated and the difficulties of building lines is increased. Without more complex regulatory schemes to overcome these difficulties, there is an inherent inconsistency between the first rule and the others. More complex regulation may overcome the inconsistency—for example, through tariff discounts for owner-shippers. But even if such schemes are legal, they may not be simple to implement and hence are by no means clearly superior to existing regulatory constraints.

Commentaries

Michael J. Piette

The paper presented by Canes and Norman is an enlargement and update of prior analysis, as they pointed out. Their paper makes a contribution to the continuing look into pipeline economics and relevant public policy issues. It seems to me the main goal of the paper is to debunk the logic and analysis used by critical observers of the pipeline industry, most particularly the U.S. Department of Justice. The paper analyzes what may be called themes woven into the DOJ analysis of pipelines.

As I follow what Canes said, I see three major areas that need to be touched on. One is the question of pipeline undersizing, which we may have beaten to death by now. The second is the theory of vertical integration as it relates to oil pipelines. The last is the workability or unworkability of various proposals offered in place of regulation, such as outright divestiture or competitive rules. I should like to make the bulk of my discussion within the topic of pipeline sizing, since DOJ, Canes, and I perhaps agree that is probably the fulcrum of most of the discussions so far.

Canes's argument attempts to critique the assumptions, logic, empirical implications, and public policy ramifications of the reasoning of DOJ in the case of the pipeline undersizing accusation. The paper goes through the assumptions given by DOJ—that pipelines have economies of scale, that other stages of industry are workably competitive, that joint ventures will collude, and so forth.

Canes and Norman concluded, using these same assumptions, that DOJ's analysis is flawed. While I would agree with Canes in questioning of some policy implications resulting from the DOJ analysis, at one point I am rather bothered by the methodology offered in the paper.

165

Canes points out that some of the DOJ's assumptions rest on questionable empirical foundations or appear inconsistent with other assumptions. I would probably agree with that, but I think there should be more analysis of which assumptions are questionable and which are inconsistent.

In the original paper from which this was derived,[1] there was more discussion of the assumptions. We need to know which assumptions are inconsistent and which assumptions are flawed empirically. I do not think there is a lack of empirical data showing that pipelines do have economies of scale, but there is a lack of empirical data showing whether or not joint ventures collude if the cost of collusion is not zero.

Canes may have backed himself, as an economist, into an awkward corner. If he states that the assumptions—in this case, DOJ's assumptions—are empirically unjustified and are not mutually consistent, then I do not feel he can take assumptions he knows to be false and inconsistent, and prove the opposite conclusion to that drawn by DOJ. It seems all he has done so far is prove that both the conclusions of DOJ and those of the authors are inconsistent with a world that does not exist.

This results partly from focusing too much on peripheral issues, such as pipeline undersizing, and a failure to focus on real issues, such as how workably competitive is the pipeline industry. To some extent, clear understanding would be facilitated by a slightly different approach. One could probably approach undersizing from the Averch-Johnson point of view, which provides a rich theoretical base and a large amount of empirical research. One could approach prorationing as a peakload pricing problem, which also has a rich theoretical base and a fairly sizable amount of empirical research.

I do not want to speculate on the outcome of such approaches, but I think that those studies would serve all of us—government, industry, and academicians—with a more complete understanding of pipeline issues and related problems.

In terms of the vertical integration of pipelines, Canes discusses some of the new contributions of literature by Klein, Crawford, and Alchian.[2] Their argument is one of property rights and quasi-rents, indicating that once a large fixed investment like the pipeline is in place, those on whom the investment depends, such as oil field owners and refineries, have an incentive to behave opportunistically. That is, when the pipeline is in place, they have some capability of holding the owner

---

[1] Michael E. Canes and Donald A. Norman, "Pipelines and Antitrust," Critique No. 005, November 1978 (Washington, D.C., American Petroleum Institute).

[2] Benjamin Klein, Robert G. Crawford, and Armen A. Alchian, "Vertical Integration, Appropriable Rents, and the Competitive Contracting Process," *Journal of Law and Economics*, vol. 21, no. 2 (October 1978), pp. 297–326.

hostage. While it does seem that vertical integration may mitigate, to some extent, the possibilities for opportunistic behavior of the parties at each stage in industry, I am not sure that is the question asked of industry by its critics.

Instead, the crux of the issue lies in accessibility to those pipelines by nonintegrated—that is, independent—shippers or field owners. It is not clear in the analysis nor in the extension that Canes has done exactly how nonintegrated entities would be treated when faced with an integrated pipeline owner. In other words, if the encompassment of a previously independent pipeline firm with production and refining stages avoids placing the oil field or refinery "hostage" to the pipeline owners, is an independent shipper equally "hostage" to an integrated pipeline, assuming some sort of vertical integration? This is not answered in the JLE article, and I am not sure it is answered here, either.

Another theme that worked its way through Canes's paper, in its most important and best discussed section, is the two proposed policy alternatives, pipeline divestiture and imposition of a set of "competitive rules."

I found myself much in agreement with Canes regarding what I foresee to be the more radical policy prescription—that is, outright divestiture. Important issues in many unclear areas, such as how firms would size their pipelines after divestiture and how independents would raise capital, are very well discussed.

The parties who prescribe "competitive rules" as a policy alternative have come up with many inconsistencies and many empirical mistakes in their assumptions. Canes does not accept these assumptions, and I think that he is correct in pointing out the free-rider problems and the changes in the risk-bearing positions.

The papers presented at this conference, particularly the paper by Canes and Norman, move us quite a way toward discovering the real significance of competition in the pipeline industry—that is, its level, its intensity, and its relevance to many of the very important public policy issues that have been raised by this forum.

Stanley Boyle

I should like to point out that I spent a total of five years working in the Justice Department and the Federal Trade Commission. My comments must be taken in that perspective.

In any situation dealing with the question of regulation, our overall goal is probably to achieve results as close to competition as possible,

167

given the cost and other considerations, which can be obtained from a noncompetitive structure.

I applaud Reynolds's efforts to achieve an approach that is an alternative to the present system, which involves the regulation of rates of return and dividends, to achieve at least a controlled quasi-monopoly situation. I use "quasi-monopoly" here advisedly because I must admit I do not see anything approaching a natural monopoly in the pipeline industry. Although many of the examples given of multiple outlets from one area to another in the shipment of crude oil and petroleum products are flawed, they do suggest that it is not clear we have a natural monopoly here.

I was particularly interested that Reynolds came up with a joint-venture rules approach. Under the joint-venture rules, as I understand them, we would try to include all shippers on a pipeline as owners, irrespective of their size and present ownership. An adjustment of share ownership would occur at some regular period, perhaps annually, and the new shares would reflect shipping throughput on the pipeline. This seems a little bit difficult, particularly after the comments of LeGrange and the others on the difficulty of finding people to participate in financing joint ventures.

Reynolds also says that he has a device for the transfer price of shares at something approaching replacement costs minus depreciation. Now, this is imaginative, but it seems to start running into problems. The first of them has already been pointed out by Reynolds in the presentation of his paper. There is no way to ensure that joint ventures, working together, will not try to capture monopoly profits. As a matter of fact, to one who had been a small shipper before, a joint venture might seem like a wonderful opportunity to join the fray and get more profits than ever before, if indeed the joint venture is obtaining monopoly profits. It does not seem to me that the great stress Reynolds puts on the annual changing of the shares of ownership, as a means of producing more competitive conditions, will actually accomplish that.

I also have a problem with "compulsory" joint ventures. It seems to me that such a situation might result in the loss of some arm's-length bargaining between firms in the industry. One might run into the kind of problem that the economist, Geoffrey Shepherd, keeps referring to, in that there would be more and more points at which various members of the industry would come into contact with one another in ownership patterns through expanded use and encouragement of joint ventures.

The fact that an individual firm can expand unilaterally does not seem to solve the problem for me. Such a firm has to make certain guarantees and throughput. It can undertake unilateral expansion, but why necessarily would it? It does not seem a general rule that firms will avail themselves of any unilateral ability to increase some portion of pipeline that will be advantageous to them.

One change from regulation to rules that really bothers me is this: Who will enforce the shifts in the ownership shares among the members of the pipeline? It seems to me at this point we add to regulation. This was a matter that Canes alluded to. We end up with more regulation than we had before. Now we have more things to look at, and more areas to take into account.

If I had 15 percent of a nice profitable pipeline, and if I were told at the end of the year that I had shipped only 12 percent of its annual total, and my share of the pipeline would therefore go down to 12 percent, I would fight to maintain my share. I would want a good justification for having my property taken away from me. The end result of this device would be to worsen regulation problems rather than improve them. I think it is an imaginative approach, but it ends up with more problems than we started with, and more regulation than we started with. If the joint-venture rules approach is the alternative, I would rather stay with what we have now.

Discussion

ROBERT J. REYNOLDS: As I said, the joint-venture rules approach could be regarded as a genuine alternative to regulation.

DR. CANES: Just one very brief comment on the thoughts of Dr. Piette. I think he said that one of the issues I did not fully address was the question of access by nonowners into pipelines. This is an issue that is raised over and over again.

It also can be raised with respect to our assertion that competition for access to a pipeline will erode whatever monopoly profits could have been earned by undersizing the line. There one has to speak in terms of incremental access both by nonowners and by owners.

In any event, we do have some data that bear upon it. We investigated four different pipelines. We looked to see what happened to the share of nonowners over time on those pipelines. I will just quote the summary figures.

In the case of the Colonial line, the first nonowner figures we have are for 1964. Less than 1 percent of total shipments were by nonowners, but by 1978, about 36 percent were nonowner shipments on that line. In the case of Plantation, 29 percent were nonowner shipments in 1950, the first year for which we have data, climbing to about 65 percent in 1963, then falling off dramatically with the opening of Colonial. The Explorer line had about 1 percent nonowner users in 1972, rising to about 29 percent in 1977. And Texoma, with only two years' data, in 1977 had 48 percent nonowner shipments, and in 1978, 52 percent nonowner shipments. In each case there is at least some indication that nonowners can gain incremental access onto those lines.

As a general proposition, how does one test whether or not nonowners are getting sufficient access onto lines owned by vertically integrated companies? One can look for direct statements by nonowners, and that test suggests that nonowners are not being denied access. We can also state that there are many nonowners using pipelines. We can cite statistics. A very interesting report has been done for the Association of Oil Pipelines by Howrey and Simon.[1] They provide a great deal of data, including a survey done by AOPL on that question.

PROFESSOR MITCHELL: Don Baker made a point relevant to this discussion. That is, in listening to this debate, theories seem to be thrown out—particularly by the instigator of a lot of this activity, the Department of Justice—and not a great deal of empirical research is done. The theories are based on pure logic or on anecdotal information.

Now, I don't know why that is the case. Maybe some would disagree that it is. It may be that data are not available. It may be that there are some numbers, some facts, that we ought to have but for one reason or another are not being made available to us.

Reynolds's paper puts forward the proposition that, if the rules of pipelines are changed to have ownership shares change with shipments, pipelines will behave better. "Better" is presumably defined in some way we can measure.

If that were the case, isn't it, in principle, a testable proposition? That is, actual pipelines do have different rules. Some pipelines change shares; some pipelines do not change shares in response to changing shipments.

MR. HARVEY: First, regarding a point that Canes made comparing the Gulf Coast prices, I don't know if Spavins will talk about that, too. But we must remember that we are in the context of Department of Energy regulations, and prior to that, Cost of Living Council regulations. Therefore, I don't think we can look at any petroleum prices on the Gulf and New York harbor since 1971 as having much relevance to anything.

It may be that if we do it very carefully, we will find some prices that are market-determined, but in late November, not only was the price in New York below the Gulf Coast price, for instance, plus transportation, but I think on November 29 the Gulf Coast price for regular gasoline was 54 cents, and the New York harbor price was 46 cents. The reason was, of course, that under Department of Energy regulations, there are very peculiar incentives. I don't think we can draw very strong conclusions from anything happening in this period.

---

[1] Howrey and Simon, "Pipelines Owned by Oil Companies Provide a Pro-Competitive and Low Cost Means of Energy Transportation to the Nation's Industries and Consumers," March 1978.

The second point—and this keeps coming up again and again—was that when there are many pipeline owners, there is a need for cooperation, and there may be a problem of collusion. If we were to go to the rules that Reynolds discussed, we might get competitive behavior when we have a large number of firms. But as has been pointed out several times, when an individual firm that owns a 25 percent interest can say no to any expansion of a line, it does not matter how many owners there are. There are no costs of collusion because the firm can veto any expansion of that pipeline beyond the size it wants, which would be the monopoly size. I don't see how the costs of collusion are relevant in a line where firms have a veto over any expansion.

Thirdly, I would like to ask whether the analysis that Canes and Norman have done regarding the competition for pipeline space under proration shows that any rents due to undersizing would be eroded regardless of the method of prorationing employed. They base their analysis on some sort of mystical proration policy where one gets space on the pipeline in proportion to one's importance in the downstream market. I would appreciate their developing their analysis in the context of historical prorationing, which I believe is usually used by pipelines. I believe that generates conclusions exactly the opposite of what Canes has suggested.

Finally, a few comments regarding the pipeline rules. There has been some attack on Reynolds's suggestions, first regarding the ownership transfer and the adjustment of shares. I don't think he is wedded specifically to the rules specified for the deepwater ports. Another rule might say that anyone buying into the pipeline owns some interest in that pipeline's capacity, and when it is appropriate to make an expansion, any of the new owners can buy an interest in that new capacity.

That effectuates a share revision. It would be like Capline, an undivided interest line, and any firm could buy as much new capacity as it wanted. There would be no free-rider problems in that case, because a firm would have to put up the money to pay for it. The throughput guarantee would be provided if the financing required it.

Another issue related to the rules and the access question, which is apparently in most pipeline company agreements, is a first refusal provision. One cannot just buy into a pipeline, if one wants to, at a negotiated price.

If one of the owners of Colonial wanted to sell to some non-owner that was willing to pay the price—that say, Murphy wanted to buy ARCO's interest—the seller would have to offer it to the other owners of Colonial at that price.

173

The football team owners have tried to get that kind of provision into all of their agreements with the Players' Association explicitly to drive down the value of player services. We are well aware that it destroys a free market for that asset, that pipeline share, when there is a right of first refusal. Perhaps eliminating that type of restriction on ownership (transfer)—and there are many restrictions like that which don't seem to be necessary—would introduce a little more competition and avoid some of the problems we are talking about.

DR. CANES: In a certain sense, I am rather happy to hear Mr. Harvey say that nothing can be concluded from the prices in the Gulf Coast and New York harbor, about the mechanisms that set them. That comment could be directed, in the first instance, at the Justice Department. It did, indeed, examine those prices and concluded that there must have been a monopoly rent being earned by the owners of Colonial pipeline. If we cannot conclude anything, then I would say they are also put on notice.

Second, there is the question whether multiple ownership of lines would cause problems for groups that sought to collude. I am speaking about the initial decision to undersize, and the question of whether a large group of owners, with varying interests, could somehow agree on a common course of action at one end of the pipeline that would maximize the joint profits of the entire group without some rather complicated cross-payment mechanism.

It is my impression that the large lines generally are joint-venture lines, sometimes with many owners. It is likely that there are differences in their relative interests, and, therefore, that it would not be simple for them to come to a common course of action.

What about pipeline agreements that allow minorities who hold on the order of 25 percent to oppose expansion? I think we have to go back to the question of how are we going to get people to come into a pipeline project? Is it not necessary to grant them some protection for their business interests in the line in order to induce them to come in?

What if, for example, one's investment becomes subject to the choices of others, who may or may not be making decisions that the first owner thinks practical as a matter of business judgment? The call for an expansion may involve financial commitments on the part of someone who would not otherwise wish to expand a line, and whose business interests will be affected.

I understand that such a reaction by a pipeline owner could have a monopoly interpretation, but even in a purely competitive situation, someone could reach the judgment that it would not be in his business interest to see an expansion take place. Yet it could be forced upon him.

As to the prorationing policies, as I understand the criticism, it is that we may weaken competition for access if a pipeline is using historical prorationing policies. Some lines do use historical prorationing, and other lines do not. We are speaking here about those that do.

Even if that is the case, we still can have entry by nonowner shippers onto the line, and one of my purposes in citing the data earlier on changes in nonowner shipments on pipelines over time was to indicate that this may be rather substantial. Even in circumstances where historical prorationing is limiting or reducing one particular form of competition for access, we still can have competition from this other source.

As to why historical prorationing might be used at all, I think it goes back to the question of the free-rider problem on oil pipelines. At least that is one reason why it might exist.

If people take some ownership interest in a line, but ownership is not particularly attractive while shipment is attractive, then there is an incentive on the part of each shipper to expand shipments, yet not to put up more ownership capital. Historical prorationing puts a limit on that, and reduces that incentive. And so, it may be a form of a protection for owners that will induce people to come into the pipelines in the first place.

HENRY M. BANTA: As to the access question, my suspicion is that the expansion of access to nonshipper owners correlates with one thing. That is the investigations of the Department of Justice. It is pretty difficult to rely on those data when the Department of Justice is an actual participant in the game. It is one of those cases of the observer's spoiling the experiment.

MR. SHAMAS: To add a practical note, let me discuss the problems related to change of ownership. The Texoma line finally ended up with ten companies. We had to borrow $100 million from Chase Manhattan.

We wanted to have an ownership reallocation at the end of five years. The credits of those ten companies varied from AAA to B, and one was so limited that the bankers said we should not consider allowing that company into the corporation. The other nine companies had to give financial guarantees for the small company, because we wanted to have a diversity of ownership. But Chase would not allow a reallocation of ownership because of its limited credit rating. Chase thought the company should not have its original proposed ownership, or

175

the capability of ever having a larger ownership, because of the potential liability of this corporation. Was this collusion? We carried a financial guarantee for the small company to get it into this common carrier pipeline.

I am chairman of the board of Texoma, even though we own only 10 percent of it. Of the Osage system, of which we own 50 percent, I am president. We had four credit ratings there. They varied from AAA to A. We were allowed by our lenders to have a 5 percent ownership reallocation at the end of five years, based on throughput, on what we actually shipped. That is in our shareholder agreement. Again, the limitation was put on by the lending agency, because they did not want some of the weaker credit risks gaining larger ownerships, because it made their loans to us a greater risk to the lenders.

Are you going to have an arrangement such as if we ride the public bus 360 days a year, we gain a certain amount of ownership? How can we change pipeline ownership annually when we have a twenty-year loan guarantee based upon the creditability or the credit rating of the owner company? That is not a very practical solution. I am not saying that it cannot be done, but it would change the whole theory of financing of major pipelines.

In regard to expansion, Texoma requires a vote of two-thirds of the owners. No single company can block the expansion. We have had an expansion already. We went from 300,000 to 400,000 barrels a day. We had to go back to the bank to borrow $8½ million dollars more to do this. It was simply a business judgment. We felt that the rate of return was there, and we thought the volumes were there, so we went ahead. We also had to guarantee another $25 million in dock facilities in Nederland, Texas, in order to handle the expansion volumes. On Osage it takes three-fourths of the owners to vote an expansion. Both of these are common carrier companies, and all of the numbers are there in the Exxon evaluation.

Now about proration. We argued for a year and a half on Texoma, trying to meet ten companies' legal staffs' opinions over what was the legal proration policy. We finally got one. It is printed in our tariff. A customer needs no historical basis. He walks up and he has the same rights as any other person. A newcomer can force someone who was there from the start to reduce his allocations or his nomination. What bothers me is that we are talking about a system that has been functioning ever since the Hepburn Act in 1906. It was aimed at the railroads, but the courts have interpreted it to include pipelines. We have tried to live up to the rules that have been placed upon us. If someone shows that we have unconscionable earnings, they should know that we invited twenty-five companies to come into Texoma, and we finally got ten. Even so, we had to give a financial guarantee for one of them.

Mr. LeGrange: I had a little confusion about what I sensed is a change in the Department of Justice's thinking on a certain aspect of Seadock, and I hoped it would be clarified by Reynolds's paper, but it was not. In fact, I am having a hard time understanding what it said at all. Others apparently understood what he said, because they commented on it and discussed it. But let me suggest that Mr. Reynolds need not communicate with economists, who are unlikely to invest money in any future pipelines. He would be wise to communicate with a businessman like myself, who might invest a few dollars in pipelines in the future. The industry will certainly need to invest in some.

The problem is the tendency to stay at the theoretical level and never get down to the kinds of hard facts that we have to look at when we make a business decision.

Let me just give two illustrations of what the problem is. Maybe they hold out some hope of how we might solve it. Looked at not from your aspect as economists, but from mine as a businessman, I thought regulation was a set of rules and procedures. They are laid out, and I know what they are. I take my risk accordingly, because I have a pattern that I am supposed to follow. Now the government says that we need a substitute for regulation, and we are going to have something called "competitive rules."

Let's look at those rules for a specific example. Look at rule number three.

It is a simple sentence, with some straightforward words in it, but look at it and try to decide whether or not to commit a company's funds to a project that has to operate under it.

Take "open ownership" for example. Why doesn't it specify in the rule how long ownership is open? Is it for a year, two years, five years, or forever? My business judgment will be tempered by how "open ownership" is defined.

That raises another obvious question: To whom is ownership open? One company comes to a project with an AAA credit rating, and another with a lower rating, whatever it happens to be. How will proper equity be worked out between one credit rating and another, to relieve the worries of the banks and the financiers who are putting up the money? I am not suggesting it cannot be handled, but before suggesting a rule like this, a little homework should be done to decide how these problems might be handled fairly and appropriately.

177

In regard to the question of price, if open ownership is open over a long period of time, how do we differentiate in price when someone wants to come in? If I read rule number three literally, it scares me, because it says someone can enter at any time and buy in at replacement costs. Under rule number three I never want to come in on the front end of a pipeline project. Why should I take the risk at the beginning, try to put together a project, and commit funds, when there may never be a project? Why not just sit back and wait, and let somebody else go through all that agony and anguish. When the project is completed, I can just walk in and say, "Give me my 20 percent, and here is my share of replacement costs."

One must carefully consider what is price and how it is determined. Price, I thought, was supposed to be a reward for risk-taking in a venture.

Price, obviously, is a consideration on rule number four. If this is going to be done annually, one needs some way to divide up the price, the credit rating, the credit risk, and the throughput guarantees. Have you really thought of all that? Is there a way to do it? Would it be satisfactory? Can you lay it out so a businessman can look at it? Then I will have an opportunity to decide what kind of risk I am taking.

I am worried about another part of rule number four. This rule is obviously concerned with successful deepwater ports. If I were a fair-sized share owner in a deepwater port, I might worry about that port ten or fifteen years from now, when it might be not growing in size but starting to contract.

As I sit there with my throughput in the year 1990 or 1995, suddenly I have a 20 percent ownership, but I am shipping 50 percent of the volume because all the other shippers are disappearing from the scene. The way I read rule number four, at that point I would be forced to bail other owners out of a bad deal simply because my firm still uses the facilities. That is the kind of thing they should have thought about in writing rule number four. If bailing out is what is intended, fine. We can abide by that. But we will put that into our business risk calculation, too, in making our decision whether or not to invest.

But if that is not really intended, we must spell out rule number four to indicate how this thing would work in real equity, and not just for a year or two. These investments are made for thirty or forty years.

DAVID BROWN: I can shed a little light on the Deepwater Port process, in the context of the regulatory burdens associated with implementing competitive rules. Mr. Reynolds said that these are constitutional rules designed to alter incentives. Indeed, it was our hope that as the rules were implemented, there would be little or no regulatory burden, and that the real problem was one of novelty, rather than burden. Mr. LeGrange has pointed out quite emphatically how novel these rules were when he was first confronted with them back in early 1977.

Despite that novelty and despite the uncertainties the Seadock proposal brought for Exxon and certain other owners, those very same rules were, in fact, implemented and are now in operation with respect to another deepwater port, Louisiana Offshore Oil Port, the LOOP Deepwater Port.

It might be useful just to run down the rules quickly and explain how generalized rules were specifically implemented in a way designed to minimize the burden involved. The first rule, open and nondiscriminatory access, is little more than a reflection of the existing regulatory burden. It cannot possibly be much of an additional burden.

In regard to rule number three, open ownership to all shippers, as Scott Harvey mentioned a possible variant of that rule would be not to keep ownership open indefinitely or perpetually on a day-to-day basis, but rather to open ownership at times of expansion. That, in fact, is what the secretary of transportation decided to do in taking our recommendations.

The LOOP shareholders' agreement had to be revised to reflect the basis upon which that expansion would take place. Indeed, the Deepwater Port License, Article 15, specifies a number of the uncertainties Mr. LeGrange mentioned. These include points such as when expansions would have to take place, whether or not the expansion would be technologically practicable, whether or not a definite and continuing need could be shown for the expansion, whether or not the expansion would be economically feasible, and whether or not the available financial arrangements would allow an expansion that would, quoting from the license, "not result in materially impairing the earning power of the investment in the pre-existing port complex." [2]

Not only was the secretary very careful not to impair the earning power of the pre-existing port complex, the shareholders' agreement, which resulted from that, very carefully insulated the initial investment from the expansion investment, should the expansion investment have any tendency to increase the risk of the overall investment.

With regard to the fourth rule, frequent ownership readjustment, this is, while part of the general theory that we have articulated, not

---

[2] U.S. Department of Transportation, "License to Own, Construct and Operate a Deepwater Port Issued to LOOP, Inc.," Article 15 (January 17, 1977), p. 10.

something that Bob Reynolds or any ivory tower theoretician has made up. This comes directly from the prevailing practice in the industry.

I don't know whether Mr. Shamas's throughput and deficiency agreements are any different, but those that I am familiar with require a shareholder readjustment based upon the throughput of the owners for a three-year period—the last three years of the first five years of operation of the port.

If the financial institutions have any problem with shifts in ownership, or shifts in creditworthiness because of the different ratings of the owners, my understanding is that the prevailing practice has been an internal ownership readjustment. Something would be done internally in terms of providing payments among the shareholders. The original signers on the debts would remain basically liable to the institutions involved.

Solutions can be worked out on these problems. In our view, while we are presenting novel requirements in some respect, in the LOOP situation there was a response to these requirements. Something was worked out, I believe, that all of those involved in the process felt they could live with. It remains to be seen exactly how LOOP operates. It is something we should all watch closely as we continue to discuss and debate the effectiveness of the competitive rules.

MR. LeGRANGE: You fingered the very problem that worries us, that when we redistribute internally, we do not get off the hook from our debt guarantee externally. That is the very issue that worries me when I look at rule number four. It doesn't do much good to swap ownership with somebody who has a terrible credit rating. The bank has not relieved me of any obligation at all.

If there has, in fact, been a lot of detailed work on competitive rules like these, then we ought not to get into forums and talk in generalities about them. We ought to take the benefit of the work that has been done, describe the research that has been done, and present the rules in a way people can understand.

Furthermore, we should not assume that if somebody was willing to invest under a set of rules, in the past, people in the future will step up and invest under those same kinds of rules. I have no idea what the situation was when those people became involved in LOOP. Obviously, they had a very strong driving force to get that port built under any conditions. And they swallowed some pretty bitter pills in taking those decisions and moving ahead. The situation at Seadock was different, and it did not proceed under the same ground rules.

STANLEY BOYLE: I would dearly love to see Mr. Reynolds come up with a set of rules that are more specific than those he has written. My natural inclination is to be somewhat suspicious of joint-venture relationships. They introduce all sorts of problems, as Mr. LeGrange pointed out. Scott Harvey mentioned the question of what happens if a firm wants to go in and expand unilaterally. What happens if it expands unilaterally, and then, for some reason or another, the through-put on the line drops substantially? If it is down to, say, 50 percent of capacity, which is not a profit maximizing level of operation, who bears the loss? A portion of the problem has been aggravated by the individual company. Even though it was willing to pay for the expansion, does it pay for its additional share of the loss, and is the rest of the loss then prorated?

MR. HARVEY: I would expect the company that initiated the expansion to be made responsible. It puts up the money to finance the expansion and takes the consequences if the expansion turns out to be unwarranted. Remember that these rules are not entirely untried or only used when mandated by the Department of Justice. We have already brought up the instance of Capline: as far as I know, some of the owners did go ahead with an expansion. Not all the owners decided to go along, and those who did financed that expansion and presumably suffered the consequences. Not all pipelines are like that, but some of them have chosen to go that route.

Regarding the financing problem, if each company goes out and raises capital on its own, the company that is a bad credit risk does not have to be protected by the company that is a good credit risk. Each company has to satisfy the bank itself and get its own credit. There are good credit risk companies and bad credit risk companies. That is a problem in the current regulatory environment. We might be able to solve it by going to the type of competitive rules we are discussing.

Most important, keep in mind with regard to competitive rules that the financing situation will be radically different. We are not talking about the consent decree's 7 percent limitation on rate of returns, which is the only reason why we have 99 percent debt financing. We do not observe that type of capital structure in the rest of American industry, and in Bob Reynolds's context of deregulation and competitive rules, we will not have the kind of financial structures we have now. Presumably the financing problem will be radically different.

MR. WOLBERT: Mr. Reynolds, it is contemplated that at the time you install the competitive rules, there will be a dismissal of the *United States v. Atlantic Refining Company et al.*?

MR. SHAMAS: Scott Harvey chose a very bad example. The Justice Department has come down 100 percent against undivided interest pipelines. Because of the frown of the Justice Department, every enlightened company in the last five years has decided it cannot go into undivided interest ventures anymore. That is why we formed joint stock companies. What you are saying is, there might be many tax advantages if companies could go into undivided interest pipelines, but we have all decided that that is too risky.

When we ten companies went into Texoma, we fought internal battles on these matters. The majority opinion of the ten legal departments was that that was the safest way to go.

DR. REYNOLDS: Many of the financing problems that have been raised here were ones I have thought about myself for some time. Because of the reasons that Scott Harvey indicated, the financing problems have always tilted me to thinking of undivided interest rather than a joint-venture form to implement these rules.

MR. SHAMAS: Don't you see the dangers of the apparent collusion in undivided interest? That is what we are always hit with.

MR. BAKER: I wanted to back up and ask you about the Department of Justice's frown, which was influencing everybody. I have done my share of bureaucratic frowning on behalf of the department, but my internal impression from the department was that there was no such neatly defined line, at least at the upper levels of the department.

MR. SHAMAS: Do you remember the original investigation of Colonial?

MR. BAKER: That was water over three dams, and they couldn't figure out what was going on.

MR. SHAMAS: Well, most of the companies in the original Colonial wanted to go to undivided interest, but it came down that that was not going to be allowed. It was going to be a joint stock company.

MR. BAKER: My point is that if someone came in to me with a pipeline proposal and said it could not be done because of what happened back in the Colonial days, I would advise him to think about it again, to go and talk to lawyers like Don Flexner and Don Kaplan in the Department of Justice, and find out, and to consider putting in a business review request.

The decision makers change, and the amount of information available to the department also changes. My general impression was that what went on in the pipeline field of the department before the Deepwater Ports was much less important than what has gone on since.

MR. SHAMAS: I think two things killed Seadock. One was overregulation. All ten of the Texoma companies evaluated Seadock, and Texoma itself evaluated it, because if it ever came into being, we thought we might lay a line over there and connect up. The minute it got to where it looked as if it would cost more than 15 cents a barrel for the use of Seadock's facilities, it became uneconomic for us. We could already beat it with the Nederland facilities.

All the things Seadock had to do to meet the regulations made it an economic investment that got to be larger than could be justified. Those two things killed it.

MR. BROWN: There is an expressed antipathy to undivided interest lines in the Deepwater Port Report. I would like to quote very quickly from that report a sentence from one of the company participants. "The Consent Decree tends to set a practical limitation on net income at 7 percent evaluation. Participating in an undivided interest venture by an established pipeline company permits the higher income from the new large-diameter operation to offset its less lucrative small-diameter trunk lines and gathering facilities, keeping the overall net income under 7 percent." [3]

The rate aggregation problem that this represents is something that we thought was unacceptable. I am very pleased to see that TAPS, which is also an undivided interest line, chose to defend its rates on a pipeline basis. Exxon Pipeline Company, for example, chose not to aggregate its investment in TAPS to its lower forty-eight investment and defend an overall rate of return on that basis.

MR. LeGRANGE: I think you put your finger on a very important point— Why are we bothering to worry about the existing regulation system? That is an area that can very easily be corrected and changed for the future.

[3] "Report of the Attorney General Pursuant to Section 7 of the Deepwater Port Act of 1974 on the Applications of LOOP, Inc., and Seadock, Inc., for Deepwater Port Licenses," November 5, 1976, p. 82.

It was our assumption that it was, in fact, being changed for the future for significant stand-alone investments. We had never assumed, in any of our internal discussions that in a venture like TAPS or Sea-dock, or even when we were in the Sohio pipeline project, we would aggregate all of our pipeline assets and in so doing lose a way of judging any indication of performance and of what we were earning on the separate ventures. We felt that those would be significant stand-alone investments, and they would meet the test on their own. To change the regulations, that area ought to be looked at, with specific reference to significant stand-alone facilities.

MR. BAKER: It seems to me a better way to deal with that issue.

MR. BANTA: Let me ask our friends from Justice a question. In trying to draw from the experience of the Deepwater Ports, it seems to me that a case for the competitive rules has to be predicated on the fact that they are relatively self-enforcing. Members can enforce them against each other, without going to some outside forum.

As I listen to the discussion here, I can envision members' lawsuits against each other. Does the peculiar structure of the Deepwater Ports administration, where the secretary of transportation is a licensing authority, make the situation unique? Does that insulate them from civil litigation, or do you see these rules as being self-enforcing in a very literal sense?

MR. BROWN: I don't think they are unique to the Deepwater Port context. The competitive rules, except for the rule about nondiscriminatory access, have been distilled in the negotiating process at DOT over several months and are embodied in the LOOP Amended and Restated Shareholders' Agreement. All the rules, all of the legal ramifications, and all of the readjustments to debt and readjustments to equity are there, though they are quite complicated and quite novel.

I did not mean to suggest that there might not be some potential effects or problems in the administration of these rules, but the problem has been approached. I believe that unless shareholders have rather bad legal advice in the operation of their ownership share, they will try to work out questionable areas among themselves, rather than taking their disputes to the secretary of transportation. The secretary would have final reviewing authority only in limited circumstances— over a decision not to expand the port within the first additional 25 percent of the plan capacity. Beyond that, even the expansion decisions would largely be self-executed.

MR. LeGrange: You ought not to leave the impression that that document you mentioned was developed in and among the participants themselves. The secretary of transportation and you yourself had a strong hand in writing that document, and I would assume you would have a strong hand in any changing of that document. I don't think that the existing owners could get together and solve all the problems and modify this instrument as they see appropriate.

MR. Baker: Still, on the question of modifying the instruments, disagreements on responsibilities are common in carrier contracts, aren't they?

MR. Brown: In the pipeline context, that would be the case. In the Deepwater Port context, because there is a regulatory body that purports to regulate all aspects of operation in the deepwater port, there is an overlay.

MR. LeGrange: Even if they had a dispute and resolved the dispute, and wanted to change this document, they could not do it just on their own.

MR. Banta: Is this a way of moving a regulatory function out of an administrative agency? If this is generalized and applied to pipelines that are not administered in a specific way by the secretary of transportation, are the disputes, of the kind handled by the ICC or the Department of Transportation, being moved into a court deciding contract law? What does that gain?

MR. Brown: Questions of capacity are not being moved, because ICC and FERC do not pay any attention to questions of capacity in their regulation. Their focus, if at all, is access.

Calvin Roush: From all of the potential problems I have heard described, it seems to me that this set of rules or regulations, or whatever you want to call them, is no more self-enforcing than what we have now.

One of the rules, number one on this list of rules proposed in the Deepwater Ports Report, is open and nondiscriminatory access. It seems to be key to the rest of the rules' working because if we do not have access to shipping, we do not have access to ownership. If we do not have access to ownership, we cannot push for expansion.

A problem with the existing regulatory framework is that it is alleged there is not open access, and that is why undersizing is profitable. Otherwise, the owners would not be able to capture the advantages of undersizing. If there is open access, those advantages, it appears to me, in theory would be so diluted that prior planning would not evaluate undersizing as a profitable undertaking.

If access cannot be enforced now, why do we think we will be able to enforce access under this revised set of rules?

MR. SHAMAS: You can ask Mr. Shenefield to give some examples of denials of access, but he doesn't give any. That is the thing that bothers me.

DR. ROUSH: The ways of excluding people may be so subtle that for some reason they are not brought out, or there may be documents that have not been discussed today, that might reveal some problems of access. But if there are these subtle means of denying access, I do not see that they are eliminated under this scheme. And if they are not eliminated, I do not see how the rest of it works.

MR. BAKER: I was head of the Antitrust Division at the time the competitive rules came about, and I was the final sign-off on those reports. They arise from the way antitrust lawyers look at a set of problems, and they have in them problems that are familiar to antitrust lawyers. Essentially, when an antitrust lawyer looks at a joint-venture problem, he asks three sets of questions. The first question is: Does the joint venture eliminate competition among the venturers? Another way of saying it is, could there be several competing joint ventures?

The second question is: Does the joint venture prevent the competitive members from competing against some other market by collateral restraints?

The third is: Assuming the joint venture is a monopoly or what is called an essential facility, is access open to it to all on reasonable and nondiscriminatory terms?

The first and the third are rather closely related, because a big joint venture is usually defended on the grounds of economies of scale or high risk. That usually means there will not be an alternative one, and then we get into the questions of access.

The compulsory access rules, I think, have consistently been recognized as involving several problems we have been discussing, mostly of the free-rider variety.

One of the leading cases on compulsory access involves the Associated Press,[4] which has been in business for seventy years. When a new

---

[4] Associated Press v. United States, 326 U.S. 1 (1945).

member comes in, does he have to pay the same as everyone has paid over the seventy years? No, but he does have to pay something as a contribution. Another area in which these problems have come up has been access to electric power systems, which are not so flexible in terms of additions.

The third area is electronic funds, transfer of services, and banking. Interestingly, in the last named area, the Department of Justice has been a very firm opponent of state statutes that would create compulsory access in single systems. Its grounds, stated throughout the documents, are that there was an enormous free-rider problem and that it dispensed with the incentive to create the joint venture, to start with.

It is true that even the first rule does create a bit of a free-rider problem, that the common carriage all come in. That is true of any common carriage rule. The common carrier has to set up the facility, so it is a minor one. But the broader risk premium of the people who take a risk at the front end is a real problem. We recognized that in the department at the time but we thought that, all things considered, the risks were worth running.

Now, your point, Mr. LeGrange, is absolutely right, that you are more than incrementally adding to the business cost, and it isn't a question of whether all will go or all won't go. Some won't go as a result of that additional risk.

The second thing is the credit premium and how you cover the cost. The third element in the access deal is the actual physical assets, and the physical assets can be much more easily handled, although they are the biggest part of the package, than the other two.

This problem that we have not solved here is a more general problem that Antitrust has not solved in other areas, as it has tried to come to grips with the application of the so-called St. Louis terminal railroad principle—the principle that requires competitors who jointly control an essential facility to allow all comers in on nondiscriminatory terms.[5]

PROFESSOR MITCHELL: I think that is very well put, and a good note on which to close this conference.

---

[5] United States v. Terminal Railroad Assn. of St. Louis, 224 U.S. 383 (1912).

Supplements

Supplement 1

TESTIMONY OF JOHN H. SHENEFIELD
Assistant Attorney General, Antitrust Division
Department of Justice
Before the Subcommittee on Antitrust and Monopoly
of the Committee on the Judiciary
United States Senate
Concerning Oil Company Ownership of Pipelines

June 28, 1978

Mr. Chairman and Members of the Subcommittee:

I welcome the opportunity to present the views of the Department of Justice on the subject of divestiture in the oil pipeline industry. My testimony today is consistent with prior statements by Department officials on the competitive problems of vertically-owned and integrated pipelines. One of those statements, the Attorney General's 1976 Deepwater Port Report, established the analytical framework that has been followed by the Department since that time. As the Committee may be aware, in the past two years, the Antitrust Division's long-held competitive concerns about the petroleum pipeline industry have turned to action on many fronts. Because of the competitive significance of oil pipelines, we have dedicated substantial resources to this area. These efforts have run the gamut from open dialogue with industry representatives to full participation in administrative hearings. In between has come continuous documentary, economic and legal study of the industry, and the communication to others of what we—in what amounts to almost a task force approach—have learned. Those views have principally taken the form of advice to federal leasing, permitting, or licensing authorities, and statements to Congressional committees, such as this one, properly concerned with the potential anticompetitive impact of petroleum pipelines.

Our efforts have provided us with a unique vantage point from which to observe the relationship of this regulated industry to those federal agencies that regulate or otherwise have attempted to license or control it. By this process we have been able to test the validity of the economic analysis we developed to explain and predict the market behavior of petroleum pipelines owned and operated by vertically integrated oil companies. My purpose today is to explain in detail how and why these conclusions point strongly toward pipeline divestiture. To put this discussion in the proper framework, I would like to begin with a general overview of the industry.

I. Industry Overview

The pipeline industry is, in many respects, invisible. Buried underground, the lines are out of public view and pipeline transportation is hardly a consumer item. This invisibility belies how important petroleum pipelines are in two major sectors of our economy—transportation and energy. I put transportation first because of the little known but surprising fact that oil pipelines regularly account for up to one-fourth of all ton-miles of freight shipped in all forms of inter-city freight transportation.[1] But as important as pipelines are as transportation networks, they take on an added significance by virtue of the identity of the sole commodity they carry—petroleum, from unprocessed to refined. In fact, pipelines have a significant role at two stages of the petroleum industry: as crude lines, they provide the key link between crude production and refining, and as product lines they connect refining centers with product markets. More oil—crude or product—is moved more miles in the United States by pipeline than by any other transport mode.[2]

A. Industry Growth. Historically, crude lines developed before product lines. By the early 1930s, for example, the railroads' share of interstate crude shipments had dwindled to 3%, but they still carried 75% of product shipments.[3] Today, crude lines' share of total crude movements is still larger than product lines' share of all product movements,[4] owing largely to the need for extensive use of trucks on surface streets to carry product to local markets.

Major advances in the development of modern, large diameter oil pipelines occurred during World War II with the construction of the Federally-sponsored Big Inch and Little Big Inch Pipelines.[5] These pipelines, at 24- and 20-inch diameters and capacities of 300,000 and 235,000 barrels/day of crude and product, respectively,[6] demonstrated

---

[1] Statement of Ulyesse J. LeGrange, ICC Proceeding, Ex Parte No. 308, May 23, 1977 (now FERC Dkt. No. RM 78-2).

[2] Senate Comm. on Energy and Natural Resources, National Energy Transportation Report, Vol. I, Publ. No. 95-15, 95th Cong., 1st Sess., 182-84 (1976) (NET).

[3] *Id.* at 169.

[4] *Id.* at 143, 184.

[5] *Id.* at 173-74.

[6] *Id.;* A. Johnson, Petroleum Pipelines and Public Policy, 1906-1959, at 325 (1967).

to the private sector the feasibility of large diameter pipelines. Despite the success of these lines, post-war movement in the industry to large diameter pipelines was relatively slow, especially for product pipelines.[7] The 1950s and 60s, however, marked the advent of the large diameter joint venture crude and product lines, such as Capline and Colonial.[8] Now, in the post-Arab embargo era, with perhaps the exception of TAPS and its progeny, growth trends have been largely confined to adding capacity to existing lines rather than building large new pipelines over different routes.[9] However, the rate at which significant geographical shifts in demand for pipeline transportation—crude or product—have occurred has traditionally been very low.

Another significant pipeline industry characteristic is the level of ownership dominance that vertically integrated petroleum companies have obtained. For example, 1975 data[10] reveal that there were 104 common carrier pipelines subject to Interstate Commerce Commission (ICC) jurisdiction. The ICC categorized ten of these lines as independents, 59 lines as affiliated with a major oil company, that is, one of the top 20 in sales, and 35 as affiliated with non-major, oil related companies. This listing included 42 pipelines which were joint venture stock companies, of which 5 were owned by non-major, oil-related companies and the remaining 37 by major oil companies. In addition, the ICC listed separately 27 undivided interest pipeline systems operated as common carriers. In some respects, these figures tend to understate the extent to which petroleum companies (majors and non-majors) dominate the pipeline industry. In 1975, the pipelines affiliated with oil companies accounted for 98.6% of all crude barrel miles and 86.9% of all product barrel miles.[11] The corresponding figures for independent pipelines were 1.4% and 13.1%.[12]

B. Industry Economics. Pipeline economics play a key role in the structure of the pipeline industry. There are two important principles involved. First, as to competition with other modes of transport, pipelines are the most efficient (*i.e.*, cheapest) overland transport mode for the movement of significant, sustained quantities of petroleum between

---

[7] NET at 181.

[8] *Id*. at 190-93; Johnson at 382-83.

[9] NET at 206.

[10] Statement of David L. Jones, Ex Parte No. 308, at 8-9 (April 28, 1977) (hereinafter cited as Jones).

[11] *Id*. at 17-18. The numbers quoted in the text are arrived at by adding together separately stated figures for major and non-major petroleum companies.

[12] *Id*.

any two given points.[13] Even where there is waterborne competition, the larger diameter lines, which include most interstate lines constructed in recent years, are equally, if not more, efficient. Second, as to competition within the pipeline transport mode itself, pipelines exhibit great economies of scale resulting from the technological fact that pipeline capacity varies with the square of the radius of the pipe. This means, for example, that a 36″ diameter pipe will carry nine times as much oil as a single 12″ diameter line, or three times as much as three separate 12″ lines. Since construction costs, for the most part, reflect a more linear progression, throughput costs steadily decline as pipeline size increases (assuming that utilization increases correspondingly). This phenomenon of declining unit costs has been observed to take place over the entire existing range of technologically feasible pipeline sizes, which are now approaching diameters exceeding 4 feet.[14] As a result, in almost all circumstances, one properly sized pipeline will be more efficient than two or more in satisfying the available transportation demand between any two given points.

These two basic economic facts—premier cost efficiency and tremendous economies of scale—are usually referred to as the "natural monopoly" characteristics of pipelines. It does not necessarily follow, however, that because all pipelines have natural monopoly *characteristics* that each and every pipeline has monopoly *power*. I will explore the requisites of monopoly or market power more fully shortly, but I would emphasize here that the natural monopoly characteristics of pipelines are generally so significant that it would not be surprising to find that, between any two given overland points with a petroleum transportation demand, a single pipeline is the lowest cost, if not exclusive, means of transport. A second pipeline, or a more expensive transportation mode, such as rail tank cars, could not hope to compete effectively, and the pipeline's monopoly is the natural result of ordinary economic and technological forces, rather than monopolizing or predatory behavior.

These basic principles have been well understood by the pipeline industry since its earliest days, when its cost efficiencies over rail transportation first emerged. Those advantages soon led to substantial abuses of market power by the old Standard Oil Trust,[15] precipitating the Hepburn Act in 1906,[16] a Congressional decision that pipelines should

---

[13] NET at 213-14.

[14] *See, e.g.,* Report of the Attorney General Pursuant to Section 7 of the Deepwater Port Act of 1974 on the Applications of LOOP, Inc. and Seadock, Inc. for Deepwater Port Licenses, at 28 (November 5, 1976) (Deepwater Port Rep.).

[15] Standard Oil Co. of New Jersey v. United States, 221 U.S. 1 (1911).

[16] 34 Stat. 584 (1906).

join railroads as subjects of regulation by the ICC. I am sure this Committee is quite familiar with that bit of history, and I won't dwell on it except to note that it makes clear that regulation under the Interstate Commerce Act was intended to curb pipeline monopoly power in two ways: (1) directly, by limiting rates to just and reasonable levels; and (2) indirectly, by imposing operational, common carrier obligations of fairness and nondiscrimination on the pipeline owner in its treatment of any and all potential pipeline shippers. I will conclude my overview with a few brief comments on the efficacy of those regulatory curbs during the past 70 years.

C. Industry Regulation. The primary and most direct means for effective control of oil pipeline abuses of their market power is rate regulation. Unfortunately, however, oil pipeline rate regulation to eliminate monopoly power has proved an elusive goal ever since the passage of the Hepburn Act. Shortly after the enactment of that legislation, the industry mounted an unsuccessful constitutional challenge to the concept of rate regulation.[17] By the time that litigation was concluded, however, Congress had, in 1913, passed the Valuation Act,[18] which required the ICC to establish a "valuation" for each railroad and pipeline subject to its jurisdiction. Traditionally, that valuation was based on inflation-adjusted cost indices, component-by-component, and periodic adjustments to reflect some approximation of current or "fair" value of the facility. This task, however, was not even *begun* for pipelines for over twenty years, the priority of valuation funds and ICC efforts going instead to the railroads until 1934.[19] During this period, there were simply no standards for just and reasonable rates. Pipeline rates were, by any measure, truly monopolistic. For example, even at the depths of the Depression, the pipeline industry as a whole was earning extraordinary before-tax returns on depreciated investment of up to 34%.[20]

In the 1940s, with initial pipeline valuations largely established, the ICC finally set rate of return standards for the industry—8% on valuation for crude lines,[21] 10% for product lines.[22] Despite the fact that these benchmarks applied industry-wide and took no account of indi-

---

17 The Pipeline Cases, 234 U.S. 548 (1914).

18 49 U.S.C. §19a.

19 A. Johnson, *supra* at 240-41.

20 L. Cookenboo, Crude Oil Pipe Lines and Competition in the Oil Industry 98 (1955).

21 Reduced Pipe Line Rates and Gathering Charges, 243 I.C.C. 115 (1940); Minnelusa Oil Corp. v. Continental Pipe Line Co., 258 I.C.C. 41 (1944).

22 Petroleum Rail Shippers' Ass'n v. Alton & Southern Railroad, 243 I.C.C. 589 (1941).

vidual pipeline risks or other pertinent factors, these standards went unchallenged, unreviewed and unchanged for well over a quarter century. It was not until the mid-1970s that the ICC—responding to complaints about rates paid by shippers on an *independent* line[23]—initiated a general, industry-wide rulemaking into the rate base and rate of return standards historically applied.[24] After a somewhat uncertain beginning, that proceeding is well underway, now before the Federal Energy Regulatory Commission (FERC), with the Department an active participant in defining the issues.

At about the same time as the ICC's development of rate of return standards, the 1941 Consent Decree between the Department and the major vertically integrated pipelines was entered.[25] That Decree, still in effect today, establishes that dividends paid by defendant pipeline companies to their shipper-owners are not unlawful rebates in violation of the Elkins Act[26] if they do not exceed 7% of the ICC's valuation figure. The wisdom and efficacy of this agreement have been the subject of considerable Congressional scrutiny.[27] Much of the criticism is justifiable, for it is clear the Decree has, for a long time, failed to operate as the Department intended it should. Since stockholders are ordinarily paid dividends only after debtholders are paid interest, the Decree has been interpreted to permit expensing of interest prior to an allowance of dividends, while permitting 7% dividends based upon an overall valuation that is independent of the overall debt-equity capital structure of the pipeline.[28]

In response to this situation, debt financing in the industry has risen sharply since 1941. Prior to 1941 pipelines were funded almost entirely from equity funds provided by their shipper-owners, and investments in oil pipelines from outside sources were rare.[29] Oil pipelines had capital stock outstanding of over $264 million in 1940, while total

---

[23] Petroleum Products, Williams Brothers Pipe Line Company, 355 I.C.C. 479 (1976).

[24] Interstate Commerce Commission, Notice of Proposed Rulemaking and Order [49 C.F.R., Chapter X], Valuation of Common Carrier Pipelines, Valuation Dkt. No. 1423 (1971 Report) and (1972 Report) (Service Date: September 3, 1974).

[25] United States v. Atlantic Refining Co., Civil Action No. 14060 (D.D.C. 1941).

[26] Elkins Act, 32 Stat. 847 (1903), *as amended,* 34 Stat. 587 (1906), 49 U.S.C. §§41-43.

[27] *See, e.g.,* Hearings Before the Subcomm. on Special Small Business Problems of the House Select Comm. on Small Business, 92d Cong., 2d Sess. (1972); Hearings on the Consent Decree Program of the Department of Justice Before the Antitrust Subcomm. of the House Comm. on the Judiciary, 85th Cong., 1st Sess., V. 9, pt. 1 (1957).

[28] United States v. Atlantic Refining Co., 360 U.S. 19 (1959).

[29] Jones at 17-19.

debt of pipeline companies reporting to the ICC that year was under $21 million, and that amount was attributable to only eight companies.[30] Today, heavy debt financing with minimal equity contribution is commonplace, with debt-equity ratios of 90:10 or higher.[31] This leveraging greatly increases the return to total capital. To take a somewhat simplified example, if a pipeline were 90% debt-financed with an 8% interest rate on its debt securities, the total return (7% of valuation in dividends plus interest charges) would be more than double the total return (7% in dividends only) to the pipeline if it had no debt.

In the pipeline regulatory proceedings we have been participating in recently, it has become increasingly clear that the pipeline industry generally—even those lines not covered by the Consent Decree—has attempted to parlay the Consent Decree dividend loophole into an ICC rate regulation loophole. For example, in the TAPS case,[32] each of the eight carriers involved ignored the ICC's 1940s 8% rate of return standard and filed initial rates based on Consent Decree methodology: 7% dividends plus actual interest expense, all highly leveraged at 80% debt and greater. This approach, and the applicability of Consent Decree standards to rate regulation, was squarely rejected by the ICC last summer, when, in response to protests by the Department and others, it suspended the initial rates filed by the TAPS carriers.[33] However, this action came only after 35 years of silence during which most vertically integrated pipelines took advantage of regulatory laxity and "double counted" interest to produce returns far in excess of the ICC's published standards. The Supreme Court recently upheld the ICC's suspension of the TAPS carriers' proposed initial rates in a unanimous opinion.[34]

The future status of the ICC fair value rate base methodology and rate of return standards is uncertain, but the signs are hopeful, owing largely to the fresh airing these questions are receiving at FERC in both the TAPS rate proceeding and the general rulemaking I alluded to, entitled *Valuation of Common Carrier Pipelines*. We have argued in both proceedings that the FERC may and should (1) abandon outmoded and cumbersome fair value rate base concepts in favor of depre-

---

[30] *Id.* at 17; House Comm. on the Judiciary, Antitrust Subcomm., Consent Decree Program of the Department of Justice, 86th Cong., 1st Sess. 176 (1959).

[31] Jones at 18-20.

[32] Trans-Alaska Pipeline System, FERC Dkt. No. OR 78-1.

[33] Order of the Interstate Commerce Commission in Investigation and Suspension Dkt. No. 9164, Trans-Alaska Pipeline System (Rate Filings) and No. 36611, Trans-Alaska Pipeline System (Rules and Regulations), June 28, 1977.

[34] Trans Alaska Pipeline Rate Cases, ___ U.S. ___, 46 U.S.L.W. 4587 (June 6, 1978).

ciated original cost rate regulation modeled after the natural gas pipeline industry, and (2) set rate of return standards that allow a fair and reasonable return (reflecting the realities of current debt markets and equity capital attraction needs), but do not compensate twice for debt, *i.e.,* interest expenses.[35] These proceedings are vitally important from a competitive perspective. To the extent rate regulation has been ineffective, the results can be likened to those that would occur if all domestic pipelines had engaged in an industry-wide price-fixing conspiracy. Showing how regulation may have duplicated those effects should thus be, in our view, a major antitrust objective.

The other method by which regulation curbs the potential monopoly power of vertically integrated pipelines is through common carrier regulation. Under the Interstate Commerce Act, it is unlawful to provide any undue preference to one shipper (such as the pipeline company's parent) over another (such as an independent).[36] But the true scope of the common carrier obligation as applied to pipelines has never been made very clear. For example, there are no regulatory decisions on the obligation of a pipeline, if any, to provide storage or terminal facilities that would make the line more generally useful to nonowner-shippers. In fact, one of the few—if not the only—operating practices addressed directly by the ICC in all its years of common carrier regulation were early claims of excessive minimum tenders.[37]

The ICC's long periods of regulatory passivity on both common carrier and rate matters appear to have been premised on a "Hear-No-Evil, See-No-Evil" approach to regulation. According to the theory, if complaints are few, the system must be working well and the public interest being served. But the interests of the shipping public are not necessarily coincident with those of the public at large, especially where, as is often the case with oil pipelines, the shipper and the carrier share a strong commonality of interest. Not unexpectedly at hearings such as this, the industry points to the absence of complaints as proof of non-discriminatory service to the shipping public. In weighing that record, however, Congress should also consider the fact that the nonowner-shipper has in the past been faced with a climate of regulatory indifference in an industry dominated by carrier-affiliated shipments. This is

---

[35] *See* Memorandum Defining and Stating the Position of the United States Department of Justice on Policy Issues Raised by This Proceeding and Statement of Suggested Procedures to Be Employed for Bringing the Proceeding to a Conclusion, at 38-54 and 66-84 (April 3, 1978) (DOJ 4/3/78 Memo).

[36] Interstate Commerce Act, Sections 1(4), 2 and 3(1), 49 U.S.C. §§1(4), 2, and 3(1).

[37] *See* Brundred Bros. v. Prairie Pipe Line Co., 68 I.C.C. 458 (1922); Reduced Pipe Line Rates, *supra* n. 21; Petroleum Rail Shippers' Ass'n, *supra* n. 22.

an additional burden to any reluctance that may exist in the nonowner-shipper to disturb a customer/supplier relationship by resort to litigation.

In reality, where shippers and carriers are truly different entities, interests cannot fail to frequently clash, and the adversary fallout provides the regulator with insight into the workings of, and the dynamic tensions within, the regulated system, as well as the basis for enlightened compromise. The absence of true adverseness, then, is a substantial disadvantage: if the regulator does not take affirmative steps to regulate effectively, it is unlikely to have the opportunities to develop the expertise ever to do so. Indeed, as I have noted, it was not until there were complaints regarding the rates of *independent* pipeline companies that the ICC, after nearly 70 years, began to awaken to its affirmative responsibilities respecting rate regulation.

To conclude this overview, I would stress that if the pipeline problems were no more than what I have already described, I could end the discussion here. Our expectation might reasonably be that efforts at rate reform will ultimately prove successful, and that common carrier regulation can be substantially improved if the FERC avoids the trap of regulatory passivity to which the ICC fell victim, or if there are some minor adjustments to the Interstate Commerce Act, or perhaps both. But the pipeline problem is unfortunately not so simple. We have taken a much closer look at the consequences of vertical integration for the pipeline's regulated environment and concluded that to a great extent industry structure predetermines the failure of regulation.

II. Circumvention of Rate Regulation

I will try to describe, as simply as I can, how vertical integration creates the incentives and opportunities for noncompetitive behavior in the regulated context. To do this requires discussion of how and when rate regulation may be circumvented, the ineffectiveness of regulatory restraints on circumvention, manifestations of this behavior, and how such conduct can be perpetuated.

A. How and When Circumvention Occurs. To understand how and when rate regulation may be circumvented, three questions must be addressed:

1) When does a pipeline have market power?
2) Is rate regulation at all effective in curbing that power?
3) When does vertical integration enable a pipeline company to circumvent the constraints of rate regulation?

1. *Market power in pipelines.* In analyzing the first question, we have formulated some general conclusions about market power in pipelines that apply across the board to independents and integrated pipelines alike. To some degree, all pipelines have the incentive to circumvent existing regulatory constraints, but only those pipelines with market power can succeed. Market power, as I have already suggested, is something more than the mere fact that pipelines possess economies of scale or inherent cost advantages. Market power is the amount of a pipeline's monopoly advantage over its competitors, measured either at the upstream or the downstream end of the pipeline, or both.

Focusing for the moment on the downstream market, the factors affecting the level of market power include:

(1) The size of the pipeline's share of the downstream petroleum market;

(2) The level of effective transportation competition either from other pipelines or from other modes of transportation;

(3) Elasticity of demand in the downstream petroleum market; and

(4) The ability of other suppliers to the downstream petroleum market to expand output.

In applying these tests, the prices and outputs must be at or near those that would obtain under competition.

Thus, in general, a pipeline will have market power downstream when (a) its throughput comprises a significant share of the downstream market, (b) it can supply the downstream market with less expensive petroleum than can other suppliers, either because of lower transportation costs than alternative modes of transportation (including other pipelines) or because local producers (or refiners) cannot expand output without increasing costs, and (c) consumers in the downstream market are willing to pay a higher price to obtain petroleum if the supply is reduced.

A good example of a pipeline-type facility that will have downstream market power when it is built is LOOP. Its input of crude oil into Gulf Coast and Midwest crude markets will be substantial; it will have significant cost efficiencies over other modes transporting much-needed imported crude into that market (such as lightering, transshipment, and small tankers)[38]; and downstream refiners would very likely be willing to pay more for imported oil if supply were reduced. On the other hand, the proposed Sohio Pipeline from Long Beach to Midland, Texas, would not have market power in the Gulf Coast/Midwest crude market, because its input of crude into that market would not be sub-

---

[38] Deepwater Port Rep. at 51-53.

stantial and the delivered cost via that pipeline of oil in that market is comparable to that for foreign sources of supply. I invite the Committee's attention to our recent Sohio Pipeline Report[39] for a fuller analysis of the somewhat unusual Sohio situation.

Likewise, a pipeline will generally have market power upstream when (a) its throughput comprises a significant share of the upstream market, (b) increased supplies of petroleum can be absorbed upstream only at a lower price, and (c) suppliers upstream will be willing to accept lower prices to sell their crude or products, either because of a lack of transportation alternatives at costs as low as the pipeline's, or because contraction of output would be difficult or unprofitable. This type of market power is ordinarily termed "monopsony" power.

Our study of pipeline returns in connection with our rate reform efforts has produced what we consider to be strong empirical evidence that many pipelines have substantial market power—power which rate regulation has failed to curb. For example, in an evaluation of pipeline profitability conducted by our Economic Policy Office last year, our economists concluded that the rates of return for many oil pipelines are significantly above what is encountered in the *most profitable* electric utility and natural gas pipeline companies.[40] This bears out our historical impression that pipelines have exhibited monopoly power all through the years of ineffective regulation.

2. *Effectiveness of rate regulation.* These observations lead to a second question, whether rate regulation is at all effective in curbing monopoly power. This question is significant because with completely ineffectual rate regulation there would be no need to use vertical integration to reap the pipeline's monopoly profits. Such profits could be obtained directly with excessive tariffs. As is suggested by my earlier remarks, however, the ICC rate standards of 8% and 10% on valuation set in the 1940s, along with the Consent Decree, did finally provide some practical bounds on the rates of return pipelines have earned.

These bounds have acted as a partial curb on pipeline monopoly power. In suggesting that rate regulation has had some effectiveness, however, I mean only that rate constraints have been sufficient to create incentives in the regulated enterprise to circumvent those limitations. It is, if anything, clear that rate regulation has *not* been truly effective.

---

[39] Report of the Antitrust Division, Department of Justice, on the Competitive Implications of the Ownership and Operation by Standard Oil Company of Ohio of a Long Beach, California-Midland, Texas Crude Oil Pipeline, June 1978 (Sohio Rep.).

[40] Statement of George A. Hay, Ex Parte No. 308, at 6-7 (May 27, 1977).

3. *Means of circumvention.* There remains the question of how the vertically integrated structure of a pipeline permits circumvention of rate regulation in a manner foreclosed to an independent. The first step in this analysis requires identification of incentives. An independent pipeline company is in the transportation business and no other. Its natural incentive would be to maximize transportation profits. If it has monopoly power, it can be expected to exercise it; if it does not, it can be expected to accommodate itself to the needs of the widest possible variety and number of potential customers, earning a fair return on each barrel of oil shipped. In general, the greater the pipeline's throughput, the greater its profitability.

On the other hand, a vertically integrated pipeline company is not a pure transportation company, in theory or in fact. The incentive of the integrated firm as a whole is to maximize *overall* profits, not just *transportation* profits. Our study of this industry, therefore, focuses on whether those differing incentives result in a distortion of the independent's transportation-profit-maximizing calculus when the pipeline is vertically integrated. The critical question that has emerged is *whether, given at least some partially effective regulatory constraint, the vertically integrated structure permits circumvention of that constraint.*

We have concluded that if a vertically integrated pipeline has market power, it ordinarily possesses the ability to use its vertically integrated structure to circumvent rate regulation. Our analysis of this point should come as no surprise to this Committee; it first emerged in the publication of the Attorney General's Deepwater Port Report in November 1976,[41] and has been repeated on numerous occasions since—including appearances here.

As we see it, the ability to avoid regulation is rooted in the ability of the shipper-owner to limit pipeline throughput below the level an independent would size or operate the line. If the vertically integrated pipeline owner is a significant seller in the downstream market and the pipeline has market power downstream, the pipeline's throughput may be such as to ensure that at least some not insubstantial amount of supply arrives by the more expensive transport modes. Where this occurs, the downstream market price would likely reflect the cost of delivery by the next least expensive transport mode, allowing the pipeline's shippers to pocket the difference in transportation costs. For example, if the pipeline's cost advantage downstream over tanker shipments into the market is 20¢/barrel and tankers are being used, then the market price for *all* oil downstream will reflect that higher transportation cost for the market's incremental supply. Those fortunate

---

[41] Deepwater Port Rep. at 3-5 and 103-09.

enough to ship over the pipeline rather than by tanker will pocket 20¢ on each barrel they sell. To the extent that those shippers are affiliated with the pipeline company, the vertically integrated firm has circumvented regulation, capturing downstream monopoly profits initially denied by the rate regulator. The magnitude of the gains to the vertically integrated firm that behaves in this manner will depend on its share of the downstream market and its ability to use the pipeline for transportation as opposed to the more expensive tankers.

There is a symmetrical analysis for the upstream petroleum market. If the vertically integrated pipeline owner is a significant buyer in the upstream market and the pipeline has market power upstream, the pipeline's throughput may be such as to ensure that at least some not insubstantial amount of petroleum leaves the upstream market by the more expensive transport modes. The market price for all upstream oil will fall to reflect the higher cost of transportation out of the market by the next least expensive transport mode, allowing the pipeline's shippers to pocket the difference in transportation costs. For example, if the pipeline's cost advantage upstream over tanker shipments from the market is 20¢/barrel and tankers are being used, then the market price for *all* oil upstream will be 20¢/barrel lower than if no tankers were used. This is because an upstream seller would not ship oil out on tankers unless he would thereby receive a higher price than he would receive selling in the upstream market. Those fortunate enough to ship over the pipeline rather than by tanker in essence pay 20¢ less for each barrel they purchase that is delivered downstream. To the extent that those shippers are affiliated with the pipeline company, the vertically integrated firm has avoided rate regulation, capturing upstream monopsony profits initially denied by the rate regulator. A similar analysis can be made showing reduced acquisition costs for vertically integrated companies bidding upstream for drilling rights, including those offered by the federal government. The pipeline owner must be a significant buyer (of petroleum or drilling rights) in the upstream market in order to gain by this strategy; the larger the owner's share of the upstream market, the greater the potential gains.

In either the upstream or downstream situation, however, it is clear that successful circumvention requires the incremental supply to be diverted to more expensive transport modes. This might not occur if the pipeline had adequate capacity to handle all transportation demand. This makes pipeline capacity questions crucial. In general, configuration, sizing, expansion, extension and various types of access decisions can all play important roles in ensuring that the incremental supply is forced into more expensive transport modes. Decisions on these matters may thus be entirely at odds with those that might be reached by an independent transportation company.

I should also emphasize that the existence of market power in a pipeline does not automatically enable a vertically integrated owner to profit from avoidance of rate regulation. If the pipeline has market power upstream, for example, but the vertically integrated owner is a net seller in the upstream market, then the owner is unlikely to gain by restricting throughput in order to drive down the price of petroleum upstream. We found this to be the case with the Sohio-Long Beach to Midland, Texas, pipeline proposal.[42] Similarly, for a pipeline that has downstream market power, if the vertically integrated pipeline owner is a net buyer in the downstream market, the owner is unlikely to gain by restricting throughput to increase the price of petroleum downstream, because it shares the burden of those increased costs. These extreme cases illustrate the fact that the pipeline owner's share of the downstream market supply or the upstream market demand will limit the extent to which the owner is likely to circumvent rate regulation in an effort to capture elsewhere the pipeline's monopoly profit potential denied by regulation.

B. Regulatory Restraints on Throughput Restriction Behavior. The pipeline industry's basic response to our throughput restriction point is that effective common carrier and prorationing obligations would totally thwart any effort to circumvent rate regulation by vertically integrated pipeline owners. These curbs, it is argued, would not only regulate improper behavior, they would eliminate any incentive to restrict throughput. We disagree. Examination of the myriad ways pipelines can govern the terms and conditions of access suggests that regulation would be incapable of completely frustrating throughput restriction behavior. The vertically integrated pipeline company has at its disposal a variety of subtle operational means short of outright access denial to implement throughput restriction for the purpose of recapturing monopoly profits elsewhere. These practices include: failing to provide common carrier terminals or storage facilities; imposing excessive minimum tenders, inconvenient routings or shipping schedules, onerous commodity specifications, and biased prorationing rules; or failing to establish through routes or joint rates.

It is not clear that these practices could be completely curbed by closer regulatory scrutiny of the common carrier obligation. It is clear, however, that most of them have never been the subject of *any* regulatory scrutiny. But at the most fundamental level, effective prorationing

---

[42] Sohio Rep. at 48.

would not alter basic throughput restriction incentives. Prorationing would merely require the monopolist to dilute its monopoly return by sharing it with others. This might be an incentive to the pipeline owners to avoid tailoring the pipeline's capacity exclusively to their needs, but it would be no incentive to forgo monopoly profits entirely by properly sizing the line. Either way, the clear object would be to divert some oil to more expensive transport modes, even if some of that oil belonged to the pipeline owners.

It should also be pointed out that the Interstate Commerce Act provides no control over the initial sizing and configuration or subsequent expansion of a pipeline. Yet, in our view, decisions on these matters would be key to throughput restriction behavior—far more so than attempted discrimination in access.

C. Manifestation of Throughput Restriction. Manifestations of throughput restriction appear in various pipeline markets. In the deepwater port situation, for example, where a critical question was the initial sizing of the facilities, our study of the plans of LOOP and Seadock clearly revealed that their initial sizing decisions were originally based almost exclusively on the needs of the owners, conservatively stated, rather than the shipping public at large.[43]

Current product transportation between the Texas-Louisiana Gulf Coast refining area and metropolitan centers in the Middle Atlantic States provides another example. In that market significant quantities of petroleum products move by tanker across the Gulf of Mexico, around the Florida peninsula and up the East Coast for delivery to cities on the Eastern Seaboard that are also served by either Colonial or Plantation Pipeline or both. Shipment by tanker over this route currently costs as much as 75% more than that on Colonial or Plantation.[44] In 1976, for example, gasoline, jet fuel, kerosene, and distillate fuel oil—products all of which could be shipped by pipeline if more capacity were available—were shipped by tanker and barge from the Gulf Coast to the Middle Atlantic States in quantities totalling 120,656,000 barrels, or about 331,000 barrels per day.[45] The current tanker/pipe-

---

[43] Deepwater Port Rep. at 19-21, 70-73, and 114.

[44] These figures were derived from data contained in Platt's Oilgram Price Service, May 25, 1978 by comparing the pipeline rates from Lake Charles, La., to Newark, N.J., with the average tanker rate of petroleum products shipped by tanker from the Gulf Coast to the New York area.

[45] These figures were derived from the Monthly Petroleum Statement, Bureau of Mines, Department of the Interior, December 1976 (now published as the DOE Energy Data Reports).

line tariff differential is about 37¢/barrel.[46] These figures provide an upper bound estimate on the misallocation of transportation services in this market in 1976 of $44,600,000 of deadweight loss—that is, unnecessary additional transportation cost. In 1973, approximately 250,000,000 barrels of petroleum products were delivered by pipeline from the Gulf Coast to the Middle Atlantic States area, much of it through Colonial Pipeline.[47] If at least the same amounts of product have been pipelined annually since then—which appears to be the case— pipeline shippers could therefore have realized as much as $93,000,000 of increased profits downstream in 1976. Even if one were to discount these amounts by two-thirds for errors that might be contained in such rough estimates, the total annual potential savings to consumers of these products (one-third of the sum of our estimate of the annual extra transportation cost and the annual potential monopoly profits) could be in excess of $45,000,000 annually. The Colonial and Plantation Pipelines continue to operate at capacity and have been prorationed for years, despite expansions.

D. Means of Perpetuating Throughput Restriction. Finally, two basic questions about the perpetuation of throughput restriction need to be considered. First, if we have accurately described the abilities and incentives in the vertically integrated pipeline industry, why haven't there been more complaints by nonowner-shippers? This is readily explained if we analyze the impact of throughput restriction on non-owner-shippers. If they get some access to the pipeline, nonowner-shippers participate in the monopoly profits represented by the transportation differential. Surely they will not be heard to complain. Potential shippers whose oil is forced into other transport modes receive a normal return, because the market price is based upon those higher transportation costs. For such shipments, how can they be expected to complain about a denial of access to monopoly (as opposed to normal) profits?

This leads us to the second question: if the pipeline industry is so profitable, either directly or indirectly through vertical integration, why haven't more firms entered the market in order to earn some of these monopoly profits? New pipelines are generally financed at high levels of debt relative to equity, and this debt is usually backed by guaranteed

---

[46] The pipeline rates used are from Lake Charles, La., to Newark, N.J., as of May 15, 1978, as listed in Platt's Oilgram Price Service, May 25, 1978. An average tanker rate, weighted in proportion to type of petroleum products shipped, was calculated from spot tanker rates from the Gulf Coast to the New York area, as listed in Platt's Oilgram Price Service, *supra.*

[47] Estimate from National Energy Transportation System Map No. 12, "Total Petroleum Products Movement," accompanying NET.

revenue expectations from the project, usually in the form of through-put commitments by potential shippers. To the extent that rate regulation is at present ineffective, potential shippers would not want to pay the excessive tariffs that independent pipeline companies would charge. This would indicate that potential shippers of any significance would insist on becoming equity owners of any new pipeline rather than give throughput commitments to third parties. The result of this form of financing is an effective barrier to entry by independents.

In addition, it might be the case that several nonowner-shippers might be interested in new capacity along a route. Usually expansion of existing facilities, if they are fairly sizable, is the most efficient way of providing new capacity. We are well acquainted, however, with the practice of freezing equity participation in existing multiple-owner pipelines. It has been the usual practice to fix ownership shares in perpetuity within a few years after operations commence on these lines.[48] In addition, few nonowner-shippers have ever obtained equity participation in joint venture stock companies over which they make shipments after the lines are built. These practices have the result of stabilizing market shares, increasing barriers to entry, and, in general, reducing competition to a much greater extent than if the joint venture were divorced from ownership by vertically integrated oil companies.

III. The Merits of a Structural Solution

That completes my summary of the most significant elements of what I will refer to hereafter as "the pipeline problem." In the remainder of my testimony, I would like to turn to the issue of whether some sort of structural relief is the most appropriate solution. I hope I have made clear my belief that the problems are substantial, and that—and this is critical to this Committee—when closely examined, the problems are rooted not only in regulatory failure but also in the structure of current ownership. If that is the case, I begin predisposed to the view that it is very unlikely that refinements to the Interstate Commerce Act are going to cure the problem. Until the basic anticompetitive incentives inherent in the industry's current structure are altered, long-overdue improvements in the effectiveness of rate regulation can be nullified for much of the industry. This is not to say, however, that rate reform is in any way secondary to structural reform. For reasons which I hope will shortly become clear, that is not the case.

A. The Competitive Rules. Having stated my predisposition for structural relief, let me briefly review what I consider the only practical

---

[48] Deepwater Port Rep. at 12-13, 32-33.

alternative to divestiture available to curb the incentives of a vertically integrated pipeline to circumvent rate regulation. I am speaking, of course, of what have come to be known as the "Competitive Rules," a proposal put forth by the Department in response to the proposed licensing of LOOP and Seadock, the joint venture deepwater ports whose stockholders are almost exclusively vertically integrated petroleum companies.

Briefly, the Rules provide that

(1) The pipeline must provide open and nondiscriminatory access to all shippers, owner and nonowner alike.

(2) Any pipeline owner or shipper providing adequate financial guarantees can unilaterally request and obtain expansion of capacity within the technological limits of the pipeline.

(3) The pipeline must provide open ownership to all shippers at a price consistent with rate base/rate of return regulation.

(4) The ownership shares of the pipeline owners should be revised frequently to permit each owner's share to equal its share of average throughput.

As a potential solution to the pipeline problem, these Rules could no doubt be incorporated into the Interstate Commerce Act and yield substantial procompetitive benefits that approximate structural relief. As we made clear in the Deepwater Port Report, the Rules are designed to create a competitive environment in which the shippers' natural incentives to seek monopoly profits are used against each other, resulting in the dissipation of those very profits and the end result that every shipper is an owner shipping at or near true economic costs.[49]

Yet at the time we proffered those Rules to the Secretary of Transportation, we made clear in our Report that we regarded them as a second-best alternative to divestiture—one that would require continuous monitoring to ensure that such regulation was achieving the goals set out for it. We also made clear that in view of Congress's decision not to prohibit oil company ownership of deepwater ports, we felt constrained to explore procompetitive alternatives to divestiture.[50]

If we have reservations about the efficacy of this approach in the context of two deepwater ports in a small geographic area, our concerns greatly multiply when consideration is given to applying the Competitive Rules to a vast, nationwide pipeline network, with its multiple points of input, output, and interconnection. The problem is not that the incentives underlying the Rules do not translate to the pipeline context;

---

[49] *Id*. at 107-08.
[50] *Id*. at 103-05.

they do completely. Rather, the problem is one of the adequacy and burdens of implementation and regulatory enforcement. Rule 4, share readjustment, may be the only one that is largely self-regulating once it is implemented.

Rule 3, open ownership, presents substantial implementation and enforcement problems regarding the financial qualifications of proposed new owners and the transfer terms involved. For example, in the context of a proposed new owner for an existing joint venture stock company pipeline, some exclusionary financial responsibility standards would be essential to avoid improper dilution in quality of issued debt security. In addition, the readjustment of shareholder debt and equity to take account of changes in risk over time as equity owners join a project presents questions of considerable financial complexity. There are also definitional or financing problems that would arise when one sought to become either an owner of only part of the pipeline system or a non-shipping owner.

Rule 2, expansion, presents implementation and enforcement problems similar to Rule 3. The adequacy of financial guarantees, the technological feasibility of expansion, the relationship of expansion throughput guarantees to initial throughput guarantees, and the relative rights of owners and nonowners requesting expansion are all complicated questions likely to emerge in any proposed expansion. In the joint venture stock company context, any effort to adequately anticipate these problems in a shareholder's agreement will require a very complex instrument.

Lastly, Rule 1, nondiscriminatory access, is probably reflective of existing law under the Interstate Commerce Act. But as I indicated earlier, that kind of regulation is, as a practical matter, a very imperfect device for controlling throughput restriction.

In sum, while laudable, most of the Competitive Rules represent substantial regulatory burdens of implementation, enforcement, or both. I do not mean to suggest that these problems make the Competitive Rules unworkable, even in a joint venture stock company context. Indeed, we have seen in our deepwater port experience that the imposition by the Secretary of Transportation of a modified version of the Rules on LOOP, a joint venture stock company, appears to have been successful. The effort involved a total revision of LOOP's basic financial documents into something much more complex. The oil industry, as represented by LOOP, responded well to the implementing challenge posed by the Competitive Rules, even though they were quite novel at the time. In general, once the Rules are in place, and if they are implemented according to objective terms, the regulatory burden will be limited to adjudication of disputes about their application. The Competitive Rules also have one advantage over divestiture. Although the Rules would be imposed on all pipelines, those being operated in a competitive fashion will be largely unaffected by the operation of the Rules: only where pipeline throughput is not optimal are shippers likely to seek expansion and equity participation. Across-the-board divestiture, on the other hand, might affect some pipelines that do not present competitive problems.

On balance, I believe the basic flaw in the Competitive Rules, even in the LOOP context, may be characterized as their tendency to treat the symptoms rather than the disease. The treatment goes on forever, at considerable expense to all, and only the approximation of a cure is effected. Divestiture, on the other hand, attacks the problem directly because the problem is clearly rooted in the structure. What might appear to be a drastic approach, therefore, is in reality a clean and decisive break with the burdensome requirement of continuous, pervasive regulatory scrutiny. It is also one which essentially transforms pipeline owners into ratepayers. I trust it is clear from my earlier remarks that this sort of adversary relationship between carrier and ratepayer provides better regulatory information. This is a substantial positive by-product of divestiture. Indeed, our experience with the industry strongly suggests that oil company ratepayers would be very capable of defending their interests well in such an environment, greatly enhancing private and public scrutiny of the pipeline industry and improving regulation in general.

B. Social Costs of Divestiture. Weighed against these benefits of divestiture, however, must be its social costs. While we are not expert in measuring general social costs, we are acquainted with claims of efficiency arising out of vertical integration—efficiencies it is argued would be lost if divorcement of oil pipelines by the oil industry were required. To us, however, those costs appear to be much smaller than might be expected. I will address that issue in two contexts: prospective and retrospective divestiture.

1. *Prospective divestiture.* The principal difference between these two forms of divestiture is the degree of their impact on the existing pipeline industry. Retrospective divestiture would essentially require *existing* vertically integrated pipeline companies to sell or transfer their assets to companies or stockholders not affiliated with the oil industry. Prospective divestiture would merely prohibit the ownership and operation of *future* pipelines by oil companies. In addition, prospective divestiture should properly prohibit the future acquisition by oil com-

210

panies of pipelines, whether existing or planned. Viewed in this light, it is clear that prospective divestiture does not contemplate the sort of disruption of existing pipeline investments that retrospective divestiture represents.

But that does not end the inquiry into the feasibility of prospective divestiture. One major argument of divestiture opponents is the claim that divestiture promotes inefficiencies in terms of significant added transaction costs between separate corporations that are not organized around a continuous and efficient flow of petroleum. This argument, however, ignores several important facts: (1) The basic transaction costs between pipelines and the other stages of the industry are already commonplace. The vertically integrated pipeline industry's almost universal response to regulation under the Interstate Commerce Act has been to form pipeline company subsidiaries.[51] This approach has kept the regulatory spotlight, however dim, focused on the pipeline subsidiary and off the parent. This means, of course, that the form, if not the substance, of pipeline company bookkeeping transactions must be preserved. (2) The historical development of the industry also belies the claim that efficiencies were responsible for vertical integration or even played a significant role in that process. Our understanding of that history is that oil companies entered the pipeline industry to take advantage of monopoly pricing opportunities and to enhance their position in the oil industry overall, not to reduce transaction costs.[52] (3) Finally, the assumed "continuous flow" of petroleum from company production to company crude line to company refinery to company product line to company retail outlet is, in most cases, likely to be simply nonexistent—a holdover dream from the days of John D. Rockefeller. Substantial crude and product markets—spot, term, or exchange—are necessary between transportation links because perfect match-ups of company demand and company supply are uncommon.[53] This is especially true in the refining and marketing of foreign-based production. In sum, the claimed inefficiencies of pipeline divestiture in terms of significant added transaction costs are largely illusory.

Another argument is that capital formation would be considerably more difficult as a result of divestiture. In essence, the argument is that the current financing barrier to entry of independents into the industry—which I described earlier—would continue. But an independent pipeline

---

[51] *E.g.,* A. Johnson, *supra* at 97-98.

[52] *Id.* at 5.

[53] *See* Senate Judiciary Comm., Petroleum Industry Competition Act of 1976 (S. 2387), S. Rep. No. 94-1005, 94th Cong., 2d Sess. 135 (1976).

company would have no intrinsic difficulty obtaining financing if the oil industry were willing to make throughput commitments to it on the same basis that they are now willingly made by oil companies to their own pipeline company subsidiaries or to joint ventures with which they are affiliated.[54] We have looked at this question at some length and have every reason to believe that once divestiture forecloses equity participation by shippers, their throughput commitments to independent carriers would be much more readily forthcoming than they are now. However, I do foresee the possibility of some industry reluctance in some circumstances, grounded in the fact that throughput guarantees—which must be assumed to have some quantifiable value—might be sought without some consideration being offered in exchange. Absent some consideration, the shipper-guarantor might feel it was placing itself at a slight disadvantage with a shipper that was not a guarantor—provided, of course, the pipeline ever gets built. To overcome this problem, some form of incentive could be made available to the guarantor. For example, a tariff-related incentive, administered in a careful, nondiscriminatory manner, could be made available to *shipper*-guarantors. This is a matter to which the Department has been giving close study in recent months, in connection with the financing plans of such disparate projects as the Northern Tier Pipeline and the Seadock project (as taken over by the Texas Deepwater Port Authority). Moreover, it is a question that we have proposed for consideration by the FERC in *Valuation of Common Carrier Pipelines*.[55]

Of course, as I indicated earlier in my testimony, prospective divestiture is only one of the alternative means of dealing with the competitive problems of shipper-ownership of pipelines. The Administration has not yet taken a position on this issue. In the Department's view, however, the benefits of prospective divestiture are great and its social costs small. In particular, we attach little credence to the claims that it won't work or would reduce efficiencies or would end the natural growth of the industry. We believe a clear and convincing case for prospective divesture can be made.

2. *Retrospective divestiture.* Because of its much clearer potential for disruptive effects, we have greater concern about across-the-board restructuring of the entire existing interstate pipeline industry by retrospective divestiture. Retrospective divestiture, therefore, requires careful analysis of its potential impact on the industry as it exists today. An

---

[54] *Id.* at 128-30.
[55] DOJ 4/3/78 Memo at 79.

important point to keep in mind, however, is that retrospective divestiture need not necessarily focus on the industry as a whole; rather, a class of pipelines likely to represent significant concentrations of market power may properly serve as the exclusive target of such structural relief. In this way, the benefits of divestiture could be largely, if not entirely, preserved, at the price of far less industry disruption, although there would be significant legislative and administrative costs with respect to both identifying the class of pipelines to be divested and supervising the divestiture.

The type of disruption and costs that divestiture might foster would be unfortunate, but I would emphasize that it would be of limited duration and that there is good reason to believe from past divestiture experience that the targets of divestiture would survive the process with minimal capital formation problems.[56] In addition, the traditional problems of identifying and collecting corporate assets for sale or transfer would be substantially lessened in this industry: most, if not all, the pipelines likely to be affected are already operated as joint ventures with their own corporate identity, or else as separate subsidiary corporations of their vertically integrated parents.

In sum, if corrective surgery is limited to the class of pipelines likely to exhibit substantial market power, we believe a strong case for retrospective divestiture can be made. In reaching this conclusion, I certainly agree with divestiture *opponents* that, before such restructuring is imposed, it is up to divestiture *proponents* to make a convincing case that there is a problem and that there is no effective, less drastic solution. I think that in terms of identifying the incentives and opportunities for regulatory circumvention, our burden has been met. What yet needs to be done is to define the class of pipelines whose divestiture would, on balance, be of social benefit. As a general matter, the major interstate crude and product trunk lines can be an initial point of focus, as they are the most likely to evidence substantial market power of the type I have discussed. It would then be necessary to study each of the relevant markets served by such pipelines in some detail—after the fashion of the Sohio Pipeline Report—on several key items: the pipeline's market power, the owners' share in upstream and downstream markets, and manifestations of throughput restriction behavior. Finally, such information would have to be drawn together in a way that would allow some general conclusions regarding the class of pipelines that would comprise a proper subject of retrospective divestiture.

---

[56] S. Rep. No. 94-1005, *supra* at 133-135.

IV. The Primacy of Rate Reform

I have saved for last one vital point that must not be lost sight of in any divestiture effort, litigative or legislative. The point is best made in the context of the story of the sale of the vertically integrated Great Lakes Pipeline Company to the independent Williams Brothers Pipeline Company in 1966. Rounding the figures slightly, Williams paid $280 million for a pipeline system whose net original cost was only $100 million and whose ICC valuation rate base was $160 million. The purchase price, therefore, reflected about $120 million of capitalized monopoly profits to Great Lakes.[57] The lesson is clear: divestiture must be preceded by general rate reform in the industry, or we run the risk of artificially high transfer prices or like distortions that would capitalize monopoly profits for those vertically integrated pipelines with monopoly power.

There is also another reason why rate reform should precede divestiture. Rate reform will prevent all pipelines, vertically integrated or independent, from obtaining monopoly profits at the pipeline level. Absent rate reform, independently-owned pipelines would have the incentive and ability, subject only to market power constraints, if any, to capture in the tariff whatever excessive profits are allowed by imperfect rate regulation. The same would be true for vertically integrated pipelines. But for some vertically integrated pipelines there might be a substantial incentive to maximize the total firm's profitability by maximizing throughput on the pipeline. To do this, the pipeline would avoid throughput restriction behavior, even if not forced to do so. For example, as our throughput restriction analysis suggests, pipelines whose parent companies are significant net upstream sellers or significant net downstream buyers may adopt this strategy. The end result is a throughput that is not distorted by the monopoly rate-setting strategy of the unregulated independent.

I trust that these points on rate reform adequately explain the approach we are taking and our current allocation of priorities in the Department with respect to oil pipelines. The energies of our pipeline experts are now largely consumed in rate reform efforts. Unless Congress were to legislate the types of reform of rate regulation we are working hard to achieve, that effort ought to take precedence over pipeline divestiture efforts. We do expect, however, to reach some definite conclusions in the coming months regarding (1) the implications of various financing possibiilties for independent pipelines and (2) identification, on the basis of market power analyses, of a class of pipelines

[57] Petroleum Products, Williams Brothers Pipe Line Company, 351 I.C.C. 102, 107-08 (1975).

that should properly be the subject of any retrospective divestiture effort. In the meantime, our efforts on the rate reform front show signs of bearing fruit. As we continue these efforts, you may be assured of our continued cooperation with Congress on efforts to implement a divestiture proposal that takes account of the primacy of rate reform.

That concludes my prepared statement of the Department's views on the competitive issues created by shipper-ownership of petroleum pipelines. I would be happy to answer any questions the Committee may have, or to elaborate on any specific point I have discussed here today.

Supplement 2

REPORT OF THE ANTITRUST DIVISION
DEPARTMENT OF JUSTICE
ON THE COMPETITIVE IMPLICATIONS OF
THE OWNERSHIP AND OPERATION BY
STANDARD OIL COMPANY OF OHIO OF A LONG BEACH,
CALIFORNIA-MIDLAND, TEXAS CRUDE OIL PIPELINE

June 1978

INTRODUCTION

This report analyzes the competitive implications of the application to the Secretary of the Interior by the Standard Oil Company of Ohio (SOHIO) for a right-of-way grant across federal lands in order to construct and operate a crude oil pipeline system from Long Beach, California, to Midland, Texas, pursuant to Section 28 of the Mineral Leasing Act, 30 U.S.C. §185. The proposed pipeline would create a transportation link between the West Coast and the midcontinent areas of the United States, connecting by pipeline for the first time these two major crude oil markets. On December 22, 1977, the Department of Interior sought the advice of the Antitrust Division of the Department of Justice as to the appropriate terms and conditions to the right of way grant "which can protect the public interest and meet the requirements of the Mineral Leasing Act that the pipeline be operated as a common carrier."

An Antitrust Division recommendation of terms and conditions to the right-of-way grant must necessarily be preceded by an analysis of whether SOHIO's ownership and operation of this pipeline is likely to have serious anticompetitive effects. Based upon our analysis of the unique circumstances faced by the proposed SOHIO pipeline, we have concluded, from a competition perspective, that there is no reason to deny SOHIO's application. If the Secretary decides to grant the application, we do suggest that some clarification of SOHIO's common carrier obligations be made incidental to the right-of-way grant. Summarized below is the reasoning and recommendations that are more fully explicated in the main body of our report.

SOHIO plans to build a port and pipeline facility designed to move substantial quantities of Alaskan crude oil from the West Coast to the

217

Midwest refining area. The crude oil supply picture for the foreseeable future is an excess of supply on the West Coast and a shortage in the Midwest. In recent years, this Midwest shortage has been alleviated by ever-increasing imports of crude, whose reduction and displacement by Alaskan crude is now a matter of Administration policy. As the SOHIO pipeline will substantially reduce transportation costs associated with moving Alaskan crude into the midcontinent area, its construction is consistent with, and a part of, that policy. None of the several West Coast to Midwest pipeline proposals is as far along in the planning stage as SOHIO's, and the denial of the SOHIO application would be unfortunate.

Given this unique set of circumstances, we have sought to evaluate the likely incentives for SOHIO, as a vertically integrated petroleum company, to misuse this particular transportation facility in an anti-competitive manner. Our conclusion that such incentives are minimal in SOHIO's case results, in part, from the existence of the very supply imbalance problems the pipeline is designed to alleviate, and other facts that suggest SOHIO will be unable to affect crude oil prices by virtue of its pipeline ownership. This conclusion also depends upon effective regulation of the pipeline's rates by the Federal Energy Regulatory Commission and the effective enforcement of workable common carrier obligations, both of which we are seeking to accomplish.

With respect to the downstream market, even at its ultimate, expanded capacity, the SOHIO pipeline will afford SOHIO only a relatively small share of Midwest crude oil supply. Moreover, the market price for Alaskan crude in the Midwest will continue, for the foreseeable future, to be set by the price of imported crude, which can be expected to remain a significant source of supply in the Midwest for many years. Under such market conditions, SOHIO should have little incentive to artificially control pipeline throughput, since to do so would not yield it any benefit in terms of the price it can expect to receive for its crude oil.

Upstream, on the West Coast, a similar conclusion is evident. Pipeline throughput restriction would decrease the value of SOHIO's vast Alaskan holdings by increasing West Coast surpluses and depressing crude prices in that area. Excessive throughput expansion to increase West Coast prices by creating an artificial shortage would probably benefit SOHIO's competitors more than SOHIO: virtually all the oil that would be drained from the West Coast to the midcontinent would have to be SOHIO's, while the remaining oil would belong to SOHIO's competitors. Moreover, the world price of crude is a limiting factor in any throughput manipulation strategy since any attempts to raise price above

218

that level would simply result in a resumption of imports to the West Coast.

Our conclusions about the SOHIO pipeline stand in sharp contrast to the situation, like that involving recent deepwater port applications, where a vertically integrated pipeline controlling a significant share of the downstream crude market can exercise substantial market power by artificially restricting pipeline throughput. For reasons that are fully detailed in the report, such conditions are either not present or are neutralized by countervailing market forces in the case of the SOHIO pipeline.

The recommendation that we have made to the Secretary is that the grant of right-of-way be conditioned upon a clarification of the common carrier obligation. We believe the entire facility—port, storage facilities and pipeline—should be bound by an express common carrier obligation. In addition, the Secretary should approve fair and equitable prorationing rules that are coextensive with the reach of the common carrier obligation.

Our report begins with a description of the SOHIO proposal, and our understanding of the relevant economic and regulatory context. The next section provides the detailed competitive analysis of these basic facts in terms of the incentives we have been able to identify. The final section recommends the clarification of the common carrier and prorationing obligations.

I. Background

A. Description of the SOHIO Proposal[1]

SOHIO, through its subsidiary, SOHIO Transportation Company of California, proposes to construct and operate a crude oil transportation system from the Port of Long Beach, California to Midland, Texas, where it would connect with existing crude oil pipelines. Under the SOHIO plan, oil would enter the pipeline through a new port facility to be constructed at Long Beach, and move from there through a pipeline system which would consist primarily of an existing, surplus natural gas pipeline which SOHIO would acquire from El Paso Natural Gas Company and Southern California Gas Company and convert to petroleum use. As proposed, the port facility would have a capacity of 700,000

---

[1] See generally Bureau of Land Management, *Final Environmental Impact Statement: Crude Oil Transportation System—Valdez, Alaska to Midland, Texas (As proposed by SOHIO Transportation Company)*, Executive Summary (1977) (EIS), and Port of Long Beach, *Overview, Sohio—West Coast to Mid-Continent Pipeline Project* (1977).

barrels per day (b/d), would be owned by the Port of Long Beach,[2] and would be operated by SOHIO under a long-term agreement. The initial capacity of the pipeline would be 500,000 b/d, or 5/7ths of the port's capacity. SOHIO estimates that the system could be constructed in 18 to 22 months.[3] SOHIO has also indicated that the capacity of the pipeline could be doubled to 1,000,000 b/d at a later date if a second natural gas pipeline, one still in use, is converted to carry crude oil.[4]

B. Economic Context[5]

1. Crude Oil Supply on the West Coast. The onset of production of substantial quantities of Alaskan North Slope crude oil has changed the sources of crude oil used to satisfy West Coast demand. Prior to North Slope production, West Coast refineries supplemented local crude production with substantial quantities of imported oil. The price of incremental quantities of petroleum in PAD V[6] reflected the world market price together with the appropriate transportation differential.[7]

---

[2] The Port of Long Beach is operated by the City of Long Beach Harbor Department, which is a part of the city itself. Long Beach is a municipality chartered under the Home Rule Provisions of the California constitution. The Port derives its authority from the charter of the City of Long Beach rather than from the provisions of the California code dealing with ports and harbor districts.

[3] Statement of Fred G. Garibaldi, Project Manager for the Long Beach to Midland Pipeline, *Hearings on S. 1868, National Crude Oil Transportation Act of 1977 before the Senate Commerce Committee*, 95th Cong., 1st sess. (September 14, 1977) (Senate Hearings).

[4] EIS at I-41. The proposed grant of right of way is limited to the initial phase of development, 500,000 b/d.

[5] See Appendix map, Crude Oil Supplies, for an overview of U.S. crude oil supply.

[6] Petroleum Administration for Defense District (PAD) V includes the states of Alaska, Arizona, California, Hawaii, Nevada, Oregon and Washington.

[7] At present the price of crude petroleum in the United States is regulated by the Department of Energy. Crude oil sources from the Alaskan North Slope, which constitute the bulk of PAD V reserves, are effectively not subject to price limitations because, while the wellhead price is regulated at the so-called "upper tier" level (presently $11.60 for North Slope crude, see 42 Fed. Reg. 41566, August 7, 1977), the delivered price (wellhead price plus transportation) is not. Since transportation costs from the North Slope of Alaska exceed the difference between the wellhead price limitation and the delivered world market price in PAD V, the North Slope wellhead price actually received by producers is determined by the delivered world price less transportation costs from the North Slope, not by price regulation. However, the price received by producers of non-North Slope PAD V crude is not subject to high transportation costs. Thus, the price received by those producers is limited by Department of Energy price regulation.

The value of the difference between the price received by all producers in the U.S. whose prices are limited by the controls and the market value of the oil is redistributed to all U.S. refiners by the complex entitlements program. The analysis in this report is not affected by the present system of wellhead price controls and entitlements.

Production of crude from the North Slope of Alaska today accounts for approximately 1.2 million b/d.[8] In several years production from the North Slope is expected to reach 2 million b/d.[9] Pursuant to Section 28(u) of the Mineral Leasing Act, 30 U.S.C. §185(u), President Carter has declined to approve the exportation of North Slope Alaskan crude oil. Therefore, any PAD V crude oil production which cannot be absorbed on the West Coast must be shipped to other U.S. markets.

Alaskan oil will not, immediately at least, eliminate the need to import some oil for use in PAD V refineries. Alaskan North Slope crude has a density of 27° API and contains 1.04 percent sulfur by weight.[10] Product demand, environmental concerns, and the present configuration of PAD V refineries contribute to the need to import into PAD V some crude feedstocks of lighter gravity or of lower sulfur content.[11]

In addition to North Slope crude production, the Elk Hills Naval Petroleum Reserve is expected to increase petroleum production for commercial consumption.[12] Present plans call for future production levels of 200–250,000 b/d but this level may be increased to 300–350,000 b/d.[13] Additional sources of crude oil may also be developed in off-shore areas adjacent to Alaska and Southern California.

---

[8] The precise capacity of TAPS is a factual issue to be resolved in the *Trans Alaska Pipeline System* Proceeding, FERC Docket No. OR 78-1.

[9] See Statement of Alton W. Whitehouse, Chairman of SOHIO, in the *Trans Alaska Pipeline System* Proceeding, FERC Docket No. OR 78-1 (November 30, 1977).

[10] See *Oil and Gas Journal,* June 7, 1976.

[11] Gravity, which is a measure of the density of crude petroleum, is measured in API degrees, according to a system established by the American Petroleum Institute. A lower API degree number indicates a higher gravity and a "heavier" crude. The mix of products that can be obtained from any given barrel of crude depends on both the gravity of the crude and the extent of refining. The output of a barrel of heavier crude will tend to yield a greater proportion of heavier products, e.g., residual fuel oil, for a given degree of refining than will a barrel of lighter crude. Alternatively, in order to produce the same proportion of lighter products, e.g., motor gasoline, as a barrel of lighter gravity crude, a barrel of heavier gravity crude must be refined more intensively. A high sulfur content, such as that found in Alaskan North Slope crude, requires special expensive processing by refineries, and may be the source of environmental concern.

[12] The Naval Petroleum Reserves Production Act of 1976, 10 U.S.C. §7420 *et seq.,* provides that the Elk Hills reserves shall be produced at the maximum efficient rate as determined by the Secretary of the Navy subject to the approval of the President. Congress apparently indicated what it thought to be the "maximum efficient rate of production" when it required that a pipeline with adequate capacity to accommodate "not less than" 350,000 b/d be constructed and operational by 1979. 10 U.S.C. §7422.

[13] See President Carter's Statement, December 21, 1977; and *Oil Daily,* December 22, 1977. The production rate for 1978 is estimated to be approximately 98,795 b/d. Department of Energy, Invitation for Bids for the Sale of Crude Oil from Naval Petroleum Reserve No. 1, as amended (December 8, 1977).

The increase in the crude oil available in PAD V may exceed the increase in consumption. As of December 2, 1977, PAD V crude oil inputs to refineries were 2,259,000 b/d.[14] At that time, total supply was 2,509,000 b/d[15], of which 250,000 b/d was excess. It has been variously estimated that the PAD V excess supply will range between 300,000 b/d and 600,000 b/d in 1978[16] increasing to between 750,000 and 900,000 b/d by 1982.[17]

At the present time, the only practical way to transport surplus crude oil from the West Coast into the "midcontinent area," PADs II and III,[18] is by tanker through the Panama Canal to ports on the Gulf Coast. This mode of transportation is costly and dependent upon use of tankers whose number is quite limited.[19] For example, SOHIO has

---

[14] API Refinery Report, *Oil and Gas Journal,* December 12, 1977. We attempted to obtain recent statistics from the Department of Energy, Bureau of Mines. However, the most recent figures that the Bureau had available were for June 1977. These figures, of course, predate North Slope production. We compared the daily average of the Bureau's statistics for the twelve months ending June 30, 1977 with the API averages for December 1977 and determined that the differences could be explained by initiation of North Slope inputs.

[15] The sources of supply, at that time, were: Alaska—868,000 b/d (of which 700,000 were from the North Slope), California—947,000 b/d, and imports—694,000 b/d. Despite the surplus, crude oil of specific gravity and sulfur content must be imported as feed stocks for PAD V refineries. See *supra.*

[16] Federal Energy Administration (FEA), *Equitable Sharing of North Slope Crude Oil* at 11-13 (April, 1977). Statement of Fred G. Garibaldi, Project Manager for the Long Beach, California to Midland, Texas pipeline, Senate Hearings (September 14, 1977).

[17] EIS at I-3. Projections of total PAD V production are unreliable due to uncertainties about the levels of production from Alaskan fields and West Coast off-shore areas. In addition, the level of surplus will be affected by the ability and willingness of West Coast refiners to modify their facilities to use North Slope crude oil. See note 11, *supra.*

[18] PAD II includes the states of Ohio, Kentucky, Tennessee, Indiana, Illinois, Michigan, Wisconsin, Missouri, Iowa, Minnesota, North Dakota, South Dakota, Nebraska, Kansas, and Oklahoma. PAD III consists of Arkansas, Alabama, Louisiana, Mississippi, New Mexico, and Texas.

[19] Under the existing law, goods transported in domestic trade must be carried by vessels registered in the United States. 46 U.S.C. §883 ("Jones Act"). All tankers registered in the U.S., which are constructed with federal assistance under the "construction-differential subsidy" (CDS) program are prohibited from being used in domestic commerce without the permission of the Secretary of Commerce. 46 U.S.C. §1101-1294. The number of non-CDS ships is apparently so small that the Maritime Administration (MarAd) determined that in order to adequately transport oil from Alaska, it would be necessary to permit the use of some super-sized CDS tankers (greater than 100,000 dead weight tons [dwt]) (*The U.S.-Flag Tanker Fleet and Domestic Carriage Requirements,* U.S. Dept. of Commerce Maritime Administration, October 21, 1976). MarAd will require their owners to repay the Secretary an amount of the CDS proportional to the period of the vessel's life during which it is used in the trade. 42 Fed. Reg. 33035 (June 29, 1977).

chartered a super-sized tanker to carry its oil from Valdez to a stationary facility off the west coast of the Isthmus of Panama. The oil will be reloaded onto smaller tankers able to traverse the Canal and deliver oil to Texas and Louisiana ports. The cost of transporting oil from California to the Gulf Coast via the canal route has been estimated to be somewhere between $1.50 and $2.60 per barrel.[20] Since Alaskan oil must compete with imported oil at the Gulf ports,[21] any West Coast producer selling in the midcontinent area must bear this additional expense.

The impact of these developments has been to change the market for crude petroleum in PAD V. Before these changes, the market price of new PAD V crude production was limited by the price of imported oil, including transportation costs to the West Coast. Under present circumstances, the price of crude on the West Coast can potentially be depressed below the delivered price of world oil on the West Coast. The lower limit of the price of PAD V crude will now equal the price of that crude in other areas of the United States less the cost of transporting that oil by the most expensive mode of transportation which must be used to ship any surplus to the midcontinent area. Therefore, any effect of the adoption of the SOHIO proposal will necessarily be restricted to a range of crude prices in PAD V within these two limits.

2. The Mid-Continent Crude Oil Supply. While PAD V will be experiencing a crude oil "glut," other areas of the United States are importing, and are expected to continue to import, a large part of their crude oil supply. The midcontinent area of the United States, PADs II and III, at present relies on imported oil to meet much of the demand. As of June 30, 1977, crude oil refinery inputs were 3,744,899 b/d in PAD II and 6,079,551 b/d in PAD III, totaling 9,824,450 b/d for this broad market area. At that time in PAD III imports averaged 2,218,356 b/d, 36 percent of total average inputs. Local production averaged 5,724,126 b/d. Imports into PAD II averaged 1,293,167 b/d, 35 percent of total average inputs, and 897,677 b/d were produced in PAD II.[22]

Once Canada implements its announced policy to end exports and restrict its production to Canadian refineries, PAD II's only access to

---

[20] See *Oil Daily*, June 27, 1977, and Testimony of Alton W. Whitehouse, Jr., Hearings, in the *Trans-Alaska Pipeline System* Proceeding, FERC Docket No. OR 78-1, Transcript at 1554-1555, 1592-1594, 1609-1610 (February 14, 1978).

[21] See discussion, *infra*.

[22] DOE Monthly Energy Reports, "Crude Petroleum, Petroleum Products, and Natural Gas Liquids," July 1976–June 1977. The sum of imports and local production for each PAD is not equal to the total refinery inputs for each PAD, as some local production is shipped to other PADs.

imported oil will be transportation from other PADs—in most cases by barge or pipeline from PAD III.

A significant supply problem is developing in the Northern Tier states of Idaho, Montana, North Dakota and Minnesota.[23] The refineries serving these areas have historically relied upon Canadian exports, a source which has been curtailed for some time and which may soon dry up entirely. The Canadian government recently indicated it would consider interim "swapping agreements,"[24] but these can at best be only a short-term solution. Once Canadian exports stop completely, many of the Northern Tier refineries will be cut off from their present source of supply. At present, there are no economically efficient alternative methods for delivering crude oil to some of these refiners.[25]

3. The Administration's Policy. The Administration has taken the position that the midcontinent deficiencies should be satisfied to the extent possible by domestic resources, especially North Slope crude oil. Consistent with this policy, the Administration has taken the following actions:

First, the Administration is on record as supporting the development of a transcontinental pipeline system. On September 14, 1977, John F. O'Leary, then Administrator of the Federal Energy Administration, told the Senate Committee on Commerce, Science and Transportation that "the Administration firmly believes that the construction of at least one, and perhaps two, of the proposed west-to-east pipelines is urgently needed in order to assure an efficient and economic means of delivering Alaskan crude oil to those areas of the country which need it."[26]

---

[23] The term "Northern Tier states" has been used in various contexts to include every state that borders on Canada or one of the Great Lakes from Washington to Ohio. In this report, the term is limited to the four states listed in the text.

[24] *Platt's Oilgram News,* December 27, 1977.

[25] Minnesota's shortage may be relieved by two new pipelines to Minneapolis area refineries. Williams Pipeline Company recently began transporting crude oil via its 18 inch 80,000 b/d pipeline from Oklahoma, (*Oil Daily,* November 8, 1977). Koch Industries Inc. has announced construction of a 24 inch pipeline to carry, initially, 130,000 b/d of crude oil from Wood River, Illinois. *Oil and Gas Journal,* October 31, 1977.

These pipelines, however, may not relieve the shortages in the other Northern Tier states. Several companies are considering using railroad tank cars, a very expensive means of transporting crude oil, to supply these refineries. For a comparison of the costs of shipping petroleum by various modes of transportation, see *National Energy Transportation Report Volume I-Current Systems and Movements,* prepared by the Congressional Research Service at the request of Senator Henry Jackson, Chairman of Senate Committee on Energy and Natural Resources and Senator Warren G. Magnuson, Chairman of Senate Committee on Commerce, Science, and Transportation, 95th Cong., 1st sess. (May 1977) (National Energy Transportation Report).

[26] Senate Hearings (September 14, 1977). Cf. Federal Energy Administration, *Equitable Sharing of North Slope Crude Oil,* April 1977.

Second, the Department of Energy recently changed its entitlements regulation to encourage the flow of Alaskan oil to the midwest rather than the "shutting in" of California crude. The program, which began on January 1, 1978, increased the entitlement value for lower tier California crude oil by $1.74, thus making California crude more attractive to West Coast refiners. This should encourage California production, thereby increasing both West Coast supplies and the incentive to ship excess crude oil eastward.[27]

Most recently, on December 21, 1977, President Carter announced that the Administration would encourage expansion of production levels for domestic fields, particularly at Prudhoe Bay and Elk Hills, in an effort to reduce imports and strengthen the dollar. At the same time, the President reaffirmed his support of the measures to expedite the construction of a transcontinental pipeline.[28]

4. SOHIO's Position. The SOHIO proposal would offer one response to the need to transport oil from the West Coast to the midcontinent area. As discussed above, the SOHIO pipeline would deliver 500,000 b/d from PAD V to PADs II and III, reducing that region's dependence on imports.

SOHIO is the largest owner of Alaskan crude oil, but apparently has no ready market for it.[29] Together with its parent, British Petroleum, Ltd. (BP),[30] SOHIO owns slightly more than half of the Prudhoe Bay reserves and slightly less than half of the capacity of TAPS. As production at Prudhoe Bay is limited by the TAPS capacity, 1.2 million b/d, SOHIO has slightly more than 600,000 b/d to market.

SOHIO has no significant crude oil production outside of the North Slope.[31] As of January 1, 1977, SOHIO had a total refining capacity of 288,000 b/d, all from refineries located in Ohio. This constitutes 49 percent of Ohio's refining capacity of 589,950 b/d.[32] SOHIO's share of PAD II refining capacity is 7 percent of the 4,125,171

---

27 *Oil and Gas Journal,* December 19, 1977.

28 *Platt's Oilgram News,* December 22, 1977. See also *New York Times,* December 22, 1977.

29 *Oil and Gas Journal,* July 18, 1977; *Washington Post,* October 17, 1977.

30 British Petroleum acquired its majority ownership of SOHIO as a result of an agreement which transferred BP's Prudhoe Bay holdings to SOHIO.

31 First National City Bank, *Energy Memo,* x:2 (1974).

32 *Oil and Gas Journal,* Annual Refining issue March 28, 1977. SOHIO's share of motor gasoline sales in Ohio was 22.3 percent in 1976. We attempted to obtain additional information on SOHIO's share from the Bureau of Mines but the Bureau informed us that this information was proprietary and could not be released to us pursuant to Section 14 of the Federal Energy Administration Act of 1974, 15 U.S.C. §773.

b/d total. This amounts to only 3 percent of the combined total refining capacity in PAD II and III (almost 11 million b/d). In terms of the gallonage of gasoline sold, SOHIO and BP together accounted for 1.75 percent of the national total in 1976, which would have made them the twelfth largest marketer. Individually they ranked fourteenth and sixteenth, respectively.[33]

SOHIO does not own any refining capacity on the West Coast and has been unable to secure buyers for all of its production.[34] Since its significant Prudhoe Bay partners all have substantial West Coast refining capacity, as a practical matter SOHIO owns most of the "excess" crude oil in PAD V. At present, SOHIO must ship its excess to PADs II and III via the Panama Canal at an additional cost of $1.50 to $2.50.[35] In addition, with TAPS at its design capacity, Exxon may also develop a surplus.[36]

5. Alternative Pipeline Proposals. In addition to the Long Beach to Midland Pipeline, four other pipeline projects have been proposed to alleviate the West Coast surplus by transporting the oil to the midcontinent area. These are (1) the ARCO-Trans Mountain Pipe Line reversal; (2) the Kitimat Project; (3) the Four Corners Pipeline reversal; and (4) the Northern Tier Pipeline.

a. General descriptions. 1. The ARCO-Trans Mountain Pipe Line Project. The Atlantic Richfield Company (ARCO) and Trans Mountain Pipe Line Company, Ltd. (Trans Mountain), formed a joint venture to develop a crude oil transportation system from Cherry Point, Washington, to Edmonton, Alberta, where it would connect with the existing pipeline system for delivery to the midwestern states. The system would be created by reversing the present southwest flow of crude oil to northeast. Initially, oil would be transported in each direction during alternate periods producing a "yo-yo" effect.[37] This conversion process would take a little more than one year and would cost $140 million. During its initial phase, this pipeline would carry 180,000 b/d eastward. With some new pipeline construction the volume could be increased to 350,000 b/d.[38]

---

[33] *1977 National Petroleum News Factbook Issue.*

[34] *Washington Post,* October 17, 1977; *Oil and Gas Journal,* July 18, 1977.

[35] See *supra.*

[36] *Oil and Gas Journal,* July 18, 1977.

[37] This is the result of Canadian government policy which requires that British Columbian refineries use oil from Canadian sources rather than from Alaska.

[38] Statement of W. M. Witten, Senate Hearings, September 14, 1977.

ARCO is also a TAPS owner and a Prudhoe Bay producer. (It owns 17.8 percent of the oil.) Unlike SOHIO, however, ARCO has sufficient capacity in its own West Coast refineries to use virtually all of its initial North Slope production.[39]

On October 5, Congress passed the Marine Mammals Protection Act, 91 Stat. 1167 (1977), which included an amendment offered by Senator Magnuson prohibiting new port facilities in the Puget Sound east of Port Angeles. The Act thus prohibits the expansion of the Cherry Point facility which would have served the Trans Mountain Pipe Line. Despite suggestions that the port facility be relocated, perhaps to Port Angeles, Trans Mountain has announced it will withdraw the application which it had filed with the Canadian National Energy Board (NBEB).[40]

2. The Kitimat Pipeline. This proposal called for the construction of a pipeline from a new port facility to be constructed at Kitimat, British Columbia, to Edmonton, Alberta, where it would connect with the Interprovincial Pipeline and other pipelines for delivery to the Midwest. The present plans estimate that the 753 miles of 36-inch pipeline would cost $750 million (Canadian) to construct, and that the initial capacity would be 500,000 b/d, expandable to 700,000 b/d within five years. The project would take two years to construct.[41]

The project was proposed by Kitimat Pipeline Ltd., a consortium of refiners, pipeline companies, and oil companies. The original owners were Koch Industries, Inc. (26 percent); Ashland Oil Canada, Ltd. a unit of Ashland Oil, Inc. (24 percent); Interprovincial Pipe Line, Ltd. (15 percent); Murphy Oil Corp. (15 percent); Hudson's Bay Oil & Gas, Ltd. (15 percent); and Farmers Union Central Exchange, Inc. (5 percent). During the summer of 1977, Conoco Pipe Line Co. replaced Hudson's Bay Oil & Gas, Ltd., as a participant in the project. Both are affiliates of Continental Oil Co. which owns refineries in Minnesota and Montana dependent upon Canadian oil. Koch Industries dropped out of the project in October 1977 to concentrate on the "Minnesota" pipeline being constructed by its subsidiary, Northern Pipe Line Co., to connect Twin Cities refiners with the Capline system which originates in Louisiana.[42] SOHIO has reportedly succeeded to Koch's 26 percent interest.[43]

---

39 *Oil and Gas Journal,* July 18, 1977; *Washington Post,* October 17, 1977.

40 *Oil Daily,* November 17, 1977.

41 *Wall Street Journal,* January 11, 1978.

42 *Oil Daily,* October 11, 1977. See note 25, *supra.*

43 *Wall Street Journal,* January 11, 1978.

Like the Trans Mountain proposal, this project must be approved by Canada's NEB. Although Kitimat Pipeline, Ltd., withdrew its application in the face of competition from the Trans Mountain Project last April, the project was revived after the Magnuson amendment became law. Recently, however, a staff report to Canada's West Coast Port Inquiry Commission questioned whether it was in the Canadian national interest to permit the construction of a port facility for imported oil, as it might deter the development of Canadian oil resources.[44]

3. Four Corners Pipeline. The only project actually under construction at present is the Four Corners Pipeline.[45] In late 1976, ARCO purchased the pipeline from its five partners (Continental Pipeline Co., Shell Pipeline Co., Standard Oil Co. of California, Toronto Pipeline Co., and Superior Oil Co.). The pipeline had been constructed to carry 70,000 b/d of crude oil from Red Mesa, Colorado to California but recent throughput has been only 3,000–3,500 b/d. The $12 million project will be ready in the spring of 1978 with an initial capacity of 28,000 b/d. Capacity could be expanded to 140,000 b/d. When the oil reaches the Four Corners area it could be sold to New Mexico refiners or transferred to the Texas-New Mexico Pipeline, which runs to the Houston area.

4. Northern Tier Pipeline.[46] This proposed pipeline would be owned and operated by the Northern Tier Pipeline Company, a consortium of seven independent companies. At this point no major oil company is a participant.[47] This proposal calls for a 42-inch pipeline from Port Angeles, Washington, southward around Puget Sound and then eastward to the intersection with Glacier Pipeline System northwest of Billings, Montana, and a 40-inch pipeline continuing on to Clearbrook, Minnesota, the intersection with Lakehead and Minnesota Pipelines. The initial capacity of the pipeline would be 700,000 b/d, eventually expanding to 933,000 b/d.

As planned, the Northern Tier Pipeline would deliver oil to the Midwest as well as to points along its route. The sponsors indicate that, assuming the pipeline becomes operational in 1980, the initial actual throughput would be 540,000 b/d, of which 125,000 b/d is planned

---

[44] *Oil Daily,* November 17, 1977.

[45] *Oil and Gas Journal,* October 31, 1977.

[46] See generally Northern Tier Pipeline Co., "Northern Tier Pipeline System Benefits to the Northern Tier States and the Nation" (April 27, 1977); "General Description of the Northern Tier Pipeline System," December 14, 1976.

[47] United States Steel Pipeline Co., a subsidiary of U.S. Steel Corp., has reportedly acquired approximately a 40 percent interest in the Northern Tier Pipeline Company. *Oil Daily,* November 17, 1977; *Platt's Oilgram News,* November 2, 1977.

for destinations along the pipeline route,[48] with 415,000 b/d planned to be shipped through to Clearbrook and on to refineries in Minnesota, Wisconsin and Illinois. When these projections are extended to 1984, estimates are 193,000 b/d along the route and 740,000 b/d to Clearbrook and beyond.

b. Overview of pipeline proposals. As described above, each of the proposed pipelines terminates in the midcontinent area. At present, the midcontinent area is supplied with crude oil via the extensive crude oil pipeline network that connects the Gulf Coast ports and indigenous producing areas with refining centers. The Gulf Coast ports receive imported oil and domestic oil produced in Alaska and shipped via the Panama Canal route. In late 1980, the Louisiana Offshore Oil Port (LOOP) is scheduled to begin receiving oil imported on very large crude carriers (VLCCs) which cannot use existing Gulf Coast ports. LOOP will be connected by pipeline to the existing pipeline network for delivery throughout the area, and especially to Chicago area refineries via Capline.

Although each of these proposed systems is able to serve various markets en route, each was, with the exception of the Four Corners Pipeline, designed primarily to serve some part of the midcontinent area. In order to be economically efficient, each system apparently must be able to deliver oil at a total cost comparable to other systems serving significant parts of this area. The analyses made by the proponents of the various pipeline proposals indicate that each would be able to deliver oil to the Chicago area at a cost competitive with existing routes.

On the other hand, these pipelines face varying obstacles to construction and are in different stages of obtaining financing and permits. For example, the Trans Mountain Pipeline was effectively blocked by the passage of the Marine Mammals Protection Act, 91 Stat. 1167 (1977). The Kitimat project must overcome the concern that it might deter the development of Canadian oil resources. The Northern Tier Pipeline must still obtain financing.

C. The Regulatory Context

Before SOHIO can begin construction it must obtain approvals, permits, and rights-of-way from various Federal, state, and local agencies. Many of these agencies are concerned with the environmental impact of the

---

[48] Fred Garibaldi told the Senate Commerce Committee that SOHIO had studies indicating that refineries along this route could use "up to 150-200 thousand barrels per day of Alaskan Oil." Senate Hearings, September 14, 1977.

proposed pipeline system. An Environmental Impact Statement (EIS) was prepared by SOHIO and the Bureau of Land Management pursuant to the National Environmental Protection Act of 1969 (NEPA), 42 U.S.C. §4321 *et seq.*, and became final in the fall of 1977. Approval of the EIS has accelerated the formal permitting process. Significant steps in the approval process are discussed briefly below.

1. Federal Permits. Since the pipeline will cross federal lands administered by the Secretary of the Interior, the Secretary has the authority to grant a right-of-way across the federal lands pursuant to the Mineral Leasing Act of 1920, as amended, 30 U.S.C. § 185(c)(2). On December 1, 1977, Secretary Andrus announced that he would grant the SOHIO project right-of-way. Although the Secretary was required by the Act to give 60 days' notice of such action to the Senate and House Committees on the Interior and Insular Affairs, both Committees waived the notice requirement, as permitted by the Act. 30 U.S.C. § 185(w). As a result, the Secretary may grant the final permits at any time,[49] and according to the staff of the Department of the Interior, such action is imminent.[50]

2. Abandonment of the Gas Pipelines. Before SOHIO can convert the existing gas pipelines to crude oil use, the owners of existing sections of the natural gas pipeline must obtain the permission of federal or state regulatory agencies to "abandon" gas transmission service. El Paso owns the Arizona and New Mexico section, while Southern California Gas Co. owns the California section.

On November 10, 1977, the Federal Energy Regulatory Commission (FERC) approved the application of El Paso to abandon interstate gas transmission on the pipeline pursuant to Section 7(b) of the Natural Gas Act. 15 U.S.C. §717f(b). (FERC Opinion No. 4.) However, El Paso was dissatisfied with some aspects of the FERC order and petitioned the Commission for reconsideration. This application was denied on May 26, 1978. (FERC Opinion No. 4-A.)[51] The application of Southern California Gas Co. to abandon its intra-state service is

---

[49] See letter of Leo Krulitz, Solicitor of the Interior Department to John Shenefield, dated December 22, 1977.

[50] Before the project may be constructed it must receive the approval of several other federal agencies in addition to that of the Secretary of the Interior. A list of these agencies and their requirements is contained in the FEA Report, *Equitable Sharing of North Slope Crude Oil* at 60-63 (April 1977). Many of these approvals are routine and are expected to follow the Secretary's decision.

[51] Staff at the Department of the Interior have indicated that they are unwilling to grant the necessary right-of-way until the FERC order becomes final.

not subject to FERC approval, but must be approved by the California Public Utilities Commission.[52]

3. State Regulation of the Port Facility. At present, construction of the port facility must await the approval of the South Coast Air Quality Management District, which is subject to the concurrence of the California Air Resources Board and the United States Environmental Protection Agency. In addition, it will need the separate approval of the California Coastal Commission.

The port's application met with initial resistance from the environmental agencies, which are concerned that the facility will increase pollution in the Los Angeles basin. The application was recently approved, however, subject to certain air quality conditions.[53] As a result, SOHIO has agreed to limit the size of the facility and to demonstrate that it has reduced other sources of pollution, as a "tradeoff" for permission to construct the port. SOHIO and the environmental agencies are currently negotiating the terms of those conditions. It may ultimately be necessary for SOHIO to reduce the size of the facility from three berths and 700,000 b/d capacity to two berths and 500,000 b/d. In either case SOHIO would be permitted to operate two berths. The third, if constructed, would be leased to another operator.

The California Coastal Commission has issued a proposed permit for the construction of the port. Its terms, which have not yet been accepted, would require the port facility be operated as a "common carrier." The permit limits the capacity of the port facility to 700,000 b/d.[54]

II. COMPETITIVE ANALYSIS OF THE SOHIO PROPOSAL

Our analysis of SOHIO's ownership and operation of the proposed Long Beach, California, to Midland, Texas pipeline indicates that, on balance, its ownership and operation of the pipeline are not reasonably likely to result in any competitive harm, assuming that the pipeline and related facilities operate as a common carrier. We have examined the competitive effects of SOHIO's proposal on the market for crude oil in the combined area of PADs II and III, in Arizona and New Mexico, and in PAD V. In none of these areas does there appear to be the opportunity and incentive for SOHIO to successfully exercise market power.

---

[52] *Oil Daily,* November 23, 1977, reported that Southern California Gas Co. has filed the necessary petition.

[53] *Wall Street Journal,* January 20, 1978.

[54] California Coastal Commission, Coastal Development Permit A 185-77 issued to the Port of Long Beach (SOHIO) (October 19, 1977).

An analysis of the competitive implications of any oil pipeline should begin with several observations. First, pipelines are generally the most cost-efficient means of transporting large volumes of petroleum with the exception of supersize ocean tankers.[55] Second, from a technological perspective all pipelines are natural monopolies. This does not, however, necessarily imply an ability to exercise monopoly power. "Natural monopoly" is an economic term used to describe an activity in which the economies of scale dictate that one firm can provide any desired level of output at a lower cost than more than one firm. This technological characteristic stems from basic engineering features of a pipeline and implies that the average cost of a unit of throughput will decline throughout the range of technologically feasible pipeline sizes.[56] Therefore, if a decision to build new pipeline capacity is warranted at all, it would be more efficient to build a single pipeline of sufficient size to satisfy the entire demand for transportation between two points than to build several smaller pipelines. In this technological sense all pipelines are natural monopolies.

These economies of scale are, however, not sufficient to endow any particular pipeline owner with monopoly power. The pipeline must also possess the ability to exercise market power in either the downstream market or the upstream market served by the pipeline.[57] Therefore, in order to assess the competitive implications of the SOHIO proposal, an examination of the opportunity and incentive for the exercise of market power in the markets served by the pipeline is required.

If market power exists, the exercise of such power by a pipeline owner will, in general, result in a distortion of the level of the pipeline's throughput. This distortion can be achieved either by direct manipula-

---

[55] Truck and rail transportation is economic only for small volumes, for short distances, or in situations where great flexibility is needed. Barge traffic is potentially limited to areas where adequate inland waterways are available and is not competitive with large pipelines. See *National Energy Transportation Reports,* Vol. I, at 209.

[56] See *National Energy Transportation Report,* Vol. I, at 209-211 (discussion of pipeline costs); and Meyer, Peck, Stenason, and Zwick, *The Economies of Competition in the Transportation Industries* (Harvard University Press: 1959) at 126-133, 225-227, 248 (concise discussion of the basic economics of pipelines). See also Pearl and Enos, "Engineering Production Functions and Technological Progress," 24 *Journal of Industrial Economics* 55 (September 1975); Cookenboo, *Crude Oil Pipelines and Competition in the Oil Industry,* (Harvard University Press, 1955): Cookenboo, "Costs of Operating Crude Oil Pipelines," *Rice Institute Pamphlet* at 35-113 (April 1954).

[57] It is not necessary for a pipeline owner to buy or sell oil in order to be able to exercise market power. A pipeline owner has the ability to exercise market power if it can affect, by means of pricing or throughput decisions, the price of oil in either the downstream or the upstream market.

tion of the pipeline's throughput or by excessive pipeline tariffs. The rates of the proposed pipeline, like all common carrier pipelines, will be regulated by the FERC.[58] The Division has serious doubts about the correctness of the methodology presently employed to regulate pipeline rates and is involved in two ongoing proceedings before the FERC in an attempt to improve the regulation of pipeline rates.[59] This analysis assumes that improved FERC regulation will deny the owner of a pipeline the ability to exploit monopoly power through excessive tariffs. Therefore we will consider only the potential of SOHIO to effect anticompetitive outcomes through the use of direct manipulation of the pipeline's throughput.

A. Impact of the SOHIO Project in PADS II and III

The terminal point of the line is Midland, Texas. Midland, Texas is at present the center of a pipeline network that transports crude oil from West Texas to the major PAD III refining centers and to interconnections with the network of crude oil pipelines that serve PADs II and III.[60] Oil from West Texas becomes part of a larger stream that includes petroleum from other parts of Texas, other domestic producing areas, and imports from the Gulf Coast. Under Phase I of the proposal, the proportion of this area's crude oil supply that would come through the SOHIO pipeline would be small—about 5 percent.[61] If Phase II expan-

---

[58] Interstate Commerce Act, 49 U.S.C. §1 *et seq.*, as amended by the Department of Energy Organization Act, Pub L. 95-91, 91 Stat 565 (August 4, 1977).

[59] See Statement of Views and Arguments of the Department of Justice filed on May 27, 1977, before the ICC in Docket No. 308 now before the FERC in Docket No. RM78-2; Petition for Administrative Review in Docket No. RM78-2, filed December 12, 1977; Joint Prehearing Brief of the Department of Justice, et al., in the Trans Alaska Pipeline Investigation, FERC Docket No. OR 78-1, December 30, 1977; and Memorandum Defining and Stating the Position of the United States Department of Justice on Policy Issues Raised in this Proceeding, and Statement of Suggested Procedures to be Employed for Bringing the Proceeding to a Conclusion, filed in Docket No. RM 78-2 on April 3, 1978.

[60] See Appendix map, Crude Oil Pipeline Capacities, 1975, Federal Energy Administration, Office of Oil and Gas.

[61] The oil delivered at Midland through the pipeline is expected to be of heavy quality and high sulfur content relative to much of the current crude supply in PADs II and III. The pipeline's share of these quality crudes will be higher than its share of total crude supply. These quality crudes are less desirable because of the additional refining needed to produce the mix of products in demand in the United States. However, there is at present little information to indicate that market power will result from these quality differentials. At present, gravity price differentials are quite small, about 2 cents per barrel per degree API for West Texas high sulfur crudes, and 3 cents per barrel per degree API for OPEC crudes. The long run determinants of gravity differentials are not completely understood. For a discussion, see C. T. Roush, Jr., "The California Crude Oil

sion is granted SOHIO's market share would be about 10 percent. These proportions will decrease as the total demand for crude oil in PADs II and III increases. (For a graphic description of the source of crude oil in PAD II and PAD III, see Appendix map, Crude Oil Supplies.)

At present approximately 36 percent of the crude oil supply to this area consists of imports.[62] These imports are available at the delivered price of international petroleum. Thus, the incremental supply of petroleum to the midcontinent is forthcoming at the world price of petroleum. It is the cost of this incremental supply that determines the market price of oil in PADs II and III. These imports will still be the incremental supply to the midcontinent throughout the range of throughput possible for this pipeline. SOHIO could not, by restricting throughput, increase the price of the delivered oil above the world price. To pursue a strategy of throughput restriction would be merely to invite greater imports of foreign crude at the world price, a result which could not benefit SOHIO in PADs II and III.[63]

In the event of an oil embargo or other event significantly limiting supply from foreign sources, the competitive check offered by imports into PAD II and PAD III would be reduced.[64] However, the standby

---

Market: An Examination of the Posted Price Structure," unpublished doctoral dissertation, University of California, Santa Barbara, 1974, Chapter 1, 6. See also Manes, "The Quality and Pricing of Crude Oil," 12 *Journal of Industrial Economics* 151-62 (1968); and note 10, *infra*. The sulfur content of the pipeline's delivered crude, while high, is approximately equal to the sulfur content of most Middle Eastern crudes. The U.S. presently imports a significantly greater volume of these crudes than will be supplied by this pipeline.

[62] Bureau of Mines, *Monthly Petroleum Statements,* June 1977. See discussion, *supra.*

[63] This situation is in sharp contrast to the factual situation presented in consideration of the Gulf Coast Deepwater Ports. The Attorney General has expressed considerable concern about the potential ability of Gulf Coast Deepwater Ports to influence the downstream price of oil by restricting throughput of imported crude. See Report of the Attorney General Pursuant to Section 7 of the Deepwater Port Act of 1974 on the Application of LOOP, Inc., and SEADOCK, Inc., for Deepwater Port Licenses 3-5, 103 (Nov. 5, 1976) (hereinafter Deepwater Port Report). That analysis made clear that the ports in question had the potential for transporting up to 63 percent of all imports into the United States by 1990, fully one-third of the entire nation's daily needs. *Id.* at 51.

[64] This event would not in itself necessarily endow the owners of this pipeline with market power in PAD II and PAD III. The share of domestic crude oil supplies in the midcontinent transported by the pipeline, at its initial expected throughput, even if imports are entirely eliminated, may still be below the level that is usually believed to be by itself indicative of market power. Any analysis of the competitive position of this pipeline in the event of a significant import restriction requires a careful analysis of the supply and demand for crude oil, under the circumstances of the import restriction, in PAD II, PAD III, and PAD V. There is no reasonably clear data with which to perform such a definitive analysis at this time.

controls authorized by the Emergency Petroleum Allocation Act of 1973, 15 U.S.C. § 751 *et seq.,* would be available to determine petroleum pricing and allocation. The price of petroleum from wellhead to final product would be subject to regulation. This would deny SOHIO the ability to use its ownership of the pipeline to manipulate throughput to gain excess profits in PAD II and PAD III.

This situation would be similar to the extensive regulation of natural gas described in the Attorney General's report pursuant to Section 19 of the Alaska Natural Gas Transportation Act of 1976. That report concluded that as long as the system of pervasive controls then in effect continued, potentially troublesome competitive problems would not arise.[65]

Furthermore, SOHIO's position in Ohio will not be enhanced by its ownership of the pipeline. The petroleum arriving in Midland will not be less costly feedstock for SOHIO's refineries in Ohio than similar oil purchased from domestic or imported sources. If SOHIO does not use a barrel of its oil as a feedstock for its refineries, it can sell that barrel at the market price. If SOHIO uses the barrel of its oil instead of selling it, it has forgone the payment it would have received, that is, the market price. Therefore, the effective cost to SOHIO of crude oil feedstock is the same, whether the oil is purchased or is its own production shipped through the pipeline. Because in the absence of downstream market power, the potential for SOHIO to reap gains downstream is not present, SOHIO would not appear to have any incentive to restrict pipeline throughput.

B. The Impact of the SOHIO Project in Arizona and New Mexico

SOHIO's ownership of this pipeline does not appear to create for SOHIO any significant power to raise price in the markets for oil along the route of the pipeline in Arizona and New Mexico. In Arizona, refining capacity as of January 1, 1977, was 4,000 b/d.[66] Extensive amounts of petroleum products are shipped into Arizona by the independent Southern Pacific Pipeline.[67] The ability of SOHIO to raise the price of crude oil in Arizona is thus limited by the small contribution of the Arizona-refined products to the total supply of petroleum products in Arizona.[68] In addition, ARCO's Four Corners Pipeline will serve the

---

[65] See *Id.* at 29-30.

[66] *Oil and Gas Journal,* March 28, 1977.

[67] See *National Energy Transportation Report,* Map No. 9.

[68] In 1976 total consumption of petroleum products in Arizona averaged around 125,000 b/d. Department of Energy and Federal Highway Administration.

same areas and, thus, may provide some competition to SOHIO for the small supply of crude oil to Arizona.[69] In New Mexico, production of crude oil was 223,000 b/d as of December 2, 1977.[70] As of January 1, 1977, refining capacity in New Mexico was 119,020 b/d.[71] The crude oil produced in New Mexico which is not refined locally is shipped to the southeastern part of the state, where it enters the pipeline network near Midland, Texas. The refineries in New Mexico near the route of the pipeline are located in this southeastern part of the state; since the flow of crude is from northwest to southeast in New Mexico, SOHIO's crude oil shipped through the Long Beach to Midland Pipeline, along with the crude shipped within New Mexico, essentially competes with crude oil in Midland, Texas. In order to be able to affect the price of crude in New Mexico, SOHIO would have to be able to affect the price of crude in Midland. Therefore, for the reasons discussed in the previous section, SOHIO could not raise the price of crude oil in New Mexico.

C. Impact of the SOHIO Project in PAD V

PAD V is the upstream market for the SOHIO pipeline. The ability of SOHIO to use the pipeline to gain an advantage in the development of new energy resources or to enhance its market power in PAD V must be considered.

An initial question is whether SOHIO might be able to use its control of pipeline access and capacity to gain an advantage in the development of new oil resources in PAD V. Oil companies would be less likely to invest in new production if oil prices are depressed. This would occur if SOHIO failed to provide sufficient capacity to allow new production to find its way to areas in need of additional supply; absent such capacity, new oil would create a surplus in PAD V, and, hence, downward pressure on crude prices.[72]

---

[69] The Four Corners Pipeline is much smaller and hence less efficient than the Long Beach to Midland Pipeline. However, the estimated acquisition cost of the existing Four Corners Pipeline is so low that ARCO should be able to recover its costs and obtain a fair rate of return at a tariff competitive with the SOHIO tariff in this region.

[70] *Oil and Gas Journal,* December 12, 1977.

[71] *Oil and Gas Journal,* March 28, 1977.

[72] This downward pressure is limited to the difference between the cost of transportation on the SOHIO pipeline and the cost of the most expensive alternative mode used to ship oil to the midcontinent. At present, shipment through the Panama Canal appears to be the most costly mode of transportation that could be expected to be used. See *supra.*

Similarly, since many of the fields which are likely sources of new production are in areas subject to upcoming federal and state lease sales (such as offshore areas and the Naval and National Petroleum Reserves), the refusal of SOHIO to guarantee that there will be sufficient throughput capacity to carry the increased production would tend to erode the value of such leases to other prospective developers. The possibility exists that SOHIO could purchase these leases at a reduced price, taking advantage of the knowledge which other developers lacked, that is, that it would eventually expand throughput.

A necessary condition for this strategy is that SOHIO have control over the economically efficient means of transporting additional oil eastward. This requires either that the alternative pipeline routes cannot be built or that existing lines be fully utilized and unable to expand. If development of new oil sources west of the Rockies results in sufficient demand for transportation, there will be strong incentives for someone to build a pipeline. Even if SOHIO refuses to expand the Long Beach-Midland Pipeline, and even if alternative existing pipelines, if any, cannot be expanded, a pipeline could probably be constructed in the time period necessary to develop the leases. Thus, this strategy would be limited; throughput restriction, apparent or real, could not exceed the amount of excess supply on the West Coast that would induce new pipeline entry.[73]

There are two variations of this strategy. To discourage other firms from bidding for leases or developing new sources, SOHIO could either actually produce a surplus of crude oil and lower the price on the West Coast or threaten to create such a surplus. If SOHIO were actually to produce a surplus on the West Coast, prices and profits on sales in PAD V would be forced downward for all PAD V sellers, including SOHIO. SOHIO would have the choice of expanding the pipeline or losing profits. Furthermore, since the existing capacity, including the port, would have to be prorated under the common carrier obligations, other sellers of crude in PAD V could demand the right to use the line to move their crude from PAD V to the midcontinent. Thus, prorationing would force SOHIO to sell additional quantities of its crude in PAD V, where it would suffer the impact of the lower price. SOHIO might in fact be particularly vulnerable since, as stated above, it depends, to a greater extent than other sellers, on the pipeline to ship its "excess" crude oil to markets in PADs II and III. Finally, in order to earn the profits on the newly discovered oil that would make this strategy attrac-

---

[73] For a discussion of the prospects for the potentially competing West Coast to midcontinent pipelines, see section D, *infra,* and "Alternative Pipeline Proposals," *supra.*

tive, SOHIO would eventually have to expand the throughput of the pipeline, a fact which could probably be deduced by SOHIO's competitors.

Moreover, if other firms were to develop the leases, they would be able to coerce SOHIO into expanding the pipeline by themselves threatening to produce a surplus. These factors, including the possibility of antitrust action,[74] suggest that other firms are not likely to view the threat that SOHIO would not expand as credible enough to affect their lease bidding and production decisions.[75]

The result of these possible anticompetitive strategies by SOHIO would be to produce conditions which would not be to SOHIO's advantage. In the absence of any advantage, it is not reasonably likely that SOHIO would have the incentives to pursue an anticompetitive strategy based upon throughput restriction. It should be emphasized, however, that this conclusion assumes effective enforcement of common carrier obligations and a prorationing system to provide nondiscriminatory access to the pipeline and related facilities, including the port. In the absence of these obligations SOHIO might be able to use its control of capacity to gain competition advantage in PAD V.[76]

Consideration should also be given to whether SOHIO's ownership and operation of the pipeline could be used to enhance its ability to raise the price of crude oil in PAD V. As a seller of crude oil in PAD V, SOHIO might find it to its advantage to ship crude oil out of PAD V[77] to the midcontinent area, where the addition of SOHIO's oil will not tend to depress the price, as we have noted above. Under such conditions, it may be to SOHIO's advantage to increase shipments of oil beyond the level which would result in a "competitive" price differential between PAD V and the midcontinent, that is, a differential equaling the cost of transportation between the two markets. SOHIO would do this if it could gain more from the increase in the price earned on its West Coast sales than the increase in its net transportation cost.[78]

SOHIO's incentives to engage in this behavior are lessened, however, to the extent that the oil shipped out of PAD V would be SOHIO's, and SOHIO's PAD V sales would be reduced. As its market share on the West Coast falls, SOHIO's ability to gain from such anticompetitive behavior will decrease. SOHIO is expected to have a significant share

---

[74] See *infra*.

[75] See generally T. C. Schelling, *The Strategy of Conflict*, Chapter 5 (1960).

[76] See *supra*.

[77] Any significant seller of crude oil in PAD V would have some incentive to limit supply by moving oil out of PAD V.

[78] Net transportation cost is the sum of the price of the oil in the upstream market and the cost of transportation, less the value of the oil in the downstream market.

of the crude oil production in PAD V, approximately 29 percent. However, since SOHIO has no readily available market for much of that production, it must ship a larger portion of its oil to the midcontinent area than other PAD V sellers.[79] In fact, it appears that SOHIO's oil constitutes virtually all of the crude oil that would be shipped from PAD V to the midcontinent area at a price differential between the two markets which is not less than the transportation cost. It is expected that as a result SOHIO would be left with only a small fraction of the remaining PAD V crude supplies—not enough to affect significantly the price of crude oil in PAD V or to give SOHIO sufficient gains to make the strategy worthwhile. Therefore, SOHIO would have at best only a limited incentive to ship excess quantities through the pipeline.

What little incentive there might be is even further reduced by the possibility of imports into PAD V. The net delivered price of imported oil would be the ceiling above with SOHIO could not raise the price of its oil in PAD V. If SOHIO tried to raise the price in PAD V above this level, buyers of crude oil would merely switch to imported oil. Thus, the price of oil in PAD V could not rise above the world price.

Finally, it should be noted that any expansion of the pipeline would be subject to regulatory review. Phase 2 expansion[80] would require a new environmental impact statement and an amended right-of-way grant, subject to the approval of the Secretary of the Interior. The FERC would have jurisdiction over the abandonment of the second natural gas pipeline. Expansion of the port facility or construction of a new facility would require the approvals of the California Coastal Commission and the state environmental agencies. Any objections to expansion, once proposed, could be addressed in the appropriate agency forum and both the public and the Justice Department would have the opportunity to make their views known.

D. Impact of the SOHIO Pipeline on the Northern Route Pipeline Proposals

The Long Beach to Midland pipeline would compete with each of the proposed northern route pipelines[81] in the transportation of Alaskan crude petroleum to the Chicago market. If the surplus of Alaskan oil were the only source of crude oil for any of these pipelines, a single pipeline would probably be sufficient to eliminate the surplus.[82] To the

---

[79] See *supra.*

[80] Expansion would require the modification of a second natural gas pipeline. See *supra.*

[81] The "northern route pipeline" proposals are described *supra,*

[82] See *supra.*

extent that any northern pipeline relies on surplus Alaskan oil for its throughput, then the Long Beach to Midland pipeline damages the economic viability of that pipeline. However, if any of these proposed pipelines could deliver imported oil to the midcontinent area at a price equal or lower than alternative sources, its success would be assured.

The Northern Tier Pipeline Company claims that the pipeline could carry, in addition to Alaskan oil, imported oil from the Persian Gulf or Indonesia to Chicago at rates competitive with pipelines originating on the Gulf Coast. This claim is somewhat "independently" supported by a study conducted by Amoco,[83] using different throughput assumptions for the comparisons. Under the Amoco assumptions, the Northern Tier Pipeline would be competitive with the Long Beach to Midland pipeline and the proposed deepwater ports in the Gulf of Mexico. Furthermore, to replace curtailed Canadian exports, low sulfur crude (as distinguished from Alaskan high sulfur crude) is required by the refineries of the Northern Tier states that would be served by each of the northern route pipelines.

Even if the Long Beach to Midland pipeline does divert "surplus" oil as a source of supply to the northern route pipelines and thus makes these projects more difficult, this is not an anticompetitive result. If there is a choice to be made among proposals, this is an issue of planning for the distribution of national energy resources, not an antitrust concern. The Long Beach-Midland Pipeline may or may not substantially damage the prospects for the construction of a northern route pipeline, but from an antitrust viewpoint, this is not an anticompetitive consequence to be remedied by independent ownership or special conditions imposed upon SOHIO.

Our analysis indicates that SOHIO would not be able to achieve any monopoly gain in either upstream or downstream markets by restricting the flow of oil from PAD V to the midcontinent area. We have also noted that SOHIO's incentive to ship excessive quantities of oil is dependent upon its position as a seller in PAD V. These conclusions are not altered by the prospect of SOHIO's participation in the Kitimat Pipeline project.[84] SOHIO's downstream position in respect to crude sales would not be significantly improved as imports would remain the incremental source of supply. SOHIO's upstream position as the owner of virtually all excess crude supply in PAD V would neutralize its incentives and opportunities to manipulate prices in that area to its advantage.

---

[83] Letter from L. D. Thomas, Vice President of Amoco Oil Company to John O'Leary, Administrator, Federal Energy Administration, dated April 22, 1977 with attachment, "Supplying the Oil Needs of the Midwest/PAD IV Area."
[84] See *supra*.

The basic conclusion of our antitrust analysis of this proposal is that SOHIO will not be able to exercise monopoly power in any relevant market by virtue of its ownership and operation of the Long Beach to Midland pipeline. The availability of an elastic supply of imported crude oil will prevent SOHIO from exercising power over price in PADs II and III. The competition from the Four Corners Pipeline, locally produced crude, and imported products, will prevent SOHIO from controlling prices in New Mexico and Arizona. Finally, on the West Coast SOHIO will be unable to manipulate throughput to its advantage.

Our analysis also indicates that if SOHIO for any reason attempts to manipulate the pipeline's capacity to gain some speculative competitive leverage for the purchase of leases for new fields in PAD V, the present legal and regulatory framework will be adequate to provide a remedy. If ownership of this pipeline, for instance, were to give BP/SOHIO a strategic advantage in lease bidding, the advantage would be general and throughout the PAD V area. A pattern in which BP/SOHIO wins a disproportionate share of new leases is readily observable.[85] The Division would advise the Secretary of the Interior not to issue leases to SOHIO and BP or initiate direct action itself.[86] These measures could effectively deal with such concentration without excessive cost, particularly in view of the small likelihood that this event will occur.

The foregoing analyses and conclusions are dependent upon the operation of the Long Beach to Midland Pipeline, the port, and other facilities as a common carrier. If regulatory scrutiny of common carrier obligations is effective, such scrutiny should be adequate in this case to protect the public interest in the efficient allocation of resources. To this end, existing law provides plenary authority for ensuring the common carrier status of the pipeline, the port and the other facilities.

The common carrier status of oil pipelines is mandated by the Interstate Commerce Act (ICA), 49 U.S.C. § 1 *et seq.* Section 1(1)(b) applies the provisions of the ICA to all common carriers engaged in

---

[85] As noted in a letter from John H. Shenefield, Assistant Attorney General, Antitrust Division, to Secretary of the Interior Andrus, January 19, 1978, the concentration of PAD V reserves is at present a matter of some concern. Thus, apart from problems that might flow from SOHIO's pipeline ownership, any notable increase in BP/SOHIO's share might tend to create a situation inconsistent with the antitrust laws and lead to a recommendation that SOHIO not be permitted to acquire further Federal oil leases in PAD V.

[86] Section 28(y) of the Mineral Leasing Act, 30 U.S.C. § 185(y), provides that "the grant of a right-of-way or permit pursuant to this section shall grant no immunity from the operation of the Federal antitrust laws."

"[t]he transportation of oil . . . by pipe line, or partly by pipe line and partly by railroad or by water, . . ." in interstate commerce. The term "common carrier" by definition in Section 1(3)(a), includes "all pipeline companies." The same section defines "transportation" to include "all instrumentalities and facilities of shipment or carriage, irrespective of ownership or of any contract, expressed or implied for use thereof, and all services in connection with the receipt, delivery, elevation, and transfer in transit, . . . storage and handling of property transported." Rates charged for such service must be "just and reasonable." Section 1(5). It is unlawful for a common carrier to "make, give, or cause any undue or unreasonable prejudice or advantage." Section 3(1). The Secretary of Energy has general jurisdiction over the "transportation of oil by pipeline," and the Federal Energy Regulatory Commission has specific jurisdiction over pipeline rate regulation under the ICA pursuant to Section 306 and Section 402(b), respectively, of the Department of Energy Organization Act, Pub. L. No. 95–91, 91 Stat. 565 (1977).

Thus, the primary consequence of federal common carrier regulation here is that SOHIO will have a statutory duty to furnish transportation upon reasonable request by any shipper, and to charge just and reasonable rates. The access and nondiscrimination duties require the carrier to provide reasonable access to non-owner shippers and prohibit it from charging higher rates to those shippers than to owners. These duties are crucial in light of the economic characteristics peculiar to pipelines, discussed above.

The Mineral Leasing Act, as amended, 30 U.S.C. §185(1), ensures that the same common carrier provision applies to all pipeline systems that transverse federal lands even if those pipelines would not otherwise be subject to the Interstate Commerce Act. The legislative history is clear that these obligations are in addition to those already imposed by the ICA. Senate Report No. 93–207.[87] Section 185(r)(1) states that "Pipelines and related facilities authorized under this section shall be constructed, operated, and maintained as common carriers."

The Secretary is directed to obtain from the applicant all such plans, contracts, agreements, or other information which he deems necessary to determine whether the right of way should be granted or renewed and the terms and conditions which should be included in the right-of-way grant to ensure that the pipeline is operated in accordance with its common carrier status. This information may include but is not limited to that concerning "(A) conditions for, and agreements among owners or operators, regarding the addition of pumping facilities, looping, or otherwise increasing the pipeline or terminal's throughput capacity, in

---

[87] 1973 *U.S. Cong. and Admin. News* 2440.

response to actual or anticipated increases in demand; (B) conditions for adding or abandoning intake, offtake or storage points or facilities; and (C) minimum shipment or purchase tenders." 30 U.S.C. §185 (r)(6). The Mineral Leasing Act also establishes the Secretary's broad power to prescribe conditions to ensure that the applicant constructs, operates, and maintains the pipeline system in accord with the requirements of the statute. 30 U.S.C. §185(f). It is manifest, therefore, that the Secretary has the power, if he deems necessary, to condition the grant or renewal of a right-of-way upon the requirement that the pipeline owner expand capacity, provide reasonable storage facilities, establish reasonable minimum tenders, a broad common carrier obligation, and the like.

If the Secretary "has reason to believe" that any owner or operator of a pipeline subject to the Mineral Leasing Act is not operating the pipeline as a common carrier, he may request the Attorney General to prosecute an appropriate proceeding before a regulatory agency or court with jurisdiction. As an alternative, the Secretary may proceed under the Act to suspend or terminate the grant of right of way for noncompliance with the section. 30 U.S.C. §185(r)(5). These procedures require due notice to the holder of the grant, a reasonable opportunity to comply and an administrative proceeding pursuant to Section 554 of the Administrative Procedure Act, 5 U.S.C. §554. 30 U.S.C. §185(o)(1).

The "Common Carrier" provision in the Draft Stipulations to the right-of-way grant for the Long Beach to Midland Pipeline ("SOHIO Draft Stipulation")[88] provides that:

A. Grantee shall construct, operate, and maintain the pipeline as a common carrier;

B. Grantee shall accept, convey, transport, or purchase, without discrimination, all oil delivered to the pipeline without regard as to whether such oil was produced on Federal or non-Federal lands; and

C. Grantee shall accept, convey, transport, or purchase, without discrimination, oil produced from Federal lands or from the resources thereon in the vicinity of the pipeline in such proportionate amounts as the Secretary may, after a full hearing with due notice thereof to Grantee and other interested parties and a proper finding of facts, determine to be reasonable.

---

[88] SOHIO Draft Stipulation A-5.

CRUDE OIL SUPPLIES
(millions of barrels per day)

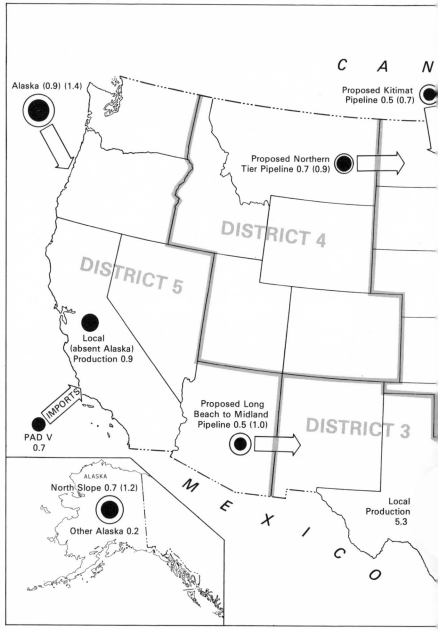

Alaska (0.9) (1.4)

C A N

Proposed Kitimat
Pipeline 0.5 (0.7)

Proposed Northern
Tier Pipeline 0.7 (0.9)

DISTRICT 4

DISTRICT 5

Local
(absent Alaska)
Production 0.9

IMPORTS

PAD V
0.7

Proposed Long
Beach to Midland
Pipeline 0.5 (1.0)

DISTRICT 3

M
E
X
I
C
O

ALASKA
North Slope 0.7 (1.2)

Other Alaska 0.2

Local
Production
5.3

SOURCE: Department of Energy, Monthly Energy Reports (July 1976–June 1977).
API Refinery Reports, *Oil and Gas Journal,* December 12, 1977.

244

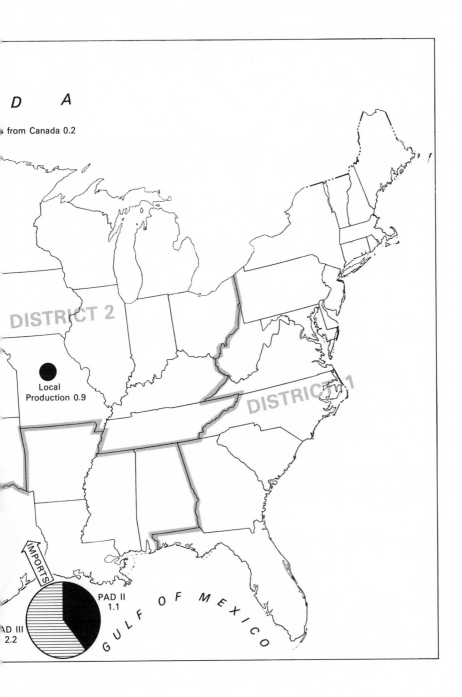

D A

from Canada 0.2

DISTRICT 2

Local
Production 0.9

DISTRICT 1

IMPORTS

PAD II
1.1

AD III
2.2

G U L F O F M E X I C O

CRUDE OIL PIPELINE CAPACITIES, 1975
(thousands of barrels daily)

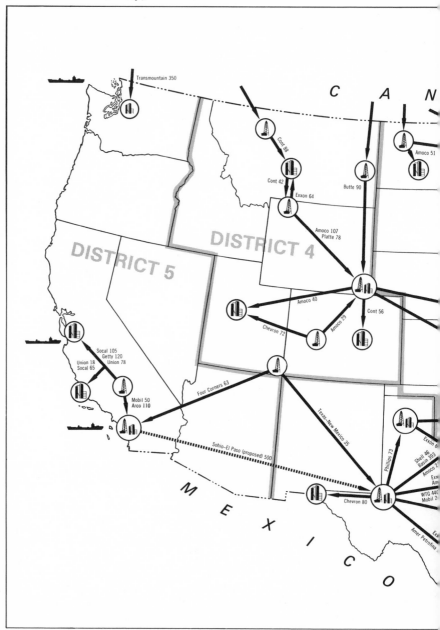

SOURCE: Federal Energy Administration, Office of Oil and Gas.

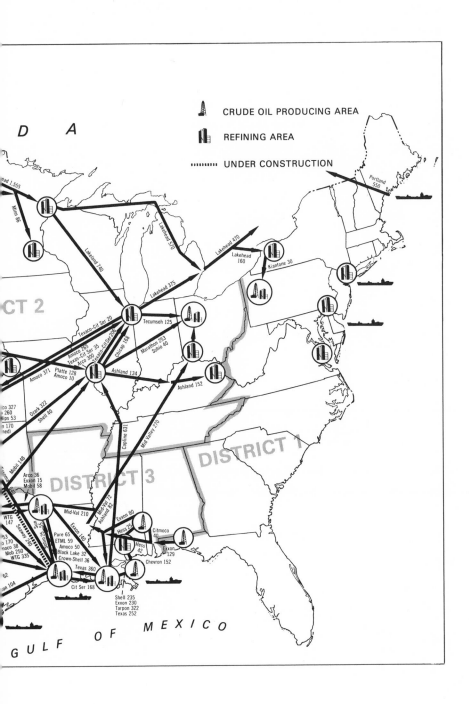

CRUDE OIL PRODUCING AREA

REFINING AREA

·········· UNDER CONSTRUCTION

D A

CT 2

DISTRICT 3

DISTRICT 1

GULF OF MEXICO

ad 1,555

Minn 86

Lakeland 740

Lakehead 510

Lakehead 375

Lakehead 470

Lakehead 160

Kiantone 30

Portland 550

Tecumseh 125

Texaco-Cit Ser 20

Amoco 285
Texas-Cit Ser 35
Arco 300
Texaco-Cit Ser 206

Chicap 168

Marathon 253
Sohio 40

Amoco 371

Platte 128
Amoco 10

Ashland 134

Ashland 152

co 327
260
ips 53
e 170
ned)

Ozark 322
Shell 40

Capline 631

Mid Valley 270

Mobil 148

Arco 36
Exxon 15
Mobil 58

WTG
147

53
170
noco 38
Mob 250
WTG 335

Mid-Val 210

Mid-Val 72
Ashland 82

Exxon 140

Exxon 80

Hess 25

Citmoco
40

Exxon
129

62

on 104

Pure 65
ETML 59
Amoco 50
Black Lake 32
Crown-Shell 36

Texas 360

Cit Ser 168

Hess
42

Chevron 152

Shell 235
Exxon 230
Tarpon 322
Texas 252

247

Product Pipeline Capacities, 1975

(thousands of barrels daily)

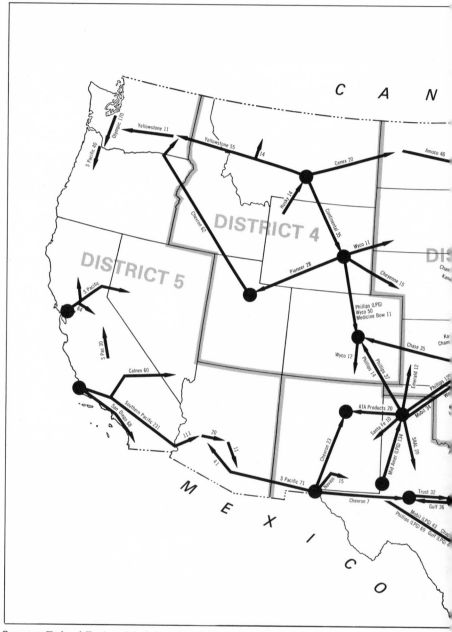

Source: Federal Energy Administration, Office of Oil and Gas.

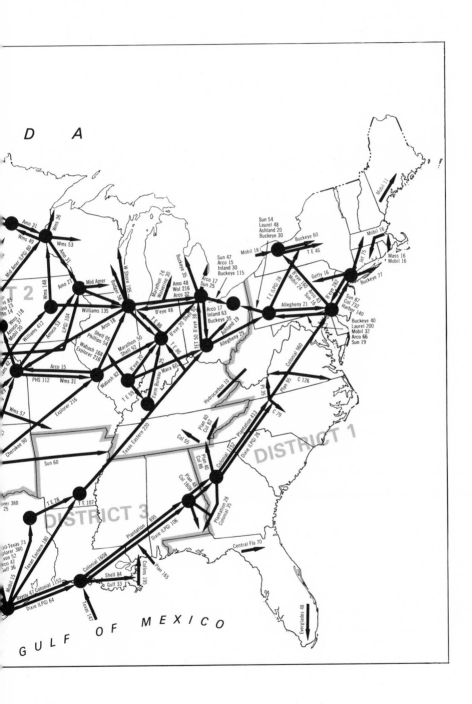

This language is adapted essentially verbatim from the Mineral Leasing Act, Sections 28(c)(1), (2)(A) and (2)(B), except that Section (c)(1) of the Act refers to "pipelines and related facilities."

The term "pipeline" is defined in the stipulations as "the line transversing Federal land for transportation of oil and related facilities as discussed in this grant [of right of way]." SOHIO Draft Stipulation A-1.G. In contrast, the term "Pipeline System" as defined in Draft Stipulation A-1.H, encompasses "all facilities, whether or not located on Federal land, used by a Grantee in connection with the construction, operation, maintenance, or termination of a pipeline." "Related facilities" is defined broadly to include, *inter alia*, terminals, including certain port facilities, storage tanks, and gathering lines. SOHIO Draft Stipulation A-1.I.

The Draft SOHIO Stipulations are apparently modeled after the stipulations and agreements entered into by the Department of the Interior and the owners of the Trans Alaska Pipeline System (TAPS) in 1974. The Common Carrier provisions used there are similar. See TAPS agreement, Section 3. However, Section 2 of the agreement defines the term "pipeline" to include the "forty-eight (48) inch diameter pipe and its Related Facilities," and the term, "Related Facilities," is defined in the TAPS stipulations as "those structures, devices, improvements and sites substantially continuous as of which is necessary for the operation or maintenance of the Oil transportation pipeline" including, *inter alia*, "(5) surge and storage tanks" and "(7) terminals."

In light of the economic characteristics of pipelines, as discussed above, we believe that the common carrier provision should be revised to eliminate any ambiguity as to whether the common carrier obligation extends to related facilities. Under the Mineral Leasing Act and the Interstate Commerce Act, related facilities are clearly included in the common carrier status. Of course, the common carrier provision does not reach ancilliary facilities not employed in the transportation of oil. An example cited in the Senate Report is a natural gas pipeline built solely to carry fuel to a pumping station.[89] The precise language of the provision will have to be carefully drafted.

The use of a more broadly defined term, such as "pipeline system," would also remove any ambiguity about the scope of the stipulation's applicability. Obviously, any common carrier conditions imposed by the Secretary under Section 28(f) of the Mineral Leasing Act, 30 U.S.C. §185(f), cannot be limited geographically to the sections of the pipeline transversing Federal lands. Transportation of oil by pipeline is a continous operation beginning with, in the case of the Long Beach–

---

[89] 1973 *Cong. & Admin. News* at 2440.

Midland Pipeline, the unloading of oil tankers at the Port of Long Beach and ending with the delivery of the oil at the Midland terminal or some intermediate offtake point. Both the Interstate Commerce Act and the Mineral Leasing Act recognize this fact by providing for common carrier regulation not only of the pipeline but also of the related facilities.

In particular, our analysis leads to the conclusion that the port facility is governed by these acts and must be in practice subject to common carrier status.[90] The port is an integral part of the entire facility, and a potential bottleneck to nondiscriminatory access to the Long Beach-to-Midland Pipeline. The California Coastal Commission, we believe, has taken the proper action in conditioning its permit for the port facility on its operation as a common carrier.[91]

Finally, the duty of the carrier to transport without discrimination all goods tendered to it from all shippers to the limits of its capacity is at the heart of existing common carrier obligations. In cases in which the carrier lacks capacity to carry all the goods tendered, it must accept a part of each shipper's tender in proportion to the total amount of goods tendered. Thus, in the case of pipeline companies, the carrier should be required to establish fair and equitable rules under which such "prorationing" would be conducted.

These rules should apply to oil carried by all parts of the facility, that is, the port, storage facilities and the pipeline. In other words, the prorationing requirement should be coextensive with the common carrier obligation.

The prorationing requirement is so crucial to the common carrier obligation that we believe the Secretary should require SOHIO to disclose its plans for prorationing capacity in the event the pipeline is fully

---

[90] The fact that the port will be owned by the Port of Long Beach, a part of the municipality of Long Beach does not "defeat" regulation under the ICA.

The issue of whether the port facility would be subject to the federal common carrier obligation was addressed on April 18, 1977, when then Assistant Attorney General Donald I. Baker advised Senators Stevenson, Magnuson and Jackson, in separate letters, that there was a "strong argument" that the port facility would be subject to common carrier status under the ICA, notwithstanding the involvement of the municipality.

It has been held, for example, that by engaging in interstate commerce by rail, a state subjects itself to the commerce power of Congress, even though the statute effectuating that power is silent as to Congressional intent to regulate the states. See, e.g., Parden v. Terminal Ry., 377 U.S. 184 (1964); California v. Taylor, 353 U.S. 533 (1957). This exercise of the commerce power does not reach state functions essential to the state's separate and independent existence. National League of Cities v. Usery, 426 U.S. 533 (1976). The Interstate Commerce Commission has long acted in accord with this principle by asserting regulatory authority over state owned railroads. California Canneries Co. v. Southern Pacific Ry. Co., 51 I.C.C. 500 (1918); United States v. Belt Line R. Co., 56 I.C.C. 121 (1919).

[91] See note 54, *supra*.

utilized. To date we understand that SOHIO has not been asked to make such disclosures, so we cannot evaluate the effect on competition that its plans might have. We do not know whether the prorationing rules SOHIO will establish will have any anticompetitive potential, but this precaution is surely warranted in any event. Prior approval of prorationing plans should be a condition of the grant of right-of-way.

Supplement 3

UNITED STATES OF AMERICA
FEDERAL ENERGY REGULATORY COMMISSION

| Treatment of Certain Production-Related Costs for Natural Gas to be Sold and Transported through the Alaska Natural Gas Transportation System | Docket No. RM79-19 |

COMMENTS OF UNITED STATES THE DEPARTMENT OF JUSTICE

John H. Shenefield
Assistant Attorney General
Antitrust Division

Donald L. Flexner
Deputy Assistant Attorney
General
Antitrust Division

Communications with respect to this Document should be addressed to:

Donald A. Kaplan
Chief
Robert Fabrikant
Assistant Chief
Energy Section
Antitrust Division
U.S. Department of Justice
Washington, D.C. 20530

Janet R. Urban
Judy Whalley
Attorneys
Antitrust Division

Thomas Spavins
Gregory Werden
Economists
Economic Policy Office
Antitrust Division

U.S. Department of Justice
P. O. Box 14141
Washington, D.C. 20044

March 19, 1979

UNITED STATES OF AMERICA
FEDERAL ENERGY REGULATORY COMMISSION

Treatment of Certain
Production-Related Costs
for Natural Gas to be
Sold and Transported   Docket No. RM79-19
through the Alaska Natural
Gas Transportation System

COMMENTS OF UNITED STATES THE DEPARTMENT OF JUSTICE

I. Introduction

The Department of Justice ("Department") hereby submits these comments on the Notice of Proposed Rulemaking and Statement of Policy ("Proposed Rule") which was issued by the Federal Energy Regulatory Commission ("Commission") on February 2, 1979 in the above captioned proceeding. These comments address the competitive impact of the Proposed Rule and express the Department's concern that adoption of the Proposed Rule, without adoption of competitive safeguards, could have a significant anticompetitive impact upon future production of Alaskan Natural Gas that would be transported to the lower forty-eight states through the Alaska Natural Gas Transportation System ("ANGTS").

The Department is the Executive Branch agency responsible for enforcing the antitrust laws and is charged with promoting the nation's policy in favor of competition embodied in those laws.[1] The state of competition in the energy sector has a critical impact on all segments of the American economy. It is therefore important that the competitive effect of governmental decision-making concerning production and transportation of energy resources be analyzed. Governmental policies which would have a significant anticompetitive effect should not be adopted if a less anticompetitive alternative is available to accomplish the government's purpose.[2]

The Department has actively participated in the decision-making process regarding ANGTS. Pursuant to Section 19 of the Alaska Natural Gas Transportation Act of 1976, the Department conducted a study of the antitrust issues relating to the production and transportation of

---

[1] See, e.g., Northern Pacific Railroad v. United States, 356 U.S. 1 (1958).

[2] See, e.g., FMC v. Aktiebolaget Svenska Amerika Linien, 390 U.S. 238 (1968); United States v. Civil Aeronautics Board, 511 F.2d 1315 (D.C. Cir. 1975), Northern Natural Gas Co. v. FPC, 399 F.2d 953 (D.C. Cir. 1968).

Alaskan natural gas. The results of that study were published in July 1977 in the *Report of the Attorney General Pursuant to Section 19 of the Alaska Natural Gas Transportation Act of 1976* ("Report"). The Report identified several potential competitive problem areas associated with the sale of Alaskan natural gas and recommended solutions to these problems. The major potential problem identified in the Report was the danger that if (1) regulation of the wellhead price of gas were relaxed, (2) the Alaskan gas producing areas were workably competitive, *and* (3) the pipeline possessed downstream market power, then producers who owned or controlled this monopoly transportation system could circumvent regulation of pipeline tariffs and reap monopoly profits in unregulated operations, through denial of access to non-owners and restriction of downstream supply. As a prophylactic measure, the Department recommended that ownership or control of the pipeline by the North Slope producers be prohibited. The recommendation was accepted by the President in the *Decision and Report to Congress on the Alaska Natural Gas Transportation System* (September 1977) ("President's *Decision*"), which was ratified by Congress.[3]

Through its participation in the process undertaken by the Commission to implement the decision of the President and Congress,[4] the Department has undertaken to ensure that to the greatest extent possible, under both the Alaska Natural Gas Transportation Act of 1976 and the antitrust laws, the production and transportation of Alaska natural gas is both competitive and efficient.

II. Issue Presented

In the Proposed Rule the Commission set forth its preliminary conclusion that the public interest, as well as the specific language of the President's *Decision,* form a basis for a definitive finding by the Commission that the conditioning facility should be the responsibility of the producers (Proposed Rule at 14-15). The Proposed Rule sets forth the further preliminary conclusion that the costs of gathering and conditioning should be borne by the producers and subsumed in the maximum wellhead price (*Id.* at 14-17). Thus, the Commission proposes to

---

[3] Joint Resolution of Congress, H.R.J. Res. 621, Pub. L. No. 95-158, 91 Stat. 1268 (1977).

[4] See Letter, May 15, 1978, Joe Sims, Deputy Assistant Attorney General, Antitrust Division, Department of Justice, to John B. Adger, Jr., Director, Alaska Gas Project Office, Federal Energy Regulatory Commission, concerning competitive impact of the application to transfer to Alaska Northwest Natural Gas Transportation Company the conditional certificate of public convenience and necessity issued to the Alcan Pipeline Company; Midwestern Gas Transmission Company v. Federal Energy Regulatory Commission, District of Columbia Circuit 78-1753, Brief for the United States of America as *Amicus Curiae.*

require that producers construct and operate, either directly or through a third party contractor, the conditioning plant that will prepare Alaskan gas for entry into ANGTS.

In its Proposed Rule, the Commission invited public comment on (1) the proposed assignment of responsibility for conditioning Prudhoe Bay gas for pipeline entry to the existing producers of Prudhoe Bay gas, and (2) the mechanisms proposed to implement that assignment.

It appears that the conditioning plant under consideration will be the only conditioning plant that will be constructed and operated to prepare gas for ANGTS. In a letter of August 19, 1977, the Commissioner of the Department of Revenue of the State of Alaska advised the Department of Energy that environmental and economic considerations militate against the construction of more than one conditioning plant on the North Slope. Thus, as all gas from the North Slope that is to be transported to the lower forty-eight states through ANGTS must be processed in this conditioning plant it will have a monopoly over this necessary step in the delivery of the gas. Further, the tariffs of the conditioning plant will not be regulated by the Commission.[5]

III. Economic Analysis

Our economic analysis is predicated upon the two key factors stated above: That there will be only one natural gas conditioning plant on the North Slope of Alaska, and that the tariffs of this plant will not be regulated by the Commission. The former implies that the conditioning plant will be a "bottleneck" and have the potential to exercise market power over North Slope gas. The latter implies that regulation will not directly interfere with the exercise of whatever market power the conditioning plant may have. These factors together raise a serious danger of anticompetitive abuse which could adversely affect the ability of the ANGTS to achieve its goals.

With a single unregulated conditioning plant, the structure of the North Slope field market for natural gas can be viewed as a monopoly. The conditioning facility will be able to set the total amount of gas which gets conditioned, and to set the price of conditioning. The conditioning plant's control over volume and price ultimately flows from its power to deny access to the market to any producer of gas.

Since the Proposed Rule will make the conditioning of the gas the responsibility of the gas producers, it is assumed here that the Prudhoe Bay producers will own the conditioning plant and be able to exercise

[5] In its Proposed Rule the Commission has not proposed to regulate tariffs for this conditioning plant, and we are not aware of any authority for it to do so.

256

its market power. By permitting these producers to monopolize conditioning services, they would also be permitted to monopolize natural gas production. The conditioning plant could simply set a price for its service so high that non-owner producers would be unwilling to use the facility. No firms other than those which own the conditioning plant would bid on new leases on the North Slope since they would know that their gas could never be sold at a profit. Any reserves found in the Naval Petroleum Reserve-Alaska, the Beaufort Sea, the Chukchi Sea, and the Arctic National Wildlife Range would be controlled by the conditioning plant owners.

The situation described above can be contrary to the public interest in four ways. First, by discouraging numerous potential bidders on new gas leases, the conditioning plant monopoly could result in a significant reduction in revenues accruing to the Federal Government and the State of Alaska. Firms which are not also plant owners know that they may be denied access to the facility, and that if access is granted it will be at a high price. They may be reluctant to bid at all for new gas leases, and if they bid, their bids may be substantially lower than would be the case if the conditioning plant operated competitively.

Second, by restricting the incentive of firms to develop the North Slope resources, a healthy diversity of expectations which is inherent in the competitive process could be lost. If discouraged by the conditioning plant monopoly, various firms may restrict their exploration activities or drop out altogether. It is possible that the conditioning plant owners may not explore for gas as widely as the discouraged firms or may have more pessimistic views as to the potential of some areas. Some of the areas would then be ignored, and whatever gas would have been found would be, thus, lost to the nation.

Third, to whatever extent North Slope gas may be of sufficiently high volume and low cost to exert market power in downstream markets, the conditioning plant monopoly allows that market power to be exercised. Fourth, if, as we have suggested, non-owners will be discouraged from exploring for gas on the North Slope, that area will become the private preserve of the Prudhoe Bay producers. This will reduce the likelihood that production and exploration of natural gas on the North Slope will be competitive. The consequence could be a higher price for gas to consumers.[6]

---

[6] This is true even with wellhead price regulation. First, absent this monopoly, prices may not achieve the regulated ceiling due to competition. Second, even if the price ceiling is a constraint, a producer may profit by restricting output if the price of unregulated gas he produces in the lower forty-eight states is pushed up as a result.

These problems cannot be eliminated by setting aside lease tracts for firms which are not owners of the conditioning plant. Such firms may not be willing to bid even the minimum amount established by the United States Geological Survey or State of Alaska, and, even if they did obtain leases, the conditioning plant would still be able to exert its monopsony power. The result will likely be a restriction in gas available to consumers. This will be the case even if the plant is forced to process all gas delivered to it since the restriction will flow from its pricing policies. Thus, an unregulated conditioning plant monopoly raises serious problems which must be addressed by the Commission.

IV. Remedy

The most effective way to exploit monopoly power, in general, is to charge an excessive price for its use. Restraints on capacity and access by vertically integrated owners will only be used to exploit monopoly power if the ability to do so directly by charging an excessive amount is circumscribed by regulation.[7] Moreover, as the charges of this facility are unregulated, a ban on producer ownership of the facility, such as was adopted for the pipeline, will not produce any competitive gain since a third party owner of the conditioning plant can simply exploit all of its monopoly power at that level to the detriment of present and future producers.

Consequently, the Department recommends that the possible anti-competitive consequences of producer ownership of the gas processing facility at Prudhoe Bay be remedied by the imposition of a set of competitive rules on the ownership of the facility. The imposition of such rules is consistent with previous antitrust approaches to competitive bottlenecks.[8]

The following set of competitive rules will be capable, if correctly implemented, of dissipating the market power of the conditioning facility.[9] The four rules which should be incorporated in any rule promulgated on the ownership of the facility are:

---

[7] See Testimony of John H. Shenefield, Assistant Attorney General, Antitrust Division, before the Subcommittee on Antitrust and Monopoly of the Committee on the Judiciary of the United States Senate, June 21, 1977, pp. 16-26.

[8] See, e.g., United States v. Terminal Railroad Ass'n, 224 U.S. 383 (1912); Associated Press v. United States, 326 U.S. 1 (1945); In the Matter of Toledo Edison Company (Davis-Besse Nuclear Power Stations, Units 1, 2, and 3, Perry Nuclear Power Plant, Units 1 and 2), 6 NRC 133 (1977). See note 10 infra.

[9] These competitive rules are a variant of those that were proposed by the Attorney General for the Deepwater Ports. ("Report of the Attorney General Pursuant to Section 7 of the Deepwater Port Act of 1971 on the Application of LOOP, Inc. and Seadock, Inc. for Deepwater Port Licenses," November 5, 1976) and adopted in a modified form in The Secretary's Decision On The Deepwater Port

1. The facility must provide nondiscriminatory access to all producers of natural gas.

2. Any owner of the facility willing to provide the necessary capital can unilaterally obtain expansion of the facility.

3. An ownership share in the facility must be available to all parties at a price determined by a Commission rule promulgated *prior* to the construction of the facility. This price should include appropriate additional payments above the direct costs of the facility to compensate the initial sponsors for the risks and burdens of their efforts. The price should also reflect the economic depreciation of the facility.

4. The ownership proportions of the processing facility must be revised on a regular basis to ensure that each user's ownership share equals his proportionate use of the facility.

These four rules, when implemented by an appropriately specific agreement, will act to eliminate any unregulated monopoly power possessed by the processing facility. Any efforts by the processing facility to charge a monopoly tariff will have no effect since each user will receive a share of the excess charge in proportion to its use. Any potential user could frustrate any effort to restrict capacity by demanding an expansion of the facility. A producer of an additional field will be able to obtain access to the facility, and obtain any output expansion to the extent that the new producer is willing to provide the capital necessary to obtain that expansion. In addition, the rules will protect the initial owners of the facility from an effort by subsequent users to gain access to the facility without providing the necessary capital to support a proportionate share of the facility. Subsequent producer-user-owners will have to provide funds for a share of the facility proportionate to their use, and will have to compensate the initial owners for the risks and burdens of their sponsorship of the project.

V. Conclusion

The Department believes that, given the facts that the North Slope conditioning facility will be a "bottleneck" and will be unregulated, producer ownership of this facility could result in serious anticompetitive abuse. This potential for abuse can be remedied by the imposition of the competitive rules set forth above. We therefore believe it essential if competition in the exploration, development, and marketing of Alaskan natural gas is to be encouraged, that the Commission incorporate the appropriate competitive rules in any rule or policy adopted concerning the Alaska Natural Gas conditioning facility.

License Application of LOOP, Inc., Department of Transportation, December 17, 1976.

Supplement 4

An Analysis of the Rates of Return on Petroleum Pipeline Investments

Exxon Pipeline Company/Exxon Company, U.S.A.

Summary

To test the hypothesis that petroleum pipeline rates of return are excessive, as alleged by the United States Department of Justice and other critics of the petroleum pipeline industry, rates of return on total capital have been calculated for the petroleum pipeline companies reporting financial data (Form P) to the Interstate Commerce Commission (ICC)[1] for the period 1973-1977. These rates of return were compared to the rates of return for overall industry, for parent oil companies having pipeline subsidiaries, and for public utilities.

To facilitate comparisons of the petroleum pipeline rates of return with those of other industries, the empirical methodology employed conformed, as nearly as practicable, to that used by *Forbes* in its "30th Annual Report on American Industry" (January 9, 1978). This methodology was also applied to data obtained from Standard & Poor's Compustat Services, Inc. Whereas the *Forbes* report considered companies with annual sales in excess of $250 million, the Compustat data base includes companies with annual sales in excess of $2.5 million. The larger Compustat data base afforded additional comparisons of the average or mean rate of return between the petroleum pipeline industry and industry in general.

The results of this analysis do not support allegations that petroleum pipelines earn excessive rates of return. Analysis of Form P data shows that the pipeline industry earned a 10.0 percent median rate of return for the study period, which was about equal to the 9.7 percent median return for overall industry (*Forbes*). If average returns are examined, the results are similar. The 9.5 percent average pipeline rate of return (Form P) is about equal to the 9.9 percent all-industry average rate of return (Compustat). On average, the pipelines have lower

---

[1] Regulatory jurisdiction over petroleum pipelines was transferred from the ICC to the Federal Energy Regulatory Commission (FERC) on October 1, 1977.

returns than their parent oil companies. Oil companies with pipeline subsidiaries earned an 11.8 percent median rate of return and an 11.9 percent weighted average return compared to the 11.0 percent median return and 6.8 percent weighted return of their pipeline subsidiaries (Form P).

Although petroleum pipeline average returns were higher than public utility average returns (9.5 percent versus 6.9 percent Compustat), pipelines are riskier investments than are utilities. This is largely due to the differences in regulatory treatment of the two and is evidenced by the significantly greater variability of pipeline rates of return as compared to that of utility rates of return.

Introduction

John H. Shenefield of the U.S. Department of Justice on October 25, 1978[2] delineated the department's activities in the area of ownership of pipeline assets by integrated oil companies. Indicating a need for divestiture, Mr. Shenefield stated:

> Divestiture of (pipeline) assets is not a popular solution for (the) owners, especially since our studies show that ownership of pipelines is often profitable for oil companies to a degree inconsistent with the minimal risks involved. Profits are not the trigger for our concern—they are obviously the necessary reward for risks taken—but profits so consistently above norms as to suggest absence of competitive forces may indicate a need for our concern.

A host of similar charges with regard to the profitability of petroleum pipelines has been made by critics of the industry in recent years. Primary among these are:

- Petroleum pipelines earn rates of return that are excessive.
- Petroleum pipelines are essentially low-risk, public-utility type investments but earn returns which far exceed the most profitable public utilities.

If these charges have any substance, then an examination of petroleum pipeline rates of return should reveal returns substantially greater than overall industry returns. In addition, if petroleum pipelines are low-risk, public-utility type investments, then the variability of petroleum pipeline and utility rates of return should be the same.

---

[2]John H. Shenefield, Assistant Attorney General, Antitrust Division, "The New Rationalism and Old Verities: Deciding Who Gets What, and Why," remarks before the Graduate School of Business, Invitation Lecture Series, University of Chicago, October 25, 1978.

The purpose of this study, then, is to test the hypothesis that petroleum pipeline profits are excessive.

Examination of Rates of Return of Petroleum Pipelines

If petroleum pipeline earnings are indeed excessive as charged by industry critics, then such pipelines should earn a rate of return significantly above the rate of return earned by overall industry. This study will empirically examine this charge by computing pipeline rates of return and comparing these returns to overall industry returns. If pipeline returns are about equal to or less than industry returns, charges of excess profits can be dismissed.

Measure of Profitability. Two of the common measures of historical profitability are return on total capital and return on stockholders' equity. These two measures of profitability are distinguished by *Forbes*:[3]

> Companies obtain their capital from two sources: stockholders and creditors. Return on Stockholders' Equity is the percentage return on the stockholders' portion of the capital. . . . The Return on Total Capital is a "basic" measure of an enterprise's profitability. It does not reflect—as the Return on Stockholders' Equity does—the effects of financing decisions upon profitability. For companies that derive all of their capital from common equity, the two profitability measures would, of course, be identical. But a company that employs debt wisely can thereby boost its Return on Stockholders' Equity well above its Return on Total Capital.

In other words, return on total capital does not distinguish whether capital is obtained from debt or equity investors. It measures what a company is earning on a dollar of capital regardless of its financial structure. For this reason, return on total capital is a better measure than return on equity for comparing profitability of companies. This is especially important when examining petroleum pipeline returns because the pipeline debt-to-equity ratios are directly influenced by the Pipe Line Consent Decree of 1941.[4]

There is another reason why return on total capital is a better measure of petroleum pipeline profitability than return on equity. Often,

---

3 "30th Annual Report on American Industry," *Forbes*, vol. 121, no. 1 (January 9, 1978), p. 38.
4 The effects of the Consent Decree on debt financing are further explained in Appendix I.

the debt of a pipeline company is guaranteed by the parent oil company which owns the pipeline company's equity stock, and therefore does not reflect the amount of debt the pipeline company would be able to raise independent of parent company support. Hence, a comparison of petroleum pipeline returns on equity, where the stockholder is responsible for both the debt and equity investment, to other company returns on equity, where the stockholder has only his equity investment at stake, would be biased.

The ratio of stockholders' equity to debt varies widely across industries and will, therefore, bias an interindustry profitability comparison based on stockholders' equity.[5] The rate of return on total capital eliminates the effect of leverage and hence is a better measure for interindustry comparisons.

Data. To facilitate comparisons of petroleum pipeline rates of return with those of other industries, the empirical methodology employed conformed, as nearly as practicable, to that used in a readily available comparative report, *Forbes'* "30th Annual Report on American Industry" (January 9, 1978). The *Forbes* report includes 1,005 public companies whose annual sales were over $250 million. *Forbes* calculated each company's return on total capital as of the beginning of the year for the period 1973-1976 and the twelve-month period ending with the latest available quarterly report at the time of publication (usually the period ending with the third quarter). *Forbes* then categorized each company by industry group and calculated the arithmetic average (mean) rate of return over this period for each company in every industry group. The companies were ranked in numerical order by their average rate of return on total capital, and *Forbes* selected the median of these five-year average rates of return as the industry-average statistic.

Essentially the same methodology used by *Forbes* was applied to financial data obtained from annual reports (Form P) filed by federally regulated petroleum pipeline companies (transporting crude petroleum and petroleum products) with the Interstate Commerce Commission (ICC)[6] for each of the years 1973-1977. Rates of return on total capital were calculated for each pipeline company for each year 1973-

---

[5] The Federal Trade Commission Division of Statistics reports that the ratio of total stockholders' equity to debt for all manufacturing was 2.46 for the fourth quarter 1977, with a range of 1.5 to 5.6 for individual industries. The ratio of total stockholders' equity to debt for petroleum pipelines was 0.14 as of December 31, 1976, as calculated from the ICC *Transport Statistics.*

[6] Regulatory jurisdiction over petroleum pipelines was transferred from the ICC to the Federal Energy Regulatory Commission (FERC) on October 1, 1977.

1977. Averages for each company and the median rate of return for the industry were derived for the five-year period ending with December 31, 1977.

This methodology was also applied to data obtained from Standard & Poor's Compustat Services, Inc. Compustat maintains a computerized base of financial statement information for approximately 2,500 publicly owned companies whose annual sales are over $2.5 million. Use of the Compustat data provided an expanded information base which afforded additional financial comparisons between the petroleum pipeline industry and other industries.

The petroleum pipeline company sample was developed from the 115 pipeline companies reporting Form P data to the ICC for the period 1973-1977. To conform the pipeline sample to the Compustat sample, the 36 companies having 1977 annual revenues less than $2.5 million were eliminated. The remaining 79 pipeline companies represent 97 percent of the total capital, as determined in this study, for all 115 pipeline companies reporting to the ICC during the study period.[7]

In conforming the methodology for the Form P pipeline data to the *Forbes* methodology, it became apparent that some pipeline companies filing Form P reports have ownership in other pipeline companies also filing Form P reports, and that some pipeline companies have long-term investments in the stocks and bonds of other nonaffiliated companies. The objective of this study is to examine rates of return on pipeline operations, and therefore adjustments were made to avoid counting the same pipeline's return on capital twice; adjustments were also made to eliminate capital and returns on long-term investments not related to the pipeline business. Thus, the rates of return calculated from the Form P data represent *pipeline* rates of return on total capital employed.

One further adjustment was required to include all debt in the total capital calculations based on the Form P and the Compustat data. For the Form P data, this adjustment entailed including both notes payable and payables to affiliated companies. Notes payable are a form of debt and as such are appropriately included in total capital. In addition, a number of the petroleum pipeline companies used payables to affiliates as a form of financing and these amounts were also included in total capital. For the Compustat data, this adjustment entailed including notes payable in total capital consistent with the treatment of Form P data.

---

[7] Appendix II provides the methodology used in the *Forbes,* Compustat, and Form P data calculations.

Results. In the 1973-1977 period, as shown in Exhibits I and II, the median[8] (of the five-year averages) rate of return on total capital was 10.0 percent for petroleum pipelines, as determined from Form P data, compared to 9.7 percent for general industry, as determined by *Forbes*. Comparing petroleum pipeline rates of return to the results obtained from the Compustat data in Exhibit II, the median rate of return on total capital was 10.0 percent for petroleum pipelines compared to 9.6 percent for general industry.

Although *Forbes* calculated only five-year averages and the medians of these averages, use of the Compustat data permitted the calculation of the means and the mean weighted averages. The mean of the five-year average rates of return is the arithmetic average of the rates of return. The weighted average rate of return is determined by dividing the total annual returns earned by all companies in the particular industry group by the total capital employed at the beginning of the year in that industry group. The mean weighted average is simply the mean of the yearly weighted averages. The weighted average is significant for comparing industry rates of return in that it takes into account the size of the individual companies in the sample as well as the return earned by those companies.

The mean rate of return for the period 1973-1977 for petroleum pipelines was 9.5 percent compared to 9.9 percent for general industry (based on Compustat). The mean weighted average for petroleum pipelines was 7.6 percent compared to 9.4 percent for general industry (as shown in Exhibit II).

The Trans-Alaska Pipeline System (TAPS) was being constructed during the study period 1973-1977, and since the investment of this undertaking is significant, the rates of return for pipelines were also calculated exclusive of the TAPS participants. While it would be preferable to eliminate the TAPS investment from each participant's total pipeline capital, all of the data required to make this adjustment is not publicly available. Instead, as a sensitivity, the TAPS participants were eliminated from the pipeline sample. Elimination of these participants resulted in a mean rate of return of 9.8 percent, a median of 10.2 percent, and a mean weighted average of 10.2 percent. Thus, even excluding TAPS participants from the study, petroleum pipelines earn about the industry average rate of return.

Another way of examining the excess profits charge is to compare the earnings of petroleum pipeline subsidiaries of integrated oil companies and their parent companies. If pipelines earn excess profits, then

---

[8]The median rate of return is the middle value in a distribution, above and below which lie an equal number of values.

EXHIBIT I

COMPARATIVE RATES OF RETURN ON TOTAL CAPITAL
PETROLEUM PIPELINES VERSUS OTHER INDUSTRIES

Industry	Medians of Five-year Averages 1973–1977 (percent)
Personal products	14.0
Health care	13.6
Energy	12.0
Finance: insurance	11.8
Electronics and electrical equipment	11.4
Aerospace and defense	11.3
Industrial equipment	11.2
Leisure	11.1
Construction: contractors	11.0
Chemicals	10.8
Finance: banks	10.4
Wholesalers	10.4
Food and drink	10.2
Forest products and packaging	10.1
Petroleum pipelines	10.0
Automotive	9.8
Multicompanies: multi-industry	9.6
Information processing	9.5
Household products	9.0
Metals: non-ferrous	8.9
Supermarkets	8.9
Multicompanies: conglomerates	8.7
Retail distribution	8.7
Apparel	8.5
Metals: steel	8.4
Construction: building materials	8.1
Finance: consumer	7.6
Utilities: natural gas	7.4
Transportation: surface	6.1
Utilities: electric and telephone	6.0
Transportation: airline	4.3
All Industry Median	9.7

SOURCES: Pipelines: Interstate Commerce Commission, Form P, 1973-1977. Other industries: *Forbes,* 30th Annual Report on American Industry, January 9, 1978, vol. 121, no. 1.

EXHIBIT II

COMPARATIVE RATES OF RETURN ON TOTAL CAPITAL:
FIVE-YEAR AVERAGES FOR 1973–1977
(percent)

	Mean	Median	Mean Weighted Average
79 petroleum pipelines	9.5	10.0	7.6
2,500 other industrial firms	9.9	9.6	9.4
171 utilities	6.9	6.6	6.7

SOURCES: 79 petroleum pipeline companies: Interstate Commerce Commission, Form P, 1973-1977. All others: Standard & Poor's Compustat Services, Inc.

their earnings should on average exceed the earnings of their parent companies. However, this study found that the petroleum pipeline subsidiaries earned lesser returns than their parent companies. Of the oil companies listed with Compustat, 23 had 100 percent ownership of a pipeline company included in this study. The rates of return were determined for the 23 parent companies and their pipeline subsidiaries and are displayed in Exhibit III. For the 1973-1977 period the parent oil companies earned a mean rate of return on total capital of 11.6 percent compared to 9.6 percent for their pipeline subsidiaries. The median rate of return was 11.8 percent for the parent oil companies and 11.0 percent for the pipeline subsidiaries, and the mean weighted average rate was 11.9 percent for the parents and 6.8 percent for the subsidiaries.

Charges that petroleum pipelines earn excessive rates of return cannot be substantiated. Tests indicate that petroleum pipeline rates of return have been about equal to overall industry rates of return in the 1973-1977 period, whether median, mean, or mean weighted average rates of return are used for comparison. Further, petroleum pipeline subsidiaries of integrated oil companies earn on average less than their parent companies, whether mean, median, or mean weighted average is considered.

Examination of Rates of Return of
Petroleum Pipelines versus Utilities

"The required rate of return on an investment opportunity depends on the riskiness of the investment. The greater the riskiness of the invest-

EXHIBIT III
COMPARATIVE RATES OF RETURN ON TOTAL CAPITAL
PARENT COMPANY VERSUS PETROLEUM PIPELINE SUBSIDIARY
(percent)

	Parent Company Average 1973–1977	Pipeline Subsidiary Average 1973–1977
Amerada-Hess	14.8	3.7
American Petrofina	14.3	9.0
Ashland Oil Inc.	11.6	10.0
Atlantic Richfield Company	10.1	3.9
Cities Service Company	8.9	13.7
Continental Oil Company	12.6	13.4
Dome Petroleum	11.8	(7.4)
Exxon Corporation	13.8	7.6
Gulf Oil Corporation	11.3	7.8
Marathon Oil Company	13.1	12.5
Mobil Corporation	11.8	11.1
Occidental Petroleum Corporation	10.6	15.2
Phillips Petroleum Company	12.6	12.2
Reserve Oil and Gas	13.7	10.2
Shell Oil Company	12.0	13.6
Standard Oil Company (California)	12.4	16.6
Standard Oil Company (Indiana)	13.2	16.7
Standard Oil Company (Ohio)	8.4	0.6
Sun Oil Company	10.8	16.7
Total Petroleum of North America	7.3	5.2
Texaco	10.1	11.0
Union Oil Company of California	10.7	6.2
United Refining Company	10.2	11.8
Mean	11.6	9.6
Median	11.8	11.0
Mean Weighted Average	11.9	6.8

SOURCES: Pipeline companies: Interstate Commerce Commission, Form P, 1973-1977. Parent companies: Standard & Poor's Compustat Services, Inc.

ment, the more the return demanded by investors."[9] Critics of the industry have argued that while petroleum pipelines are essentially low-risk, public-utility type investments, they earn returns which far exceed the most profitable of public utilities. The validity of this con-

[9] E. Solomon and J. J. Pringle, *Introduction to Financial Management* (Santa Monica, Calif.: Goodyear Publishing Company, 1977), p. 332.

269

tention must be determined in two steps: first, by examining the profitability of the petroleum pipelines compared to utilities, and second, by investigating the risks associated with investment in one as opposed to the other.

Exhibit I based on *Forbes* data for public utilities and on the Form P calculations for petroleum pipelines indicates that the median rate of return on total capital was 10.0 percent for petroleum pipelines compared to 7.4 percent for natural gas utilities and 6.0 percent for electric and telephone companies over the period 1973-1977.

Exhibit II shows this same relationship based on the Compustat data for utilities and on the Form P calculations for petroleum pipelines. The 79 petroleum pipeline companies earned a mean rate of return of 9.5 percent, a median rate of return of 10.0 percent, and a mean weighted average rate of return of 7.6 percent, compared to 6.9 percent, 6.6 percent, and 6.7 percent, respectively, for 171 gas, electric, and telephone utility companies, again over the 1973-1977 period.

There are a number of important distinctions between petroleum pipelines and the utilities industries which subject pipelines to a greater risk than utilities. First, utilities are regulated legal monopolies, meaning that they are awarded a market franchise by a government agency. Petroleum pipelines are not protected from competitive entry. They cannot sign long-term agreements to guarantee capacity to a shipper or to supply a particular refiner or marketer. Second, regulators establish a protective mantle over utilities which reasonably assures them an adequate rate of return. Petroleum pipelines have no assurance of earning a reasonable return on invested capital. Consequently, any rational investor considering investment in petroleum pipelines will demand a return premium over that which he could obtain from investments in utilities.

In addition, the 1941 Pipe Line Consent Decree has, for all practical purposes, established a ceiling on the allowed rates of return for petroleum pipelines, but there is no floor or minimum rate of return. In other words, a limit is effectively set as to how profitable a petroleum pipeline can become, but a company which is losing money for one reason or another is "allowed" to continue to lose it. The lower downside potential of petroleum pipeline returns is clearly illustrated by the fact that of the 79 petroleum pipeline companies in the study, ten had five-year average returns on total capital of less than 4 percent, whereas out of the 171 utilities considered in the analysis of the Compustat data, not one had five-year average returns on total capital of less than 4 percent. In other words, 12.7 percent of the petroleum pipeline companies earned less than the least profitable utility company.

The dispersion of returns provides a measure of the degree of risk involved in investment or business decisions.[10] To examine the dispersion of petroleum pipeline and utility rates of return, the statistical variances of the rates were calculated for the two industries. Variance provides a measure of the dispersion or spread of the rates of return about the mean or average value and is indicative of the risk associated with investing capital in a particular industry or group. The variance clarifies the usefulness of the mean as a measure of anticipated rate of return. If the rates of return are tightly distributed about the average value, an investor can anticipate that the average industry return is a good indicator of individual company returns. However, if the returns are widely dispersed about the average industry value, an individual investor faces the risk of greater variability in individual company returns and cannot reasonably anticipate a return approximately equal to the average industry returns. The bar-graph distribution of rates of return in Exhibit IV illustrates the dispersion of petroleum pipeline rates of return and that of utility rates of return.

As shown in Exhibit V, the 79 petroleum pipelines had a variance of 23.7, while the Compustat sample of 171 utilities had a variance of 2.0. To examine the significance of this difference in variance, a statistical test[11] was employed. At the 99 percent level of significance, pipeline rates of return have significantly greater variability than utility rates of return. Investments in petroleum pipelines, therefore, are subject to greater risk than are investments in utilities companies.

Conclusion

This study has tested the hypotheses that:

- Petroleum pipeline rates of return are excessive.
- Petroleum pipelines are essentially low-risk, public-utility type investments but earn returns which far exceed the most profitable public utilities.

The results of the study clearly indicate that:

- Petroleum pipeline rates of return are not excessive. In fact, pipelines earn about the average for industry as a whole and earn less than the parent oil companies, whether median, mean, or mean weighted average is used for comparison.
- Although petroleum pipelines have historically earned higher returns on average than public utilities, pipelines display a substan-

---

[10] Ibid., pp. 309-310.
[11] For a detailed outline of the test, see Appendix III.

EXHIBIT IV

COMPARISON OF DISTRIBUTION OF RETURNS BETWEEN
PETROLEUM PIPELINES AND UTILITIES
(five-year averages for 1973–1977)

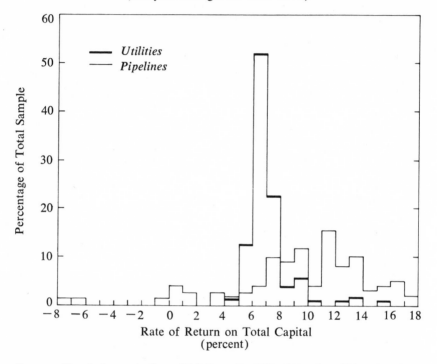

SOURCE: 79 petroleum pipelines: ICC Form P, 1973–1977. 171 utilities: Standard
& Poor's Compustat Services, Inc.

EXHIBIT V

COMPARISON OF VARIABILITY OF RETURNS BETWEEN PETROLEUM
PIPELINES AND UTILITIES: FIVE-YEAR AVERAGES FOR 1973–1977
(percent)

	Mean	Variance	Standard Deviation
79 petroleum pipelines	9.5	23.7	4.9
171 utilities	6.9	2.0	1.4

SOURCES: 79 petroleum pipeline companies: Interstate Commerce Commission,
Form P, 1973-1977. 171 utilities: Standard & Poor's Compustat Services, Inc.

tially greater variability in returns, and are subject to greater risks than utilities. While some pipelines have earned returns in excess of the most profitable public utility, nearly 13 percent of the pipeline sample here studied earned returns which were less than the *least* profitable public utility.

In summary, it has been demonstrated that charges of excess profits in petroleum pipeline companies cannot be substantiated.

Appendix I: Effects of the Pipeline Consent Decree on Pipeline Debt-to-Equity Ratios

The Pipe Line Consent Decree of 1941 [12] led to the increased use of debt financing by petroleum pipeline companies. Since this Decree was entered into, pipeline companies have financed heavily with debt in order to earn a reasonable rate of return within the limits imposed by the Decree.

The Consent Decree resulted from a Department of Justice action which named twenty major oil companies and fifty-nine petroleum pipeline subsidiaries as defendants. The suit alleged that shipper-owners were receiving rebates in the form of dividends from their pipeline subsidiaries in violation of the Interstate Commerce Act and the Elkins Act. The final judgment provided that:

> No defendant common carrier shall credit, give, grant, or pay, directly or indirectly . . . to any shipper-owner . . . any earnings, dividends, sums of money or other valuable considerations derived from transportation or other common-carrier services which in the aggregate is in excess of its share of seven percentum (7%) of the valuation [13] of such common-carrier's property, if such common carrier shall have transported during said calendar year any crude oil or gasoline, or other petroleum products for said shipper-owner, but shall be permitted . . . to credit, give, grant, or pay said percentum.

In essence, the Decree limits the annual dividend which a shipper-owned pipeline company may pay to its parent company to 7 percent of the valuation of the common-carrier's property as determined by the

---

[12] United States v. Atlantic Refining Co., Civil Action No. 14060, U.S. District Court, D.C. (1941).

[13] Valuation is the rate base determined annually by the ICC (now FERC) for each petroleum pipeline company falling within its jurisdiction. It is defined as the latest final value of a common-carrier's property owned and used for common-carrier purposes, and essentially consists of a weighted average of original cost and reproduction cost new, less depreciation.

ILLUSTRATIVE EXAMPLE OF EFFECTS OF THE CONSENT DECREE ON
DEBT/EQUITY RATIOS UNDER ICC VALUATION APPROACH FOR
HYPOTHETICAL PIPELINE

	Zero Debt	50 percent Debt	90 percent Debt
1. Revenue	230.0	275.0	311.0
2. Operating cost	50.0	50.0	50.0
3. Book depreciation (at 4%)	40.0	40.0	40.0
4. Interest (at 9%)	—	45.0	81.0
5. Income before taxes	140.0	140.0	140.0
6. Taxes (at 50%)	70.0	70.0	70.0
7. Net income	70.0	70.0	70.0
8. After-tax interest	—	22.5	40.5
9. Delevered net income	70.0	92.5	110.5
10. Rate of return on valuation	7.00%	7.00%	7.00%
11. Rate of return on total capital	7.00%	9.25%	11.05%

NOTE: All cases assume valuation = investment = $1,000.

FERC (formerly the ICC). Earnings in excess of 7 percent must be transferred to a surplus account and may be used only for maintaining normal and reasonable working capital requirements, for retiring of debt (arising out of the construction of common-carrier facilities) outstanding at the time of the Decree (1941), and for expansion of common-carrier facilities. However, the value of facilities constructed with such funds must be excluded from the rate base upon which future earnings may be generated. This provision effectuates a powerful disincentive to setting tariffs at a level which generates a return on valuation in excess of 7 percent.

The illustrative example above shows how the Consent Decree, as a matter of sound financial planning, strongly encourages a high debt-to-equity ratio. The example presents data for a hypothetical pipeline. Assuming that the pipeline costs $1,000 and that that cost is the same as the ICC valuation, the Consent Decree would limit dividends to $70.

Three cases are considered: financing with zero debt; with 50 percent debt; and with 90 percent debt. In the first case there is no interest charge and hence return on total capital and return on valuation are the same and equal to 7 percent. In the second case the annual interest charge, at 9 percent, on the $500 debt amounts to $45. Under the

Consent Decree, interest expense is allowed to be included as a cost in the tariff.[14] Removing the after-tax effects of interest from net income yields a delevered net income of $92.50 and a return on total capital of 9.25 percent. The return on valuation is, as in the first case, 7 percent.

Similarly, in the third case the annual interest charge is $81. Adding the tax savings to the net income yields a delevered return of $110.50 and a return on total capital of 11.05 percent. Again, the return on valuation is 7 percent.

In order to earn a reasonable return on total capital in this regulatory environment, a petroleum pipeline company can utilize substantial amounts of debt.

It should be noted, however, that for the same level of pipeline company net income, the rate of return on total capital will generally be different from the rate of return on valuation. This can occur because the asset bases are different (that is, valuation is not equal to total capital), and/or because the return elements are computed differently. Recognition of these computation differences is important because regulated firms are allowed to earn a rate of return on total capital which numerically exceeds the corresponding rate of return on the valuation rate base.

Appendix II: Empirical Methodology [15]

Forbes
total returns = net income including minority interests
     (+) after-tax interest expense
total capital = stockholders' equity including minority interests
     (+) long-term debt
     (+) current maturities of long-term debt
     (+) accumulated deferred taxes

Compustat Data
total returns = net income including minority interests
     (+) after-tax interest expense
total capital = stockholders' equity including minority interests
     (+) long-term debt
     (+) debt in current liabilities
     (+) accumulated deferred taxes

---

[14] This interpretation was endorsed by the U.S. Supreme Court in the Arapahoe case, United States v. Atlantic Refining Co., 360 U.S. 19 (1959).

[15] Net income is after-tax net income. A 5$ percent tax rate is assumed for interest expense. A 7½ percent tax rate is assumed on earnings of affiliated companies because of the 85 percent dividends exclusion. Stockholders' equity includes both common and preferred.

Form P Pipeline Data

total returns $=$ net income including minority interests

$(-)$ after-tax earnings from affiliated companies and other long-term investments

$(+)$ after-tax interest expense, including debt expense amortization

total capital $=$ stockholders' equity including minority interests

$(-)$ investments in affiliated companies and other long-term investments

$(+)$ long-term debt

$(+)$ debt in current liabilities including payables to affiliated companies

$(+)$ accumulated deferred taxes

Appendix III: Statistical Test of Sample Variances

Definitions

n_p = number of petroleum pipelines' rates of return included in the sample = 79.

n_u = number of utility companies' rates of return included in the sample = 171.

s_p^2 = sample variance of petroleum pipelines' rates of return = 23.72 percent.

s_u^2 = sample variance of utility companies' rates of return = 1.96 percent.

σ_p^2 = the unknown population variance of petroleum pipelines' rates of return.

σ_u^2 = the unknown population variance of utility companies' rates of return.

Statistical Test

Ho: $\qquad \sigma_p^2 \leq \sigma_u^2$

Ha: $\qquad \sigma_p^2 > \sigma_u^2$

Test statistics: $F = \dfrac{s_p^2}{s_u^2}$

$\qquad\qquad\qquad$ F is known to have an F distribution with $n_p - 1, n_u - 1$ degrees of freedom.

Criteria: If $F > F_{.01, 78, 170}$, reject Ho and conclude that $\sigma_p^2 > \sigma_u^2$; otherwise accept Ho.

Since the F distribution is such that $F_{.01, 60, 120} > F_{.01, 78, 170}$, if $F > F_{.01, 60, 120}$, Ho will be rejected.

Results

$$F_{.01,\ 60,\ 120} = 1.66$$

$$F = \frac{23.72}{1.96} = 12.10$$

By the established criteria, since 12.10 > 1.66, Ho is rejected. In fact, Ho would be rejected even at the one-tenth of one percent level of significance. The risk of variability of rates of return is greater for petroleum pipelines than for utilities.

APPENDIX IV
SUPPLEMENTARY PETROLEUM PIPELINE DATA
(thousands of dollars)

	1972	1973	1974	1975	1976	1977
Allegheny Pipeline						
Net income	612	982	891	1,379	1,610	1,449
AFIT interest expense	—	—	—	—	—	—
Affiliate and other net income	—	—	—	—	—	—
Total returns	612	982	891	1,379	1,610	1,449
Shareholders equity	5,981	6,963	6,817	8,196	9,806	11,255
Long-term debt	4,370	9,794	10,556	9,501	5,900	4,069
Affiliate and other investments	—	—	—	—	—	—
Other capital	215	191	1,457	2,052	2,651	3,155
Total capital	10,566	16,948	18,830	19,749	18,357	18,479
Return on total capital	9.29	5.26	7.32	8.15	7.89
Amdel Pipeline						
Net income			(315)	59	1,031	392
AFIT interest expense			—	—	—	13
Affiliate and other net income			—	—	—	—
Total returns			(315)	59	1,031	405
Shareholders equity		15,019	14,704	14,763	15,152	15,134
Long-term debt		—	—	—	—	—
Affiliate and other investments		—	—	—	—	—
Other capital		1	652	1,835	5,297	4,771
Total capital		15,020	15,356	16,598	20,449	19,905
Return on total capital			(2.10)	0.38	6.21	1.98

	1972	1973	1974	1975	1976	1977
American Petrofina Pipeline						
Net income			335	972	896	1,751
AFIT interest expense			293	285	278	269
Affiliate and other net income			—	—	(47)	(172)
Total returns			628	1,257	1,127	1,848
Shareholders equity		3,928	3,928	4,050	3,981	4,032
Long-term debt		9,877	9,618	9,334	9,031	8,699
Affiliate and other investments		—	—	(220)	(181)	(163)
Other capital		238	375	352	301	313
Total capital		14,043	13,921	13,516	13,132	12,881
Return on total capital			4.47	9.03	8.34	14.07
Amoco Pipeline						
Net income	19,746	20,494	20,334	27,224	33,154	36,876
AFIT interest expense	1,276	1,406	1,319	1,092	848	775
Affiliate and other net income	(3,787)	(5,702)	(5,260)	(7,166)	(10,083)	(9,506)
Total returns	17,235	16,198	16,393	21,150	23,919	28,145
Shareholders equity	113,525	117,201	122,156	131,809	144,855	159,463
Long-term debt	55,606	50,098	46,500	40,300	27,800	51,591
Affiliate and other investments	(53,405)	(52,884)	(51,663)	(50,303)	(48,958)	(47,855)
Other capital	1,668	1,731	7,421	8,798	14,497	15,292
Total capital	117,394	116,146	124,414	130,604	138,194	178,491
Return on total capital		13.80	14.11	17.00	18.31	20.37

(Table continues on the next page)

279

APPENDIX IV (continued)

	1972	1973	1974	1975	1976	1977
Arco Pipeline						
Net income	17,551	19,211	11,595	15,160	18,955	9,622
AFIT interest expense	169	111	55	4	5	31,103
Affiliate and other net income	(1,421)	(1,201)	(973)	(829)	(596)	(191)
Total returns	16,299	18,121	10,677	14,335	18,364	40,534
Shareholders equity	148,226	151,187	154,782	161,943	158,779	168,401
Long-term debt	89,600	71,200	290,000	787,744	1,290,000	1,655,000
Affiliate and other investments	(28,243)	(22,836)	(17,358)	(11,952)	(6,426)	(9,563)
Other capital	15,531	49,136	51,565	168,386	284,185	250,472
Total capital	225,114	248,687	478,989	1,106,121	1,726,538	2,064,310
Return on total capital		8.05	4.29	2.99	1.66	2.35
Ashland Pipeline						
Net income	4,469	3,487	4,501	4,412	12,291	9,421
AFIT interest expense	1,388	1,348	1,550	1,590	1,555	1,491
Affiliate and other net income	(80)	(80)	(348)	(3,004)	(4,129)	(2,023)
Total returns	5,777	4,755	5,703	2,998	9,717	8,889
Shareholders equity	27,074	15,561	5,280	4,691	9,982	16,993
Long-term debt	37,933	40,933	42,933	42,243	40,348	38,670
Affiliate and other investments	(2,418)	(2,883)	(891)	(4,051)	(6,429)	(6,184)
Other capital	1,700	4,392	20,967	22,851	21,884	18,232
Total capital	64,289	58,003	68,289	65,734	65,785	67,711
Return on total capital		7.40	9.83	4.39	14.78	13.51

	1972	1973	1974	1975	1976	1977
Badger Pipeline						
Net income	1,188	1,062	872	1,274	1,366	1,360
AFIT interest expense	262	289	283	319	341	333
Affiliate and other net income	—	—	—	—	—	—
Total returns	1,450	1,351	1,155	1,593	1,707	1,693
Shareholders equity	1,986	1,953	2,013	1,945	1,952	2,006
Long-term debt	9,000	8,580	9,360	7,740	9,620	6,900
Affiliate and other investments	—	—	—	—	—	—
Other capital	496	591	569	2,906	740	3,115
Total capital	11,482	11,124	11,942	12,591	12,312	12,021
Return on total capital		11.77	10.38	13.34	13.56	13.75
Belle Fourche Pipeline						
Net income	1,070	1,119	1,018	1,132	1,317	1,354
AFIT interest expense	239	246	239	281	269	250
Affiliate and other net income	—	—	—	—	—	—
Total returns	1,309	1,365	1,257	1,413	1,586	1,604
Shareholders equity	8,905	10,156	11,135	13,460	14,410	15,843
Long-term debt	2,000	—	—	—	—	5,163
Affiliate and other investments	—	—	—	—	—	—
Other capital	4,284	4,308	5,264	5,919	5,797	488
Total capital	15,189	14,464	16,399	19,379	20,207	21,494
Return on total capital		8.99	8.69	8.62	8.18	7.94
BP Pipeline						
Net income						(22,539)

(Table continues on the next page)

APPENDIX IV (continued)

	1972	1973	1974	1975	1976	1977
AFIT interest expense						23,745
Affiliate and other net income						—
Total returns						1,206
Shareholders equity					186,793	195,754
Long-term debt					866,824	1,099,637
Affiliate and other investments					(160)	(2,483)
Other capital					120,793	114,589
Total capital					1,174,250	1,407,497
Return on total capital						0.10
Black Lake Pipeline						
Net income	458	760	505	481	588	655
AFIT interest expense	272	233	224	215	175	194
Affiliate and other net income	—	—	—	—	—	—
Total returns	730	993	729	696	763	849
Shareholders equity	987	1,030	967	1,027	1,033	1,033
Long-term debt	7,330	7,040	6,750	6,460	6,170	5,880
Affiliate and other investments	—	—	—	—	—	—
Other capital	290	290	425	539	632	720
Total capital	8,607	8,360	8,142	8,026	7,835	7,633
Return on total capital		11.54	8.72	8.55	9.51	10.84
Buckeye Pipeline						
Net income	12,390	15,251	15,214	20,037	22,713	27,381
AFIT interest expense	1,609	1,519	1,261	2,239	3,467	3,610

	1972	1973	1974	1975	1976	1977
Affiliate and other net income	(284)	(316)	(206)	(126)	28	(315)
Total returns	13,715	16,454	16,269	22,150	26,208	30,676
Shareholders equity	107,146	106,745	114,819	114,963	129,626	132,719
Long-term debt	50,944	46,371	76,436	79,611	85,012	92,729
Affiliate and other investments	(5,170)	(4,950)	(4,493)	(5,194)	(4,824)	(7,333)
Other capital	3,664	7,573	3,014	13,026	3,021	3,287
Total capital	156,584	155,739	189,776	202,406	212,835	221,402
Return on total capital		10.51	10.45	11.67	12.95	14.41
Butte Pipeline						
Net income	747	1,012	1,420	1,649	1,223	1,338
AFIT interest expense	198	276	330	222	168	144
Affiliate and other net income	—	—	—	—	—	—
Total returns	945	1,288	1,750	1,871	1,391	1,482
Shareholders equity	3,286	3,285	3,405	3,481	3,589	3,712
Long-term debt	6,200	5,600	—	4,400	3,800	3,200
Affiliate and other investments	—	—	—	—	—	—
Other capital	602	600	5,633	660	705	754
Total capital	10,088	9,485	9,038	8,541	8,094	7,666
Return on total capital		12.77	18.45	20.70	16.29	18.31
Calnev Pipeline						
Net income	2,088	2,561	1,564	1,917	2,310	2,326
AFIT interest expense	59	82	143	48	—	—
Affiliate and other net income	—	—	—	—	(23)	(126)
Total returns	2,147	2,643	1,707	1,965	2,287	2,200

(Table continues on the next page)

APPENDIX IV (continued)

	1972	1973	1974	1975	1976	1977
Shareholders equity	11,599	14,160	13,728	15,645	16,981	18,307
Long-term debt	1,020	4,660	2,200	—	—	—
Affiliate and other investments	—	—	—	—	(1,741)	(3,910)
Other capital	468	438	506	597	3,201	3,461
Total capital	13,087	19,258	16,434	16,242	18,441	17,858
Return on total capital		20.20	8.86	11.96	14.08	11.93
Chase Pipeline						
Net income				1,298	1,556	1,660
AFIT interest expense				—	—	—
Affiliate and other net income						
Total returns				1,298	1,556	1,660
Shareholders equity			15,964	14,935	15,246	16,766
Long-term debt			—	—	—	—
Affiliate and other investments						
Other capital			139	31	124	126
Total capital			16,103	14,966	15,370	16,892
Return on total capital				8.06	10.40	10.80
Cherokee Pipeline						
Net income	1,006	273	509			
AFIT interest expense	157	186	179			
Affiliate and other net income	—	—	—			
Total returns	1,163	459	688			
Shareholders equity	5,287	5,560				

	1972	1973	1974	1975	1976	1977
Long-term debt	5,250	4,450				
Affiliate and other investments	—	—				
Other capital	1,100	848				
Total capital	11,637	10,858				
Return on total capital		3.94	6.34			
Chevron Pipeline						
Net income	6,412	6,266	9,546	7,908	8,818	8,617
AFIT interest expense	1,396	1,313	1,497	1,556	1,525	1,908
Affiliate and other net income	—	—	—	—	—	—
Total returns	7,808	7,579	11,043	9,464	10,343	10,525
Shareholders equity	9,403	10,168	10,062	9,870	9,112	8,629
Long-term debt	41,416	37,142	34,568	32,894	50,220	48,546
Affiliate and other investments	(3)	(3)	(3)	(2)	(2)	(2)
Other capital	1,733	11,200	17,221	16,842	1,851	2,868
Total capital	52,549	58,507	61,848	59,604	61,181	60,041
Return on total capital		14.42	18.87	15.30	17.35	17.20
Chicap Pipeline						
Net income	(255)	1,417	1,816	2,092	3,399	3,243
AFIT interest expense	766	772	744	738	774	727
Affiliate and other net income	—	—	—	—	—	—
Total returns	511	2,189	2,560	2,830	4,173	3,970
Shareholders equity	1,508	2,924	3,341	3,612	3,707	3,870
Long-term debt	22,140	21,300	20,460	19,572	18,684	17,796

(Table continues on the next page)

APPENDIX IV (continued)

	1972	1973	1974	1975	1976	1977
Affiliate and other investments	—	—	—	—	—	—
Other capital	1,228	1,714	3,279	6,086	5,667	6,026
Total capital	24,876	25,938	27,080	29,270	28,058	27,692
Return on total capital		8.80	9.87	10.45	14.26	14.15
Cities Service Pipeline						
Net income	2,102	1,083	4,660	3,188	3,968	3,298
AFIT interest expense	45	160	226	235	245	229
Affiliate and other net income	—	—	—	—	—	—
Total returns	2,147	1,243	4,886	3,423	4,213	3,527
Shareholders equity	17,193	16,926	21,586	21,624	20,731	22,454
Long-term debt	4,400	4,200	4,000	6,800	6,600	6,400
Affiliate and other investments	—	—	—	—	—	—
Other capital	598	548	333	469	655	724
Total capital	22,191	21,674	25,919	28,893	27,986	29,578
Return on total capital		5.60	22.54	13.21	14.58	12.60
Collins Pipeline						
Net income	1,068	943	728	698	678	616
AFIT interest expense	368	419	394	360	326	289
Affiliate and other net income	—	—	—	—	—	—
Total returns	1,436	1,362	1,122	1,058	1,004	905
Shareholders equity	1,736	2,679	3,007	3,080	3,458	4,074
Long-term debt	9,680	8,750	7,917	7,083	6,250	5,833
Affiliate and other investments	—	—	—	—	—	—

	1972	1973	1974	1975	1976	1977
Other capital	897	929	1,196	1,376	1,518	1,705
Total capital	12,313	12,358	12,120	11,539	11,226	11,612
Return on total capital		11.06	9.08	8.73	8.70	8.06
Colonial Pipeline						
Net income	30,066	34,105	26,914	42,590	59,634	53,404
AFIT interest expense	12,213	12,280	11,970	11,641	12,120	14,627
Affiliate and other net income			—	—	—	—
Total returns	42,279	46,385	38,884	54,231	71,754	68,031
Shareholders equity	49,932	52,213	48,779	49,502	49,124	48,816
Long-term debt	397,808	385,870	373,932	361,994	420,056	520,818
Affiliate and other investments	(167)	(167)	(227)	(227)	(227)	(227)
Other capital	12,093	12,208	21,709	31,337	68,865	55,728
Total capital	459,666	450,124	444,193	442,606	537,818	625,135
Return on total capital		10.09	8.64	12.21	16.21	12.65
Continental Pipeline						
Net income	8,723	9,665	8,132	10,434	10,067	11,344
AFIT interest expense	1,404	1,289	1,126	1,198	1,122	1,053
Affiliate and other net income	(3,570)	(4,109)	(3,124)	(4,184)	(1,434)	(3,942)
Total returns	6,557	6,845	6,134	7,448	9,755	8,455
Shareholders equity	47,154	56,819	58,597	62,070	62,137	58,480
Long-term debt	39,940	31,032	33,274	31,416	29,908	24,880
Affiliate and other investments	(39,012)	(39,817)	(33,293)	(35,976)	(31,642)	(28,837)
Other capital	3,376	3,738	2,355	1,881	2,596	1,964

(Table continues on the next page)

287

APPENDIX IV   (continued)

	1972	1973	1974	1975	1976	1977
Total capital	51,458	51,772	60,933	59,391	62,999	56,487
Return on total capital		13.30	11.85	12.22	16.43	13.42
Cook Inlet Pipeline						
Net income	5,773	4,313	1,500	3,228	2,595	2,610
AFIT interest expense	673	702	642	278	153	147
Affiliate and other net income	—	—	—	—	—	—
Total returns	6,446	5,015	2,142	3,506	2,748	2,757
Shareholders equity	17,393	21,706	23,206	26,434	26,129	26,299
Long-term debt	15,900	10,200	6,050	2,000	3,200	1,600
Affiliate and other investments	—	—	—	—	—	—
Other capital	5,228	3,777	3,919	3,911	2,434	3,791
Total capital	38,521	35,683	33,175	32,345	31,763	31,690
Return on total capital		13.02	6.00	10.57	8.50	8.68
Diamond Shamrock Pipeline						
Net income	369	327	423	343	1,021	1,791
AFIT interest expense	604	598	591	583	575	565
Affiliate and other net income	—	—	—	—	—	—
Total returns	973	925	1,014	926	1,596	2,356
Shareholders equity	—	—	—	—	—	—
Long-term debt	13,247	13,078	12,894	12,692	12,473	12,234
Affiliate and other investments	—	—	—	—	—	—
Other capital	2,314	2,085	2,299	2,304	5,827	4,972
Total capital	15,561	15,163	15,193	14,996	18,300	17,206
Return on total capital		5.94	6.69	6.09	10.64	12.87

	1972	1973	1974	1975	1976	1977
Dixie Pipeline						
Net income	3,707	3,408	1,844	3,864	5,595	4,285
AFIT interest expense	997	1,004	1,076	894	794	802
Affiliate and other net income	—	—	—	—	—	—
Total returns	4,704	4,412	2,920	4,758	6,389	5,087
Shareholders equity	4,886	4,682	4,762	4,888	5,190	4,687
Long-term debt	36,900	35,550	34,400	25,900	24,500	33,142
Affiliate and other investments	—	—	—	—	—	—
Other capital	1,400	1,400	1,929	7,402	11,060	2,973
Total capital	43,186	41,632	41,091	38,190	40,750	40,802
Return on total capital		10.22	7.01	11.58	16.73	12.48
Dome Pipeline						
Net income	54	(657)	(2,184)	(1,757)	393	2,522
AFIT interest expense	18	365	1,446	1,469	1,772	1,081
Affiliate and other net income	—	—	—	—	—	—
Total returns	72	(292)	(738)	(288)	2,165	3,603
Shareholders equity	71	(586)	(2,770)	(4,527)	(4,135)	(1,613)
Long-term debt	—	—	15,845	13,845	11,813	37,930
Affiliate and other investments	(1)	(1)	—	(1)	(1)	—
Other capital	489	21,332	10,503	17,874	22,824	31,078
Total capital	559	20,745	23,578	27,191	30,501	67,395
Return on total capital		(52.24)	(3.56)	(1.22)	7.96	11.81
Eureka Pipeline						
Net income	(140)	172	(93)	(63)	(364)	(212)

(Table continues on the next page)

289

APPENDIX IV (continued)

	1972	1973	1974	1975	1976	1977
AFIT interest expense	—	—	—	—	—	—
Affiliate and other net income	(30)	(29)	(36)	(28)	(27)	(26)
Total returns	(170)	143	(129)	(91)	(391)	(238)
Shareholders equity	3,240	3,353	3,258	3,134	2,711	2,454
Long-term debt	—	—	—	—	—	—
Affiliate and other investments	(785)	(915)	(615)	(825)	(707)	(656)
Other capital	—	—	—	—	—	—
Total capital	2,455	2,438	2,643	2,309	2,004	1,798
Return on total capital		5.82	(5.29)	(3.44)	(16.93)	(11.88)
Explorer Pipeline						
Net income	(13,090)	(9,954)	(5,342)	(7,526)	(9,795)	5,744
AFIT interest expense	5,912	8,159	8,234	8,167	8,394	8,372
Affiliate and other net income	—	—	—	—	—	—
Total returns	(7,178)	(1,795)	2,892	641	(1,401)	14,116
Shareholders equity	8,569	(1,385)	(6,727)	(14,253)	(24,048)	(18,304)
Long-term debt	197,280	197,280	197,280	197,280	197,280	193,230
Affiliate and other investments	—	—	—	—	—	—
Other capital	5,086	10,998	14,748	22,569	28,808	25,567
Total capital	210,935	206,893	205,301	205,596	202,040	200,493
Return on total capital		(.85)	1.40	.31	(.68)	6.99
Exxon Pipeline						
Net income	29,446	32,296	26,519	47,210	67,592	109,690
AFIT interest expense	2,873	3,681	5,379	6,429	7,113	36,092

	1972	1973	1974	1975	1976	1977
Affiliate and other net income	(6,759)	(6,714)	(4,044)	(6,628)	(10,374)	(12,023)
Total returns	25,560	29,263	27,854	47,011	64,331	133,759
Shareholders equity	105,465	101,761	105,284	132,994	184,586	276,776
Long-term debt	223,673	264,449	508,000	956,351	1,355,327	1,585,520
Affiliate and other investments	(12,131)	(12,246)	(18,056)	(18,215)	(20,075)	(21,424)
Other capital	5,761	1,172	54,548	32,394	175,186	224,072
Total capital	322,768	355,136	649,776	1,106,524	1,695,024	2,064,944
Return on total capital		9.07	7.84	7.23	5.81	7.89
Gulf Central Pipeline						
Net income	(9,865)	(6,258)	(6,752)	(2,530)	(767)	3,205
AFIT interest expense	3,328	2,051	2,762	2,283	1,879	1,784
Affiliate and other net income	—	—	—	—	—	—
Total returns	(6,537)	(4,207)	(3,990)	(247)	1,112	4,989
Shareholders equity	(3,754)	52,929	46,177	43,647	42,879	46,085
Long-term debt	97,275	46,500	44,111	41,188	41,707	39,975
Affiliate and other investments	—	—	—	—	—	—
Other capital	2,080	2,032	9,924	12,909	12,037	14,615
Total capital	95,601	101,461	100,212	97,744	96,623	100,675
Return on total capital		(4.40)	(3.93)	(.25)	1.14	5.16
Gulf Refining Pipeline						
Net income	9,572	7,092	5,843	7,221	9,553	14,291
AFIT interest expense	2	—	2	—	—	2
Affiliate and other net income	—	—	—	—	—	—
Total returns	9,574	7,092	5,845	7,221	9,553	14,293

(Table continues on the next page)

APPENDIX IV (continued)

	1972	1973	1974	1975	1976	1977
Shareholders equity	97,799	98,141	103,984	111,205	120,758	110,049
Long-term debt	—	—	—	—	—	—
Affiliate and other investments						
Other capital	878	635	1,659	6,842	11,436	16,345
Total capital	98,677	98,776	105,643	118,047	132,194	126,394
Return on total capital		7.19	5.92	6.84	8.09	10.81
Hess Pipeline						
Net income	249	1,212	1,293	(282)	492	312
AFIT interest expense	77	116	223	272	236	224
Affiliate and other net income	—	—	—	—	—	—
Total returns	326	1,328	1,516	(10)	728	536
Shareholders equity	7,426	8,638	9,931	9,649	6,033	6,346
Long-term debt	2,865	2,865	5,945	5,487	5,334	5,334
Affiliate and other investments	—	—	—	—	—	—
Other capital	7,393	11,377	18,870	14,750	14,034	12,741
Total capital	17,684	22,880	34,746	29,886	25,401	24,421
Return on total capital		7.51	6.63	(.03)	2.44	2.11
Hydrocarbon Transportation Inc. Pipeline						
Net income	4,310	3,831	3,429	4,862	6,162	7,839
AFIT interest expense	487	641	915	1,182	1,087	998
Affiliate and other net income	—	—	—	—	—	—
Total returns	4,797	4,472	4,344	6,044	7,249	8,837
Shareholders equity	53,457	41,788	30,417	32,631	37,160	38,225

	1972	1973	1974	1975	1976	1977
Long-term debt	14,100	17,800	26,000	24,200	22,400	20,600
Affiliate and other investments	—	—	—	—	—	—
Other capital	5,317	5,634	7,030	6,964	5,805	10,763
Total capital	72,874	65,222	63,447	63,795	65,365	69,588
Return on total capital		6.14	6.66	9.53	11.36	13.52
Jayhawk Pipeline						
Net income	592	911	611	561	481	387
AFIT interest expense	109	102	90	64	37	17
Affiliate and other net income	—	—	—	—	—	—
Total returns	701	1,013	701	625	518	404
Shareholders equity	10,282	11,193	11,204	11,165	11,645	11,432
Long-term debt	2,804	2,569	1,734	899	339	94
Affiliate and other investments	—	—	—	—	—	—
Other capital	715	835	1,014	1,149	991	741
Total capital	13,801	14,597	13,952	13,213	12,975	12,267
Return on total capital		7.34	4.80	4.48	3.92	3.11
Jet Lines Pipeline						
Net income	96	152	160	175	316	712
AFIT interest expense	92	97	173	126	102	85
Affiliate and other net income	—	—	—	—	—	—
Total returns	188	249	333	301	418	797
Shareholders equity	1,480	1,632	1,792	1,966	2,283	2,995
Long-term debt	3,605	3,510	2,990	2,640	1,930	3,425

(Table continues on the next page)

293

APPENDIX IV (continued)

	1972	1973	1974	1975	1976	1977
Affiliate and other investments						
Other capital	628	696	1,041	829	1,020	683
Total capital	5,713	5,838	5,823	5,435	5,233	7,103
Return on total capital		4.36	5.70	5.17	7.69	15.23
Kaneb Pipeline						
Net income	2,610	2,562	1,368	2,576	2,578	2,894
AFIT interest expense	621	664	631	594	556	511
Affiliate and other net income	(88)	(95)	(271)	(121)	(110)	—
Total returns	3,143	3,131	1,728	3,049	3,024	3,405
Shareholders equity	16,884	17,546	17,047	17,114	18,001	17,996
Long-term debt	17,753	16,710	15,695	14,470	13,245	12,020
Affiliate and other investments	(1,895)	(1,895)	(1,780)	(1,471)	—	—
Other capital	1,008	1,136	1,324	1,876	2,558	2,662
Total capital	33,750	33,497	32,286	31,989	33,804	32,678
Return on total capital		9.28	5.16	9.44	9.45	10.07
Kaw Pipeline						
Net income	218	250	423	449	426	424
AFIT interest expense						
Affiliate and other net income						
Total returns	218	250	423	449	426	424
Shareholders equity	4,898	4,898	4,892	4,878	4,866	4,862
Long-term debt						
Affiliate and other investments						

	1972	1973	1974	1975	1976	1977
Other capital	31	46	43	128	70	104
Total capital	4,929	4,944	4,935	5,006	4,936	4,966
Return on total capital		5.07	8.56	9.10	8.51	8.59
Kiantone Pipeline						
Net income		1,048	881	653	366	787
AFIT interest expense		398	370	342	313	285
Affiliate and other net income	—	—	—	—	—	—
Total returns		1,446	1,251	995	679	1,072
Shareholders equity	1,955	2,491	1,306	1,347	1,714	2,501
Long-term debt	7,650	7,083	6,517	5,950	5,383	4,816
Affiliate and other investments	—	—	—	—	—	—
Other capital	570	593	586	579	1,908	921
Total capital	10,175	10,167	8,409	7,876	9,005	8,238
Return on total capital		14.21	12.30	11.83	8.62	11.90
Lakehead Pipeline						
Net income	20,846	25,700	14,210	19,913	21,003	23,575
AFIT interest expense	4,492	5,381	5,997	5,260	4,511	4,250
Affiliate and other net income	—	—	—	—	—	—
Total returns	25,338	31,081	20,207	25,173	25,514	27,825
Shareholders equity	107,415	118,114	68,787	72,699	77,702	82,277
Long-term debt	142,000	151,000	133,350	127,462	119,819	115,022
Affiliate and other investments	(108)	(107)	(173)	(143)	(111)	(100)
Other capital	7,000	7,009	79,286	69,425	76,049	79,044
Total capital	256,307	276,016	281,250	269,443	273,459	276,243
Return on total capital		12.13	7.32	8.95	9.47	10.18

(Table continues on the next page)

APPENDIX IV (continued)

	1972	1973	1974	1975	1976	1977
Laurel Pipeline						
Net income	549	1,571	1,317	1,396	1,701	2,764
AFIT interest expense	671	620	568	522	468	418
Affiliate and other net income	—	—	—	—	—	—
Total returns	1,220	2,191	1,885	1,918	2,169	3,182
Shareholders equity	6,555	7,570	7,776	7,505	7,873	8,415
Long-term debt	24,000	22,000	20,000	18,000	16,000	14,000
Affiliate and other investments	—	—	—	—	—	—
Other capital	2,046	2,073	2,532	3,049	3,609	4,090
Total capital	32,601	31,643	30,308	28,554	27,482	26,505
Return on total capital		6.72	5.96	6.33	7.60	11.58
Mapco Pipeline						
Net income	10,905	16,367	41,295	12,523	13,513	15,828
AFIT interest expense	3,849	4,946	6,915	4	3	2
Affiliate and other net income	(285)	(266)	(20,524)	—	—	—
Total returns	14,469	21,047	27,686	12,527	13,516	15,830
Shareholders equity	83,263	96,838	126,891	135,495	143,235	157,877
Long-term debt	101,899	141,192	180,000			
Affiliate and other investments	(23,078)	(33,512)	(102,203)	(8)	(8)	(8)
Other capital	9,008	7,169	44,122	17,489	19,550	22,657
Total capital	171,092	211,687	248,810	152,976	162,777	180,526
Return on total capital		12.30	13.08	5.03	8.84	9.72
Marathon Pipeline						
Net income	5,133	5,907	7,091	10,747	15,884	14,007

	1972	1973	1974	1975	1976	1977
AFIT interest expense	1,619	1,965	1,975	1,918	2,003	5,347
Affiliate and other net income	(12)	(12)	(16)	(16)	(12)	(18)
Total returns	6,740	7,860	9,050	12,649	17,875	19,336
Shareholders equity	17,958	23,006	27,332	32,278	35,208	64,605
Long-term debt	63,285	62,288	61,232	60,176	106,653	134,917
Affiliate and other investments	(208)	(208)	(208)	(208)	(208)	(28)
Other capital	1,415	2,596	7,189	12,776	12,743	9,835
Total capital	82,450	87,682	95,545	105,022	154,396	209,329
Return on total capital	8.17	9.53	10.32	13.24	17.02	12.52
Michigan-Ohio Pipeline						
Net income	(2)	42	301	208	331	576
AFIT interest expense	1	18	3	74	178	175
Affiliate and other net income						
Total returns	(1)	60	304	282	509	751
Shareholders equity	4,119	4,160	4,462	4,420	4,750	5,327
Long-term debt	—	—	—	—	—	—
Affiliate and other investments	—	—	—	—	—	—
Other capital	138	4,192	16	4,231	3,916	4,481
Total capital	4,257	8,352	4,478	8,651	8,666	9,808
Return on total capital	(0.02)	1.41	3.64	6.30	5.88	8.67
Mid Valley Pipeline						
Net income	3,792	3,711	3,264	3,836	6,666	5,681
AFIT interest expense	141	104	144	395	434	377

(Table continues on the next page)

APPENDIX IV (continued)

	1972	1973	1974	1975	1976	1977
Affiliate and other net income	—	—	—	—	—	—
Total returns	3,933	3,815	3,408	4,231	7,100	6,058
Shareholders equity	31,067	31,211	30,364	32,326	32,342	33,320
Long-term debt	—	1,467	9,167	8,942	8,225	8,352
Affiliate and other investments	—	—	—	—	—	—
Other capital	2,701	483	2,028	4,363	5,597	5,743
Total capital	33,768	33,161	41,559	45,631	46,164	47,415
Return on total capital		11.30	10.28	10.18	15.56	13.12
Minnesota Pipeline						
Net income	2,230	2,341	2,076	2,869	2,453	2,054
AFIT interest expense	14	37	276	124	61	—
Affiliate and other net income	—	—	—	—	—	—
Total returns	2,244	2,378	2,352	2,993	2,514	2,054
Shareholders equity	12,282	13,920	12,267	14,061	16,514	12,918
Long-term debt	—	4,375	3,750	1,125	—	—
Affiliate and other investments	—	—	—	—	—	—
Other capital	395	826	3,769	4,127	3,770	5,990
Total capital	12,677	19,121	19,786	19,313	20,284	18,908
Return on total capital		18.76	12.30	15.13	13.02	10.13
Mobil Pipeline						
Net income	15,815	18,609	12,726	23,971	27,939	31,329
AFIT interest expense	2,422	2,730	3,930	3,190	3,058	3,053
Affiliate and other net income	(4,282)	(4,908)	(4,431)	(6,672)	(9,240)	(8,869)

	1972	1973	1974	1975	1976	1977
Total returns	13,955	16,431	12,225	20,489	21,757	25,513
Shareholders equity	91,234	99,283	98,372	93,324	107,898	111,920
Long-term debt	130,689	127,241	123,793	110,000	110,000	110,000
Affiliate and other investments	(49,589)	(49,502)	(49,967)	(51,845)	(44,453)	(43,226)
Other capital	3,845	4,235	7,318	5,068	6,600	10,516
Total capital	176,179	181,257	179,516	156,547	180,045	189,210
Return on total capital		9.33	6.74	11.41	13.90	14.17
Olympic Pipeline						
Net income	2,050	2,529	3,660	3,914	4,229	3,690
AFIT interest expense	606	775	1,241	1,201	1,175	1,446
Affiliate and other net income	—	—	—	—	—	—
Total returns	2,656	3,304	4,901	5,115	5,404	5,136
Shareholders equity	2,975	3,054	3,865	2,879	2,833	2,674
Long-term debt	18,118	17,053	15,987	14,921	38,855	37,789
Affiliate and other investments	—	—	—	—	—	—
Other capital	9,475	19,834	28,565	30,115	4,395	5,129
Total capital	30,568	39,941	48,417	47,915	46,083	45,592
Return on total capital		10.81	12.27	10.56	11.28	11.15
Osage Pipeline						
Net income					1,061	2,196
AFIT interest expense					477	637
Affiliate and other net income					—	—
Total returns					1,538	2,833

(Table continues on the next page)

APPENDIX IV (continued)

	1972	1973	1974	1975	1976	1977
Shareholders equity				1,998	2,159	2,315
Long-term debt				—	13,540	13,080
Affiliate and other investments					—	—
Other capital				13,289	2,215	2,880
Total capital				15,287	17,914	18,275
Return on total capital					10.06	15.81
Owensboro-Ashland Pipeline						
Net income				2,670	1,936	1,698
AFIT interest expense				1,527	1,546	1,518
Affiliate and other net income				—	—	—
Total returns				4,197	3,482	3,216
Shareholders equity			10,567	1,244	1,252	2,950
Long-term debt			38,500	38,362	37,728	37,395
Affiliate and other investments			(11,949)	72		
Other capital			2,916	4,736	7,939	9,943
Total capital			40,034	44,414	46,919	50,188
Return on total capital				10.48	7.84	6.85
Pasco Pipeline						
Net income			33	416	289	
AFIT interest expense			—	—	—	
Affiliate and other net income			(178)	(136)	(144)	
Total returns			(145)	280	145	
Shareholders equity		4,502	4,535	4,951	5,239	

	1972	1973	1974	1975	1976	1977
Long-term debt		—	—	—	—	—
Affiliate and other investments		(1,024)	(1,125)	(1,174)	(1,217)	
Other capital		4,571	8,361	7,628	8,405	
Total capital		8,049	11,771	11,405	12,427	
Return on total capital			(1.80)	2.38	1.27	
Phillips Pipeline						
Net income	6,572	6,075	4,263	6,153	12,020	11,425
AFIT interest expense	53	49	45	44	38	42
Affiliate and other net income	—	—	—	—	—	—
Total returns	6,625	6,124	4,308	6,197	12,058	11,467
Shareholders equity	50,795	51,869	52,733	53,886	58,306	68,855
Long-term debt	1,352	4,246	4,133	16,013	25,484	18,846
Affiliate and other investments	—	—	—	—	—	—
Other capital	117	122	130	137	145	1,526
Total capital	52,264	56,237	56,996	70,036	83,935	89,227
Return on total capital		11.72	7.66	10.87	17.22	13.66
Phillips Petroleum Products Pipeline						
Net income	924	1,245	606	437	1,710	
AFIT interest expense	15	13	11	9	8	
Affiliate and other net income	—	—	—	—	—	
Total returns	939	1,258	617	446	1,718	
Shareholders equity	4,213	4,616	4,616	4,616	4,702	
Long-term debt	2,977	2,385	2,831	3,381	3,249	

(Table continues on the next page)

301

APPENDIX IV (continued)

	1972	1973	1974	1975	1976	1977
Affiliate and other investments	—	—	—	—	—	
Other capital	85	106	208	248	276	
Total capital	7,275	7,107	7,655	8,245	8,227	
Return on total capital		17.29	8.68	5.83	20.84	
Pioneer Pipeline						
Net income	684	681	722	755	723	781
AFIT interest expense	7	—	—	—	3	1
Affiliate and other net income	—	—	—	—	—	—
Total returns	691	681	722	755	726	782
Shareholders equity	4,839	5,079	5,349	5,609	5,772	5,999
Long-term debt	—	—	—	—	—	—
Affiliate and other investments	—	—	—	—	—	—
Other capital	—	103	79	89	163	894
Total capital	4,839	5,182	5,428	5,698	5,935	6,893
Return on total capital		14.07	13.93	13.91	12.74	13.18
Plantation Pipeline						
Net income	11,576	11,734	6,830	12,156	19,975	23,837
AFIT interest expense	4,445	4,357	4,524	4,573	4,544	4,466
Affiliate and other net income	—	—	—	—	—	—
Total returns	16,021	16,091	11,354	16,729	24,519	28,303
Shareholders equity	23,629	22,740	23,322	23,239	24,344	25,104
Long-term debt	144,950	142,158	139,588	137,752	135,654	138,089
Affiliate and other investments	—	—	—	—	—	—

	1972	1973	1974	1975	1976	1977
Other capital	1,656	1,150	5,662	9,678	13,119	18,107
Total capital	170,235	166,048	168,572	170,669	173,117	181,300
Return on total capital		9.45	6.84	9.92	14.37	16.35
Platte Pipeline						
Net income	3,003	3,626	3,658	3,612	3,671	3,090
AFIT interest expense	376	391	430	355	307	250
Affiliate and other net income	—	—	—	—	—	—
Total returns	3,379	4,017	4,088	3,967	3,978	3,340
Shareholders equity	13,421	13,313	13,364	13,401	13,433	13,358
Long-term debt	14,620	11,960	11,500	3,540	6,280	—
Affiliate and other investments	—	—	—	—	—	—
Other capital	2,204	2,200	2,202	8,440	2,810	7,468
Total capital	30,245	27,473	27,066	25,381	22,523	20,826
Return on total capital		13.28	14.88	14.66	15.67	14.83
Portal Pipeline						
Net income	862	967	1,096	1,359	1,606	1,622
AFIT interest expense	190	172	153	134	115	96
Affiliate and other net income	(36)	(217)	(222)	(214)	(224)	(258)
Total returns	1,016	922	1,027	1,279	1,497	1,460
Shareholders equity	13,147	13,949	13,554	11,282	11,880	11,972
Long-term debt	6,388	5,636	4,885	4,133	3,382	2,630
Affiliate and other investments	(4,320)	(5,297)	(4,060)	(4,041)	(4,034)	(4,083)
Other capital	772	772	2,037	1,885	1,731	1,554

(Table continues on the next page)

APPENDIX IV   (continued)

	1972	1973	1974	1975	1976	1977
Total capital	15,987	15,060	16,416	13,259	12,959	12,073
Return on total capital		5.77	6.82	7.79	11.29	11.27
Portland Pipeline						
Net income	2,256	2,212	1,290	1,849	2,501	2,091
AFIT interest expense	324	284	257	234	212	195
Affiliate and other net income	—	—	—	—	—	—
Total returns	2,580	2,496	1,547	2,083	2,713	2,286
Shareholders equity	21,631	22,532	21,729	21,473	21,375	20,706
Long-term debt	11,700	10,725	9,750	8,775	7,800	6,825
Affiliate and other investments	—	—	(1)	(1)	(1)	(1)
Other capital	1,191	975	1,258	1,365	1,455	1,229
Total capital	34,522	34,232	32,736	31,612	30,629	28,759
Return on total capital		7.23	4.52	6.36	8.58	7.46
Powder River Pipeline						
Net income	412	162	171	937	1,343	775
AFIT interest expense	38	300	286	271	257	243
Affiliate and other net income	—	—	—	—	—	—
Total returns	450	462	457	1,208	1,600	1,018
Shareholders equity	(279)	(117)	54	991	1,434	1,909
Long-term debt	7,600	7,200	6,800	6,400	6,000	5,600
Affiliate and other investments	—	—	—	—	—	—
Other capital	400	555	764	1,132	1,335	1,522
Total capital	7,721	7,638	7,618	8,523	8,769	9,031
Return on total capital		5.98	5.98	15.86	18.77	11.61

	1972	1973	1974	1975	1976	1977
Pure Transportation Pipeline						
Net income	3,601	3,149	2,686	2,455	2,139	2,417
AFIT interest expense	563	627	669	551	481	506
Affiliate and other net income	—	—	—	—	—	—
Total returns	4,164	3,776	3,355	3,006	2,620	2,923
Shareholders equity	29,533	32,309	34,196	34,169	34,378	35,306
Long-term debt	18,162	16,586	15,010	13,457	14,113	12,919
Affiliate and other investments	—	—	—	—	—	—
Other capital	1,757	1,684	2,322	1,890	2,171	2,612
Total capital	49,452	50,579	51,528	49,516	50,662	50,837
Return on total capital		7.64	6.63	5.83	5.29	5.77
Santa Fe Pipeline						
Net income	(528)	427	1,550	1,296	1,646	3,099
AFIT interest expense	1,846	1,853	1,822	1,759	1,713	1,588
Affiliate and other net income	(180)	(180)	(180)	(180)	(180)	(45)
Total returns	1,138	2,100	3,192	2,875	3,179	4,642
Shareholders equity	(2,273)	(1,846)	(296)	1,000	2,646	5,745
Long-term debt	42,765	42,765	40,665	38,215	36,215	29,883
Affiliate and other investments	(4,000)	(4,000)	(4,000)	(4,000)	(4,000)	—
Other capital	57	29	143	2,828	5,006	8,859
Total capital	36,549	36,948	36,512	38,043	39,867	44,487
Return on total capital		5.75	8.64	7.87	8.36	11.64
Shamrock Pipeline						
Net income	503	612	322	339	755	1,051

(Table continues on the next page)

APPENDIX IV (continued)

	1972	1973	1974	1975	1976	1977
AFIT interest expense	40	19	5	—	—	—
Affiliate and other net income	—	—	—	—	—	—
Total returns	543	631	327	339	755	1,051
Shareholders equity	7,366	7,978	8,300	8,638	9,393	10,444
Long-term debt	—	—	—	—	—	—
Affiliate and other investments	—	—	—	—	—	—
Other capital	495	—	101	163	237	439
Total capital	7,861	7,978	8,401	8,801	9,630	10,883
Return on total capital		8.03	4.10	4.04	8.58	10.91
Shell Pipeline						
Net income	9,777	8,940	9,520	19,752	27,671	25,194
AFIT interest expense	3,549	3,569	4,515	4,135	3,704	4,115
Affiliate and other net income	(1,016)	(2,353)	(2,842)	(4,426)	(7,252)	(8,422)
Total returns	12,310	10,156	11,193	19,461	24,123	20,887
Shareholders equity	55,179	49,120	51,640	54,392	34,063	32,257
Long-term debt	102,149	120,656	99,072	97,009	95,086	93,163
Affiliate and other investments	(49,441)	(45,847)	(43,956)	(40,880)	(39,979)	(39,449)
Other capital	3,102	3,471	20,344	20,001	40,276	36,152
Total capital	110,989	127,400	127,100	130,522	129,446	122,123
Return on total capital		9.15	8.79	15.31	18.48	16.14
Sohio Pipeline						
Net income	2,769	2,525	2,136	1,913	2,273	(77,007)
AFIT interest expense	—	—	—	—	—	58,471

	1972	1973	1974	1975	1976	1977
Affiliate and other net income					34	73
Total returns	2,769	2,525	2,136	1,913	2,307	(18,463)
Shareholders equity	38,455	40,980	43,116	45,029	47,302	(29,704)
Long-term debt	20,000	—	394,500	1,548,988	2,466,892	3,059,067
Affiliate and other investments	(277)	(316)	(408)	(425)	(396)	(5,198)
Other capital	94,602	119,349	112,316	21,935	176,278	41,900
Total capital	152,780	160,013	549,524	1,615,527	2,690,076	3,066,065
Return on total capital		1.65	1.33	.35	.14	(0.69)
Southcap Pipeline						
Net income	2,172	2,523	1,247	3,827	4,135	4,855
AFIT interest expense	777	815	911	1,021	1,400	1,450
Affiliate and other net income	—	—	—	—	—	—
Total returns	2,949	3,338	2,158	4,848	5,535	6,305
Shareholders equity	3,785	4,278	4,075	5,872	4,207	4,625
Long-term debt	22,496	21,370	33,344	35,718	34,092	32,466
Affiliate and other investments	—	—	—	—	—	—
Other capital	1,143	5,258	1,954	2,707	3,401	6,482
Total capital	27,424	30,906	39,373	44,297	41,700	43,573
Return on total capital		12.17	6.98	12.31	12.50	15.12
Southern Pacific Pipeline						
Net income	13,356	14,329	13,317	15,972	17,891	21,497
AFIT interest expense	239	—		—	—	7
Affiliate and other net income	(6)	(6)	(6)	(6)	(6)	(3,726)

(Table continues on the next page)

APPENDIX IV (continued)

	1972	1973	1974	1975	1976	1977
Total returns	13,589	14,323	13,311	15,966	17,885	17,778
Shareholders equity	134,414	140,243	152,088	111,527	108,268	111,764
Long-term debt	—	—	—	—	—	—
Affiliate and other investments	(36,187)	(32,088)	(35,780)	(29,700)	(28,549)	(27,070)
Other capital	—	—	40	25,999	24,898	27,167
Total capital	98,227	108,155	116,348	107,826	104,617	111,861
Return on total capital		14.58	12.31	13.72	16.59	16.99
Sun Pipeline						
Net income	6,888	13,509	4,246	8,616	11,725	15,929
AFIT interest expense	281	85	114	68	50	83
Affiliate and other net income	(9)	(9)	(114)	(4,465)	(3,956)	(3,916)
Total returns	7,160	13,585	4,246	4,219	7,819	12,096
Shareholders equity	44,016	68,075	70,672	71,132	77,181	98,714
Long-term debt	3,130	3,764	2,838	1,345	2,289	1,550
Affiliate and other investments	(1,491)	(25,802)	(28,299)	(27,212)	(31,453)	(50,428)
Other capital	2,389	2,632	4,219	4,536	5,798	11,973
Total capital	48,044	48,669	49,430	49,801	53,815	61,809
Return on total capital		28.28	8.72	8.54	15.70	22.48
Tecumseh Pipeline						
Net income	397	838	362	466	696	1,088
AFIT interest expense	101	91	79	69	58	47
Affiliate and other net income	—	—	—	—	—	—
Total returns	498	929	441	535	754	1,135

	1972	1973	1974	1975	1976	1977
Shareholders equity	3,642	3,980	4,219	4,313	4,267	4,266
Long-term debt	4,240	3,680	3,120	2,560	2,000	1,440
Affiliate and other investments	—	—	—	—	—	—
Other capital	565	583	715	1,703	1,106	1,287
Total capital	8,447	8,243	8,054	8,576	7,373	6,993
Return on total capital	11.00		5.35	6.64	8.79	15.39
Texaco–Cities Service Pipeline						
Net income	372	76	467	883	1,657	3,483
AFIT interest expense	—	1	—	—	1	20
Affiliate and other net income	—	—	—	—	—	—
Total returns	372	77	467	883	1,658	3,503
Shareholders equity	14,857	14,801	15,268	15,102	14,944	14,956
Long-term debt	—	—	—	—	—	—
Affiliate and other investments	(17)	(16)	(12)	(11)	(11)	(9)
Other capital	1,001	1,138	3,893	1,388	2,160	2,946
Total capital	15,841	15,923	19,149	16,479	17,093	17,893
Return on total capital		.49	2.93	4.61	10.06	20.49
Texas Eastern Trans. Pipeline						
Net income	6,222	9,104	11,461	18,000	18,726	23,419
AFIT interest expense	2,364	2,569	3,364	2,877	2,803	2,555
Affiliate and other net income	—	—	—	—	—	—
Total returns	8,586	11,673	14,825	20,877	21,529	25,974
Shareholders equity	—	—	—	—	—	—

(Table continues on the next page)

APPENDIX IV (continued)

	1972	1973	1974	1975	1976	1977
Long-term debt	107,183	117,353	124,291	163,799	176,606	194,710
Affiliate and other investments	—	—	—	—	—	—
Other capital	474	414	11,820	14,740	18,491	22,258
Total capital	107,657	117,767	136,111	178,539	195,097	216,968
Return on total capital		10.84	12.59	15.34	12.06	13.31
Texas–New Mexico Pipeline						
Net income	3,733	3,769	2,881	4,133	5,116	4,813
AFIT interest expense	286	278	288	220	173	136
Affiliate and other net income	—	—	—	—	—	—
Total returns	4,019	4,047	3,169	4,353	5,289	4,949
Shareholders equity	16,818	17,107	16,897	16,891	16,756	16,680
Long-term debt	8,250	6,750	5,250	3,750	2,250	750
Affiliate and other investments	—	—	—	—	—	—
Other capital	3,492	3,781	4,470	5,254	4,850	5,112
Total capital	28,560	27,638	26,617	25,895	23,856	22,542
Return on total capital		14.17	11.47	16.35	20.42	20.75
Texas Pipeline						
Net income	10,756	8,636	11,612	19,420	22,899	18,154
AFIT interest expense	1,132	1,826	2,024	1,306	1,207	1,434
Affiliate and other net income	—	—	—	—	—	—
Total returns	11,888	10,462	13,636	20,726	24,106	19,588
Shareholders equity	104,220	107,307	110,818	113,747	117,136	125,241
Long-term debt	8,313	5,850	4,950	4,050	3,150	2,250

	1972	1973	1974	1975	1976	1977
Affiliate and other investments	(6)	(6)	(6)	(8)	(8)	(7)
Other capital	47,322	40,057	42,137	41,312	51,056	58,837
Total capital	159,849	153,208	157,899	159,101	171,334	186,321
Return on total capital		6.54	8.90	13.13	15.15	11.43
Texoma Pipeline						
Net income					6,397	9,155
AFIT interest expense					4,794	4,886
Affiliate and other net income					—	—
Total returns				10,199	11,191	14,041
Shareholders equity					13,605	15,606
Long-term debt				104,500	106,000	101,125
Affiliate and other investments					—	—
Other capital				8,284	(8,302)	(12,258)
Total capital				122,983	127,907	128,989
Return on total capital					9.10	10.98
Wesco Pipeline						
Net income	100	309	1,028	652	438	726
AFIT interest expense	8	—	—	—	—	—
Affiliate and other net income	—				—	—
Total returns	108	309	1,028	652	438	726
Shareholders equity	960	2,097	3,125	3,776	4,214	4,940
Long-term debt	—	—	—	—	—	—
Affiliate and other investments	(1,100)	—	—	—	(1,919)	(1,850)

(Table continues on the next page)

APPENDIX IV   (continued)

	1972	1973	1974	1975	1976	1977
Other capital	1,982	5,162	7,433	5,098	5,694	9,617
Total capital	1,842	7,259	10,558	8,874	7,989	12,707
Return on total capital		16.78	14.16	6.18	4.94	9.09
West Shore Pipeline						
Net income	1,136	1,220	1,820	1,733	1,954	1,822
AFIT interest expense	345	382	411	519	477	599
Affiliate and other net income	—	—	—	—	—	—
Total returns	1,481	1,602	2,231	2,252	2,431	2,421
Shareholders equity	2,380	2,413	2,627	2,358	2,377	2,417
Long-term debt	13,200	13,050	17,400	18,100	17,000	16,200
Affiliate and other investments	—	—	—	—	—	—
Other capital	928	935	1,508	2,087	2,640	2,916
Total capital	16,508	16,398	21,535	22,545	22,017	21,533
Return on total capital		9.70	13.61	10.46	10.78	11.00
West Texas Gulf Pipeline						
Net income	2,365	2,234	2,188	1,740	2,936	2,927
AFIT interest expense	5	1	—	—	—	—
Affiliate and other net income	—	—	—	—	—	—
Total returns	2,370	2,235	2,188	1,740	2,936	2,927
Shareholders equity	17,567	16,111	15,224	13,478	12,314	11,552
Long-term debt	—	—	—	—	—	—
Affiliate and other investments	—	—	—	—	—	—
Other capital	469	816	913	1,330	1,294	1,263

	1972	1973	1974	1975	1976	1977
Total capital	18,036	16,927	16,137	14,808	13,608	12,815
Return on total capital		12.39	12.93	10.78	19.83	21.51
Western Oil Transp. Pipeline						
Net income	1,405	1,776	1,691	2,585	2,128	2,130
AFIT interest expense	—	—	—	—	—	—
Affiliate and other net income	—	—	—	—	—	—
Total returns	1,405	1,776	1,691	2,585	2,128	2,130
Shareholders equity	240	216	657	742	120	250
Long-term debt	—	—	—	—	—	—
Affiliate and other investments	—	—	—	—	—	—
Other capital	9,804	11,389	13,265	15,811	17,495	19,682
Total capital	10,044	11,605	13,922	16,553	17,615	19,932
Return on total capital		17.68	14.57	18.57	12.86	12.09
Williams Brothers Pipeline						
Net income	14,885	16,616	14,456	16,366	18,473	21,800
AFIT interest expense	6,358	6,058	5,758	5,467	5,196	4,819
Affiliate and other net income	(231)	(185)	(2,157)	(2,157)	(1,687)	—
Total returns	21,012	22,489	18,057	19,636	21,982	26,619
Shareholders equity	122,592	131,316	139,509	140,852	146,476	139,950
Long-term debt	197,924	187,324	176,724	164,674	151,203	137,212
Affiliate and other investments	(4,095)	(3,615)	(28,360)	(24,101)	—	—
Other capital	10,723	10,658	13,947	35,111	40,322	45,479
Total capital	327,144	325,683	301,820	316,536	338,001	322,641
Return on total capital		6.87	5.54	6.51	6.94	7.88

(Table continues on the next page)

APPENDIX IV (continued)

	1972	1973	1974	1975	1976	1977
Wolverine Pipeline						
Net income	415	804	3,123	923	5,006	5,383
AFIT interest expense	705	1,308	1,232	1,311	1,450	1,648
Affiliate and other net income	—	—	—	—	—	—
Total returns	1,120	2,112	4,355	2,234	6,456	7,031
Shareholders equity	2,321	2,406	3,325	3,413	4,243	2,782
Long-term debt	33,600	33,167	32,733	32,300	36,627	39,953
Affiliate and other investments	—	—	—	—	—	—
Other capital	661	651	7,033	10,799	7,971	6,972
Total capital	36,582	36,224	43,091	46,512	48,841	49,707
Return on total capital		5.77	12.02	5.18	13.88	14.40
Wyco Pipeline						
Net income	1,538	1,489	1,085	1,573	1,507	2,042
AFIT interest expense	285	360	438	405	484	437
Affiliate and other net income	—	—	—	—	—	—
Total returns	1,823	1,849	1,523	1,978	1,991	2,479
Shareholders equity	2,294	2,523	2,260	2,171	2,190	2,273
Long-term debt	5,975	10,150	10,150	10,000	9,000	8,250
Affiliate and other investments	—	—	—	—	—	—
Other capital	5,014	450	351	1,191	1,302	1,531
Total capital	13,283	13,123	12,761	13,362	12,492	12,054
Return on total capital		13.92	11.61	15.50	14.90	19.84
Yellowstone Pipeline						
Net income	1,445	1,382	1,544	1,525	1,627	1,762

	1972	1973	1974	1975	1976	1977
AFIT interest expense	229	252	250	232	205	197
Affiliate and other net income	—	—	—	—	—	—
Total returns	1,674	1,634	1,794	1,757	1,832	1,959
Shareholders equity	4,230	4,264	4,379	4,404	4,421	4,346
Long-term debt	8,166	7,453	6,740	6,027	5,314	4,600
Affiliate and other investments	—	—	—	—	—	—
Other capital	2,463	2,577	2,664	2,615	3,179	3,336
Total capital	14,859	14,294	13,783	13,046	12,914	12,282
Return on total capital		11.00	12.55	12.75	14.04	15.17

SOURCE: Interstate Commerce Commission, Form P.

315

Supplement 5
Oil Pipelines: Industry Structure

S. Morris Livingston

I. Introduction

This paper is limited to an analysis of the industry structure of crude oil and refined products pipelines, and the conclusions stemming therefrom. In considering industry structure, the major fact to be recognized and explained is the pervasiveness of vertical integration. Why are pipelines usually built by the prospective users and what are the economic effects? What would be gained, or lost, if shippers were prohibited from owning the lines they use?

Section II describes this vertical integration, and Section III examines the reasons why it has occurred. Included in this analysis is a description of some of the numerous instances in which both crude oil and refined products pipelines exist solely to serve a single refinery. There is also an explanation of why vertical integration takes the form of joint ownership of many of the major pipelines.

Starting with the erroneous assumption that both crude oil and refined products pipelines are dominated by a few large integrated oil companies, critics allege that these owners have an unfair advantage over smaller competitors in that they can discriminate against non-owner shippers in the use of the lines, and can charge tariff rates that are excessive. Because this paper is limited to an analysis of industry structure, it does not examine the evidence as to whether owner-shippers do or do not discriminate against non-owner shippers. Neither does it go into the questions of whether tariff rates are more or less than adequate. Section IV, however, indicates how the location of refineries with respect to existing products pipelines limits the possibilities for such discrimination.

Section V examines the theory, posed by the U.S. Department of Justice, that pipeline owners can benefit by restricting capacity or throughput, thus forcing some crude oil or refined products to move by less efficient modes of transportation, causing downstream prices to be higher by this incremental cost, and increasing their downstream profits. It questions the logic of the theory and cites the evidence that pipeline owners have not acted, and should not be expected to act, in the alleged manner.

Because pipelines are usually built and owned by those that expect to use them, some of the usual measures of industry structure, such as the concentration of ownership and the ease of entry, are less significant for pipelines than for many industries. There is also the qualification that national concentration ratios are relatively meaningless because each pipeline is in no position to compete with all other pipelines. These measures, and the extent of competition among pipelines, are, however, considered briefly in Section VI.

II. Extent of Vertical Integration

In January 1974 there were sixty-five refining companies, accounting for over 96 percent of the operating crude oil distillation capacity of all U.S. refineries, that owned or were part owners of crude oil pipelines that supply their refineries. There were fifty refining companies, accounting for over 94 percent of the refining capacity, that owned or were part owners of refined products pipelines.[1]

This vertical integration is not limited to a few large petroleum companies. The larger refiners are large owners of pipelines, but they are also large users of those lines, and their aggregate ownership is not out of line with their own requirements. The percentages in Table 1 are necessarily approximations, because of the difficulty of getting complete data on the many lines not reporting to the Interstate Commerce Commission, and they will vary slightly from year to year, but they are accurate enough to demonstrate the pervasiveness of vertical integration.

All of the forty-two largest refining companies, accounting for 94 percent of total U.S. crude running capacity in January 1974, own crude oil pipelines, and all but three own refined products lines. Even among

---

[1] These numbers are approximations, and may be slightly understated, because of the difficulty of getting information about some of the smaller firms. The numbers will also change slightly from year to year. The situation on January 1, 1974, is given here because the difficult and time consuming research into pipeline ownership, going beyond the readily available data for lines regulated by the Interstate Commerce Commission, was undertaken at a time when much of the necessary information was not yet available for a later date.

the smaller refiners, twenty-three own crude oil lines and eleven own products lines, and the smaller refiners as a group have roughly the same share of pipeline ownership as of refining capacity.

The pervasiveness of this general pattern of vertical integration can be emphasized by considering the apparent exceptions, both the pipelines that are not owned by refiners and the refiners that do not own pipelines.

Table 1 shows that about 12 percent of the miles of crude oil lines are owned by firms that are not also refiners, but most of this share is not an exception to the rule that the lines are usually built by those that expect to use them. It includes numerous small lines built by crude oil producers. They are vertically integrated but the motive is the outlet for the crude rather than supply for a refinery.[2] There are also those lines, such as Buckeye and Williams, that were built originally by refiners but subsequently became independent.[3] Other exceptions are the 2,500-mile Lake Head Pipeline, which has supplied some crude oil to U.S. refineries but was built by Canadian interests primarily to transport oil from Western Canada to refineries in Eastern Canada, and the 514-mile Portland Pipe Line, which transports crude oil from a tanker terminal at Portland, Maine to refineries in Canada. There are also a number of crude oil gathering lines built and operated by firms engaged in buying oil at the well and reselling it. Thus they are shipper owners even if neither refiners nor producers.

The almost 30 percent of the miles of refined products lines that are not owned by refiners is also only in small part an exception to the rule that the lines are usually built by the prospective users. Almost half of this is in the Williams and Buckeye lines that were built by refiners but subsequently became independent. Another major segment transports products of gas processing plants, rather than refineries, and these numerous lines are frequently owned by either the producers or the users

---

[2] Of course most refiners are also crude oil producers. There is no need to distinguish whether the primary motive for pipeline construction was assured low cost transportation of their crude oil production to market or of purchased oil to their refineries.

[3] The 2,452-mile Buckeye Pipe Line, now owned by the Pennsylvania Co., was originally part of the Standard Oil Trust and became independent when the trust was dissolved in 1911. It handles both crude oil and refined products.

Williams Brothers bought in 1965 what had been The Great Lakes Pipe Line, which had been built by refiners prior to World War II. At that time Williams transported only refined products but at the end of 1973 it reported 7,299 miles of products lines and eighty-two miles of crude oil lines.

A number of gathering lines built originally by large refiners have been sold in recent years to independent operators.

TABLE 1
OWNERSHIP OF UNITED STATES OIL PIPE LINES[a]

Owners	Percent of Refining Capacity [d]	Percent of Miles [b] of Crude Oil Lines [c] [d]	Percent of Miles [b] of Refined Products Lines [d]
8 largest refiners	56.5	44.8	36.8
10 next largest refiners	24.5	24.4	20.4
10 next largest refiners	8.3	6.1	4.8
20 next largest refiners	6.2	7.2	6.7
20 next largest refiners	2.6	3.4	0.6
58 smaller refiners [d]	1.9	2.0	0.9
Non-refiners		12.1	29.8
Totals	100.0	100.0	100.0

[a] Includes both lines regulated by the Interstate Commerce Commission (now the Federal Energy Regulatory Commission) and those not reporting to the ICC. Excludes ammonia and coal slurry lines.

[b] The only available measure of all pipelines is miles of pipe. A rough calculation, based largely on data for lines reporting to the Interstate Commerce Commission, suggests that, if allowance could be made for differences in pipe diameter and therefore throughput capacity, the shares of the larger firms would more closely approximate but would not exceed their shares of refining capacity.

[c] The available data do not provide a meaningful division between gathering lines and trunk lines. Conceptually, gathering lines use relatively small diameter pipe, are connected to the lease tanks at the producing properties, and transport the oil relatively short distances to connections with a trunk line. Frequently, however, large diameter pipes extending considerable distances are reported as gathering lines. In other instances relatively short and small diameter pipes are reported as trunk lines.

[d] The shares of refining capacity, and the total number of refining companies, are as of January 1974. By January 1977, there were eighteen more refining companies, and the share of refining capacity owned by the eight largest had dropped to 52.7 percent. Where available, and this includes all ICC regulated lines, the ownership of pipelines is also at the end of 1973, corresponding to January 1, 1974 for refining capacity. For non-ICC lines, however, it was necessary to use a wide variety of sources, some of which could not be dated so precisely. Changes in the ownership of such lines within a year or two are not enough to change the total picture significantly.

of the natural gas liquids.[4] There are also, however, a few notable exceptions to the general rule that the lines are usually built by those that

---

[4] By ICC definition, data reported to and published by the Interstate Commerce Commission combine lines transporting products of refineries and those transporting products of gas processing plants. Sources of data on non-ICC lines also frequently make it difficult to distinguish these two categories.

expect to use them. The possible reasons for these exceptions will be discussed in Section III as part of the analysis of the reasons for the general pattern of vertical integration.

Refiners who do not own pipelines can be largely explained by the location of their refineries. For example, the largest refiner that owns *neither crude oil nor refined products pipelines* is Earth Resources Co. which has a 31,500-barrel-per-day refinery (Delta Refining Co.), at Memphis, Tennessee.[5] This plant is supplied with either Gulf Coast or imported crude oil via the Mississippi waterway by Earth Resources' wholly owned Valley Towing Service, which also transports oil for other customers. Thus Delta is vertically integrated in terms of water rather than pipeline transportation.

The only existing crude oil line to which this refinery could conceivably be connected is the Capline System, which passes about twenty-five miles east of Memphis on its way north from St. James, Louisiana, to Patoka, Illinois. This line has a throughput of about one million barrels per day, with individual shipments of particular types of crude amounting to tanker loads of roughly half a million barrels. The economics of tapping this stream, short of its destination at Patoka, are such that not even the owners, with much larger refineries, have attempted to do so. For example, Ashland, which is the largest owner, considered this possibility and rejected it in favor of taking delivery at Patoka and building a much longer line, than would otherwise have been necessary, south and east from Patoka to its refinery at Catlettsburg, Kentucky.

The bulk of the output of the Delta refinery is sold within a range such that truck transport is most economical, much of it to retail outlets which the company owns. The only nearby refined products line runs from Oklahoma *to* Memphis, and thus would be of no use in transporting products *from* the refinery at Memphis.

The second largest refiner owning neither crude oil nor refined products lines is Rock Island Refining Corp., with a 29,500-barrel-per-day plant at Indianapolis.[6] This plant is close, and connected, to the large diameter Marathon crude oil line which was already there before the refinery was built. Hence there was no need to build any more than a short connection to the Marathon line. Here again there are ample markets for its output of refined products within truck range of the

---

[5] The size ranking used here is in terms of operating crude running capacity on January 1, 1974, as reported by the Bureau of Mines. In this ranking, Earth Resources is number 43. In 1977 this company built a second refinery near Fairbanks, Alaska, which is connected to the Trans-Alaska Pipeline.
[6] Rock Island Oil & Refining Co., which does own pipelines, is a subsidiary of Koch Industries and unrelated to Rock Island Refining Corp. which owns the refinery at Indianapolis.

refinery. If it should wish to ship products to more distant markets, there are several refined products pipelines passing near Indianapolis that are owned by refiners, such as Marathon, who, because of their size and the distance to markets they serve, had a much greater interest in making sure that these lines were built.

The third largest refiner without either crude oil or refined products pipelines is Good Hope Refineries, a subsidiary of Gasland, Inc. (now Good Hope Industries), which had a capacity of 29,450 b/d in January 1974. This refinery is located at Good Hope, Louisiana, on the Mississippi River, and can thus be supplied with either Gulf Coast or imported crude oil by tanker or barge. It is very close, and connected, to a nonrefiner owned GATX terminal capable of handling tanker shipments. The parent markets refined products in New England and New York, largely through company owned retail service stations. Products are transported from the refinery by tanker to deep water terminals and thence by company owned barges to truck terminal facilities and by truck to retail outlets.[7]

The fourth largest refiner without either crude oil or refined products pipelines may be San Joaquin Oil Co., which ranks forty-ninth in order of refining capacity. Lacking readily available information about this firm, beyond the Bureau of Mines report that it has a 27,000 b/d refinery at Bakersfield, California, no attempt was made to determine whether it owns any pipelines. It is one of nine refineries, with aggregate capacity of 143,000 b/d, located near Bakersfield because of local crude oil production. It may own a crude oil line from this production or may rely on any one of several lines in the vicinity owned by both refiners and crude oil producers. San Joaquin's output is not enough to warrant its construction of a refined products line. The only products line out of Bakersfield is the nonrefiner owned Southern Pacific Pipe Line to Fresno, which presumably serves all refineries near Bakersfield desiring to be connected.

The fifth largest is VGS Corp., which operates three small asphalt plants in Mississippi, under the name Southland Oil Co., with aggregate capacity of 21,000 b/d, located close to crude oil production. Following a complaint by the Federal Trade Commission issued September 18, 1973, VGS accepted a consent decree prohibiting it from acquiring any crude oil line in Mississippi or Alabama. As an asphalt producer, it would have no need for a refined products line.

---

[7] This was the situation in January 1974. The refinery has since been expanded to 70,000 b/d. It still receives the bulk of its crude oil supply by tanker via the GATX terminal. A substantial part of the increased output now moves via a Shell pipeline to a connection with the Plantation Pipeline.

There are roughly sixty still smaller refining companies that do not own *crude oil pipelines*, many of them so small that the aggregate refining capacity of this group on January 1, 1974, was only about 2 percent of the U.S. total.[8] Some of them, like U.S. Oil & Refining Co. at Tacoma, Washington, are supplied by tanker or barge. Some, like Caribou Four Corners near Salt Lake City, are supplied by pipelines built by other refiners to serve their plants. Many are located close to local crude oil production, sometimes in situations where it is more economical to truck the oil to the refinery.

Many of the still smaller refiners have had no need to invest in *refined products pipelines* because they are located close to local markets that are large enough to take their total output. They have a transportation cost advantage over larger refiners that must reach these markets from a greater distance. There are also some that are primarily producers of asphalt or lubricants, not readily transported by pipeline. These explain most of the approximately seventy-one smaller refiners, accounting for less than 5 percent of total refining capacity, that have not been identified as owners of products lines. As noted later, however, there are also a few that have found it convenient to utilize pipelines owned by others rather than build their own.

Refining companies having only limited ownership of crude oil pipelines, relative to their size as refiners, can also usually be explained by their location. For example, Murphy Oil Co. owns a 92,500 b/d refinery at Meraux, Louisiana, utilizing imported crude brought in by tanker, or Gulf Coast crude transported by barge. It has a 37,000 b/d refinery at Superior, Wisconsin, which was built to refine Canadian crude oil supplied by Lakehead Pipe Line (subsidiary of Interprovincial Pipe Line) whose large diameter line from Western to Eastern Canada runs close to Superior.[9] Murphy has substantial interests in crude oil gathering lines in Montana and offshore Louisiana, and owns 20 percent of the Butte Pipe Line running from Eastern Montana to connections with lines running east from southeastern Wyoming. Ownership of these lines, however, is related more to Murphy's crude oil production in the areas served than to its refinery requirements.

Two other examples of limited ownership of crude oil lines are Crown Central Petroleum Co. and Texas City Refining Co., the latter a subsidiary of two eastern farmer cooperatives. Their refineries are both

---

[8] The number of such firms is necessarily an approximation both because of the difficulty of getting information about some of them and because both entries and exits can change the number from year to year.

[9] The Interprovincial line was undertaken by Canadian crude oil producers to provide an outlet for their crude. Its availability to Murphy was largely fortuitous.

323

located on the Gulf Coast near Houston and process primarily imported crude oil brought in by tanker. Both own small crude oil gathering systems in Texas, and Crown Central has minority interests in two trunk lines transporting oil from interior Texas to the Gulf. Both, however, rely primarily on tankers rather than pipelines.

This pervasive vertical integration calls for explanation. There must be good reasons why pipelines are usually built and owned by those that expect to use them. The fact that this shipper ownership is true of small firms as well as large suggests that it does not occur because of any attempt to monopolize. It also suggests that there are economic efficiencies that would be lost if there were forced divestiture.

Pipelines as Plant Facilities. One interesting and significant aspect of vertical integration is the number of refiner-owned pipelines that exist to serve only one refinery. They were built by the refiner to assure low cost transportation of crude oil to that refinery, or low cost transportation of products from that refinery to the markets it serves. Frequently the location of these lines is necessarily such that they are of no use to any other shipper. The owner might welcome additional traffic, with the cost savings resulting from larger volume, but this is not possible. Because such lines are as much a part of the refinery complex as the individual processing units, or the storage tanks at the site, they are sometimes referred to as plant facilities. In a variation of this situation, a pipeline may be built to serve two refineries, may be owned jointly by the two refinery owners, and again be of little or no use to any other refiner.

Appendix A describes in some detail only fifteen of the many possible examples of refiner ownership of lines that, in varying degree, might be called plant facilities. For ten of these refiners there are maps showing the relationship of the pipelines to the refinery. These fifteen examples are merely illustrative and are by no means a catalog of such situations.

The chief interest in such situations is that they shed some light on the reasons for vertical integration to be discussed in Section III. They suggest that a refinery's needs for transportation facilities may, at least at times, be met more efficiently within the firm than by arms length market transactions. Furthermore, it seems reasonably clear that a pipeline serving a single refinery was not built as a defensive move to avoid discrimination in the use of a competitor owned line.

It is equally true that, if a pipeline is so located that it can serve only an affiliated refinery, it cannot be guilty of discriminating against non-owner shippers. For the same reason, there is no reason to question whether what the refiner charges itself for transportation is more or less than adequate.

It may also be noted that competition among such pipelines is a meaningless concept. The competition is in the markets for the refined products, and the pipeline is merely one of the costs of the supply for the refinery, or one of the costs of getting the refined products to market.

It is also clear that forcing the divestiture of such lines, or prohibiting the users from building additional ones, would serve no useful purpose.

III. Reasons for Vertical Integration

There are good reasons for the pattern of vertical integration. The cost of moving crude oil or refined products by pipeline, or by water where this alternative is available, is usually so much less than the cost of rail or truck transportation that a refiner must have the lower cost in order to be competitive. Thus the refiner, or alternatively the crude oil producer, has a far more urgent need to see that the pipeline is built when and where needed than an independent entrepreneur whose only incentive is the prospective return on the pipeline investment.

The risks incurred in the construction of a pipeline can be so great as to deter an investor whose sole interest is the potential return on the pipeline investment. These risks are inherent in the peculiarities of pipeline transportation, and can be demonstrated by a number of historical examples. The prospective users are in a better position to evaluate the risks, and can afford to accept them because of their urgent need for low cost transportation. Because of this need, they could afford to pay, to an independent investor, tariff rates high enough to compensate for the risks incurred. This, however, is not a feasible alternative, primarily because of the limited return on the pipeline investment imposed by government regulation. Neither is it reasonable to expect some of the prospective users to guarantee the investment in a line in which they have no investment interest.

Furthermore, there are transaction costs involved in an arms length negotiation between the prospective user, or users, and an independent entrepreneur that can be avoided by vertical integration. Also, plans for pipeline transportation can be better coordinated, with the related plans for refinery construction, crude oil supplies and marketing investment, within a firm than by arms length transactions with an unaffiliated investor.

Refiners Must Have Low Cost Transportation. With very minor exceptions, a refinery has to receive its crude oil by pipeline, or by water where this alternative is available, in order to be competitive.[10] The difference between the cost of moving oil by pipeline, or by water where this alternative is feasible, and the cost of rail or truck transportation is so great as to be prohibitive.[11]

This point can be illustrated by some comparative costs of getting crude oil to Standard of Indiana's Whiting refinery, located just east of Chicago, in 1952. An important source of oil for that refinery was West Texas. The tariff rate from West Texas to Whiting, charged by what was then the Service Pipe Line Co., was 38 cents per barrel, plus a gathering charge of 7.5 cents and a "tender deduction" of 1.3 cents (to allow for evaporation and other losses in transit). The rail freight from Levelland, Texas, was $1.60 per barrel. The cost of trucking oil to the railroad, which would correspond to the pipeline gathering charge, would vary with circumstances but typically would be several times the gathering charge.

In the same year the pipeline tariff rate from the big East Texas field, excluding gathering charge and tender deduction, was 35 cents per barrel, while the rail freight was $1.54. The pipeline rate from Cushing, Oklahoma was 28 cents while the rail freight was $1.48. The pipeline rate from Wyoming, another important source of crude for this refinery, was 50 cents while the rail freight was $1.55.[12]

The rates quoted in the two preceding paragraphs varied somewhat among pipelines, and they also change over time. It is also conceivable that the rail rate might have been reduced somewhat if there had been any reasonable hope that this reduction would attract a substantial volume of traffic.[13] Any such differences, however, are not sufficient to invalidate the conclusion that pipeline transportation costs are far below rail rates.

There is a similar disparity between rail rates and refined product pipeline transportation costs, wherever there is sufficient volume to warrant the pipeline investment. This is illustrated in Table 2, using rates

---

[10] A very small refiner, located in a producing area, may survive on receipts of crude oil by truck from nearby wells, particularly if the amount of local production is too small to warrant a pipeline and the refiner thus has a monopsony.

[11] Historically, particularly in the 1930s, some crude oil was shipped considerable distances by rail. This, however, was a temporary expedient until the pipelines could be built.

[12] There are somewhat similar economies in water transportation as compared with railroads. For example, the barge movement from Baton Rouge to Cincinnati, at a typical three mils per ton mile in 1952, would be 60 cents per barrel. The normal rail freight from Baton Rouge to Cincinnati was $2.14.

[13] Under ICC rate setting procedures, a railroad may be allowed to quote a somewhat lower rate than would otherwise be charged if it can demonstrate that the lower rate would result in sufficient additional traffic.

TABLE 2

COMPARISON OF PIPE LINE AND RAIL RATES FOR GASOLINE FROM
TULSA, OKLA., TO INDICATED DESTINATIONS, END OF 1953
(dollars per barrel)

Destination	Pipe Line [a]	Rail plus Tax	Excess of Rail over Pipe Line
St. Louis, Mo.	.53	1.34	.81
Chicago, Ill.	.63	1.66	1.03
Omaha, Neb.	.42	1.34	.92
Fargo, N. D.	.80	2.14	1.34
Mason City, Iowa	.52	1.57	1.05
Minneapolis, Minn.	.63	1.66	1.03

[a] Includes $.05 per barrel terminalling charge in order to make pipe line rate comparable to the service included in the rail rate.

charged by what was then the Great Lakes, now the Williams line. Here again, more recent data would not alter the picture. In his statement before the Interstate Commerce Commission in Ex Parte 308 on May 27, 1977, Vernon T. Jones compared the average of 0.116-cent-per-barrel mile charged by Williams Pipe Line with an average of 0.840-cent-per-barrel mile for a seventy-five-mile haul by three truck lines.

The magnitude and significance of these transportation cost differences can be indicated by some simple comparisons. The average price of crude oil at the well in 1952 was $2.53. The average of the total refining cost of all refiners in that year, including depreciation and other overhead, was less than $1.00 per barrel. Thus it is understandable that refiners, or alternatively the crude oil producers, took the initiative to see that pipelines got built where and as needed.

The Trans-Alaska Pipeline System is a recent example of the compelling motive to see that a pipeline was built. In this instance the interest was primarily as crude oil producers rather than as refiners. The large proved reserves in the Prudhoe field, and the potential as yet undiscovered oil on the North Slope, were virtually worthless until there was economical means of transporting that oil to market. Therefore the owners of that potential production undertook the enormously difficult and costly task of building that line. The risks they incurred, and the major environmental and bureaucratic obstacles they had to overcome, are a matter of record. It is significant that the ownership shares in this line are approximately the same as the ownership shares in the Prudhoe

Bay Unit. The owners were willing to share the risks and cost of the line in proportion to their expected use.[14] They were induced to undertake this project more by their need for low cost transportation than by any prospective return on the pipeline investment.

There is not the same urgency on the part of an independent entrepreneur. He, and the financial institutions that will provide most of the investment, can afford to be much more cautious in appraising the risks relative to the prospective return. Furthermore, as explained later, neither is in as good a position to evaluate those risks as the prospective user or users of the line.

Risks in Pipe Line Investment. The risks incurred in the construction of a pipeline can be so great as to deter any investor whose sole interest is the potential return on the pipeline investment.

These risks are inherent in the peculiarities of pipeline transportation. Consider first *crude oil lines*. A crude oil line is built to transport only one commodity in only one direction, from a specific source or sources of crude oil to a specific refinery or refineries, sometimes only one and frequently only a few. This is quite different from the general transportation services offered by a railroad, which can hope that the declining traffic of any one of its customers will be more than offset by the growth of other industries and consumers along the road. There is always the risk that the volume of production to which the pipeline is connected may prove to be far less than anticipated, or that the refineries to which it is connected may find it necessary, or more economical, to turn to other sources of supply.

These risks are enhanced by the peculiarities of pipeline investment. Once a line is built this investment is largely sunk, with relatively little salvage value. The economies of scale dictate that the line should be built large enough to handle all of the prospective volume, which may be quite uncertain at the time it is built. Otherwise there is a threat of future competition from a larger diameter line with substantially lower costs. If, however, the line is built substantially larger than necessary to handle the volume of traffic that actually materializes, it will have incurred the high fixed costs related to that unnecessary investment.[15]

---

[14] Jane Atwood and Paul Kobrin, *Integration and Joint Ventures in Pipe Lines,* API Research Study #005. The authors made a similar point with regard to several pipelines transporting crude oil from fields offshore Louisiana.

[15] The larger part of the total cost of owning and operating a pipeline, including depreciation, property taxes, maintenance, repairs, interest on borrowed capital and the cost of equity capital, is determined by the size of the investment. Those expenses that are related to the volume of traffic, rather than the investment, are a minor part of the total.

These risks are not merely hypothetical. A number of examples could be cited, but a few should be sufficient.

One historical example of risks that actually materialized is the Ajax Pipe Line. This line was built in 1930 by the Standard Oil Co. (Ohio), the Pure Oil Co. and the Standard Oil Co. (New Jersey). It ran from Glenn Pool, Oklahoma, to Wood River, Illinois, whence there were existing common carrier pipelines running north and east. The first two owners were interested in low cost transportation of Mid-Continent crude oil to their refineries in Ohio. The third wanted an outlet for its crude oil production. In a related agreement Sohio contracted to buy from a subsidiary of Standard of New Jersey 28,000 barrels per day of crude oil for five years. This approximated Sohio's total refinery requirements at the time and was almost half of the capacity of the line.

Within seven years, however, the rapidly expanding production in Illinois, which went from 4,475,000 barrels in 1936 to 147,647,000 barrels in 1940, provided a much closer and much cheaper source of oil for Ohio refineries. Traffic on the Ajax line dropped from 60,500 b/d in 1937 to 7,400 b/d in 1940.

With the decline of flush production in Illinois, and the advent of World War II, Ajax again did a satisfactory business. Subsequently, however, its traffic was again reduced to far below capacity. One factor in this decline was the building of the Interprovincial Pipeline from western Canada to Sarnia, Ontario. A substantial part of the crude oil consumption at Sarnia had been supplied through the Ajax and connecting pipelines. Another factor was the competition of larger, and therefore far more economical lines. As compared with the two parallel ten-inch pipes operated by Ajax at that time, the Ozark Pipe Line was built with a single twenty-two-inch pipe. Both lines transported crude oil from Oklahoma to Wood River, Illinois. Also the Mid Valley Pipeline built a twenty-two-inch line from East Texas to Lima, Ohio. As a result of these developments, Ajax was liquidated in 1954. The fact that some of the facilities were subsequently converted to transportation of refined products may have increased their salvage value.

Another historical example is the pipeline that Standard Oil Co. (Ohio) built in the 1930s from producing areas in central Michigan to its refinery at Toledo, Ohio. Contrary to expectations, crude oil production in Michigan peaked in 1939 at 23,462,000 barrels and then declined. By 1950 it was down to 15,826,000 barrels, which was less than the crude running capacity of refineries located close to the crude oil production. In that year Mid-West Refineries acquired the line and managed to make some use of the sunk investment by reversing the flow in part of the line in order to supply its refinery at Alma, Michigan with Mid-Continent crude via a connection with another line at Toledo.

Another example is the crude oil line built by Standard Oil Co. (Indiana) from Wyoming to Freeman, Missouri, which was abandoned as a crude oil carrier for several years in the 1930s. This reflected the decline in Wyoming production from 44,785,000 barrels in 1923 to 11,227,000 barrels in 1933. With additional discoveries in Wyoming, and the increased demand for Wyoming crude, the line was reactivated, but the growth in Wyoming production to 82,618,000 barrels in 1953 and 115,572,000 barrels in 1958 created another problem. The much larger actual and prospective volume justified the building of the Platte Pipe Line east from Wyoming in 1952 with much larger diameter pipe. The economies of scale achieved by this competitive line made Standard's line obsolete. Standard abandoned it in 1955 and replaced it by a large diameter line following a somewhat different route.

A more recent example is the Four Corners Pipe Line, completed late in 1957 and built to transport crude oil, from the producing area where Utah, Colorado, Arizona, and New Mexico meet, to refineries in the Los Angeles area. The five owners included both the owners of refineries near Los Angeles and owners of crude oil production in the Four Corners area.

Prior to construction of this line, production in the Four Corners area was limited to the few thousand barrels per day that could be consumed by small local refineries. Lacking low cost transportation to larger markets, there was no incentive to explore for and develop what were believed to be very sizable reserves. In 1957 the Oil & Gas Journal estimated that the Utah portion of the area had "upwards of 500 million barrels of undeveloped reserves," and predicted that the activity induced by construction of the line would boost production in the area to 110,000 barrels per day within two years.

Actual production proved to be far less than anticipated. Furthermore, a competing line was built, also by prospective users, southeast to a connection with existing lines to Gulf Coast refineries.[16] Traffic through the Four Corners line peaked in the second full year of opera-

---

[16] This was an extension of the Texas–New Mexico Pipe Line, running from the Four Corners area to Jal, New Mexico, and completed shortly after the Four Corners Pipe Line. This also was a sixteen-inch line, and the predicted capacity was only moderately less than the Four Corners line. It appears from the map that it was not drawing from precisely the same fields as the Four Corners line. Furthermore, available data suggest that only about 12,000 b/d of its 1973 traffic originated in the Four Corners area. Nevertheless the possibility of shifting some Four Corners production from Los Angeles to Houston presumably contributed to the poor showing of the Four Corners line.

tion at just under 97,000 barrels per day. From that point it declined to less than 65,000 b/d in 1962, and to just over 23,000 b/d in 1975. By that time the major new source of crude oil on the North Slope made it evident that the future flow of oil would be east from the West Coast rather than west to Los Angeles. When Atlantic Richfield bought out the other owners in September 1976, the line was reportedly moving 3,000 to 3,500 b/d. It was taken out of service in June 1977.

Builders of *refined products pipelines* incur similar risks. A products line is designed to provide low cost transportation, from frequently only one and usually only a few refineries, to the markets they are expected to supply. Thus it is dependent on both the future output of those refineries and their ability or willingness to continue to serve the markets reached. The economies of scale require that the line be built as large as the anticipated traffic will warrant. If it is larger than the volumes actually achieved, however, it suffers the high fixed costs associated with the unnecessary investment.

Considering the long life span necessary to justify the pipeline investment, various developments can alter the anticipated traffic rather drastically. Shifts in the sources of crude oil can make it less desirable for particular refiners to sell in particular markets, while increasing the competitive advantage of other refiners reaching those markets. New transportation facilities may cause a refiner to shift more of its output to markets reached by those facilities, or may put it at a competitive disadvantage in markets it had expected to serve. For a variety of reasons a refiner's marketing strategy may be altered by withdrawing from some markets and concentrating in others.

For example, when the Laurel Pipe Line was constructed in 1959, it appeared that refineries in the Philadelphia area could process low cost imported crude oil, ship the products as far west as Cleveland and still be competitive with Midwest refiners using domestic crude. Use of this cheap foreign crude was growing rapidly, and the refiners who were also producers of the foreign crude had strong incentives to import as much as possible.[17] Therefore, the owners of three of the Philadelphia refineries undertook to build the Laurel line, west from Philadelphia, including almost 100 miles from the Pennsylvania-Ohio border to Cleveland.[18]

---

[17] This is not the place for an economic analysis of these incentives, which unquestionably existed. Briefly, they reflected the desire to produce the foreign oil as fast as possible, because of the risk of expropriation, and the imperfections in the markets for this oil.

[18] At that time Gulf owned 40 percent, Sinclair 35 percent and Texaco 25 percent. The economies of scale achieved by the Laurel line, with pipe ranging from twenty-four inches at Philadelphia to 10 inches at Cleveland, resulted in the abandonment of an existing 6-inch line from Philadelphia to Cleveland.

Shortly after the line was completed, however, the institution of mandatory import controls greatly restricted the use of foreign crude. This drastically altered the economics of at least the extension of the line into Ohio. Laurel reported losses of almost $6 million in its first six years of operation, more than wiping out the stockholders' equity. The major increase in current liabilities suggests that the owners were required to make deficiency payments, in the form of advance payments for future shipments, to satisfy bond holders and keep the line solvent.

A more recent example of a products line that did not fulfill the expectation of the builders is the Explorer Pipeline. Plans for Explorer resulted from a disagreement among the companies considering the proposed Gateway pipeline from the Gulf Coast to Chicago, with the Explorer group favoring a western route via Tulsa and the Match group favoring an eastern route, possibly incorporating an existing line owned by Texas Eastern Transmission Company.

The route favored by the Match group was never built and some of the Match group joined the Explorer project. Explorer was completed in the summer of 1972. The ownership shares, as reported at the end of 1973, were as follows:

Gulf Oil Corporation	26.7%
Shell Oil Company	26.0
Texaco	16.0
Sun Oil Company	9.4
Continental Oil Company	7.7
Cities Service Company	6.8
Phillips Petroleum Company	4.5
Apco Oil Company	2.9
	100.0%

SOURCE: Explorer's Form P report to the ICC.

Here again the line was built by those that expected to use it. It includes 50 miles of twelve-inch pipe running from Lake Charles, Louisiana, where both Cities Service and Continental have refineries, to Port Arthur, Texas, where both Gulf and Texaco have refineries. From Port Arthur, 546 miles of twenty-eight-inch pipe run via Houston, where Shell has a refinery, to Tulsa, Oklahoma, where Sun has a refinery. From Tulsa there are 653 miles of twenty-four-inch pipe to Hammond, Indiana, near Chicago. From Greenville, Texas, there is an 85 mile twelve-inch spur line to Dallas and Fort Worth and there are shorter spurs in the St. Louis and Chicago areas. The Phillips refinery at Sweeny, Texas, is some distance southwest of Houston. Phillips' 1970

Annual Report stated, however, that Explorer was connected to its terminal at Houston, which was supplied from the Sweeny refinery. Apco's two Oklahoma refineries are also some distance removed, but are connected to existing pipelines that provide a connection to Explorer. Statements in Apco's 1970 and 1971 Annual Reports make it clear that Apco was interested in Explorer, not as an investment in pipeline transportation, but as a means of supplying its markets with products from Gulf Coast refineries, and of reaching from its refineries into markets not previously available.[19]

At the end of 1972, Explorer reported gross investment in carrier property amounting to $213,411,000. Total assets, net of depreciation charges, were $216,383,000. This and the slightly smaller total in 1976 were financed as follows:[20]

	Thousands of Dollars	
	1972	1976
Current liabilities	$ 10,505	32,782
Long-term debt	197,280	197,280
Other non-current liabilities	30	240
Capital stock	21,920	21,920
Retained income (loss)	(13,350)	(45,968)
	$ 216,383	207,755

The capacity of Explorer was variously reported as initially around 300,000 b/d and expanded to 380,000 b/d in 1973.[21] Actual throughput in 1973 was only 218,000 b/d. Beginning in 1974 the line began batching crude oil in an attempt to make better use of capacity. Throughput in 1975, however, was only 212,000 b/d including 35,000 b/d of crude.

---

[19] While Explorer was built by those that intended to use it, access to the line was not restricted to the owners. Explorer reported that in 1973, the first full year of operation, there were twenty-two shippers, including fourteen non-owners. (Data submitted in response to a questionnaire circulated by the Association of Oil Pipelines.)

[20] Data for 1971 through 1975 in this and the following text table are from ICC Transport Statistics in the United States, Part 6, Pipe Lines. Data for 1976 are from a Dun & Bradstreet report on Explorer.

[21] Continental Oil Co., 1970 Annual Report, initial capacity 295,000 b/d. Continental Oil Co., 1973 Annual Report, capacity expanded from 290,000 b/d to 380,000 b/d in 1973. Phillips Petroleum Co., 1970 Annual Report, initial capacity 300,000 b/d. Shell Oil Co., 1968 Annual Report, initial design capacity 400,000 b/d. Shell Oil Co., 1970 Annual Report, ultimate design capacity 800,000 b/d.

The record through 1976 illustrates the effect of the high fixed charges associated with unused capacity. From the line's inception, the volumes handled and the losses incurred were as follows:

	Thousand Trunk Line Barrels per Year		Thousand Dollars
	Refined Products	Crude Oil	Net Loss
1971	0		260
1972 (part year)	39,767		13,090
1973	79,479		9,954
1974	80,813	1,410	5,343
1975	64,536	9,420	7,526
1976	n. a.	n. a.	9,795

At the end of six years none of the funded debt had been retired and the accumulated loss was more than twice the initial stockholders' equity. Without the backing of the principal users the line would have been bankrupt. It remained solvent because of the large deficiency payments that the owners were required to make in accordance with their commitment to the purchasers of the debt. Since these are treated as advance payments for future transportation, they appear as a large part of the almost $33 million of current liabilities reported by the line at the end of 1976.[22]

There are at least two reasons why Explorer's traffic fell so far short of the expectations when it was built, and both illustrate the risks inherent in pipeline investment. The eastern route that had been favored by some members of the original "Match" group was, in effect, built when Texas Eastern Transmission Co. expanded its product line from the Gulf and extended it into Chicago. This competing line was more advantageous for some shippers, depending on the location of both their refineries and the markets served. There was also a major expansion of large diameter, low cost crude oil line capacity from the Gulf to the upper middle east. The alternative of shipping crude to northern refineries, from which the refined products would be distributed, became, under some circumstances, more advantageous than refining the oil on the Gulf and shipping the products north.

---

[22] According to the Howrey & Simon report for AOPL, the total of deficiency payments through mid-1977 was $41.6 million. See, Howrey & Simon, *Pipelines Owned by Oil Companies Provide a Pro-Competitive and Low Cost Means of Energy Transportation to the Nation's Industries and Consumers,* March, 1978, p. 11. In the ICC hearings on Ex Parte 308, Glenn A. Walsh, vice president of Explorer, stated that shipments by non-owners amounted to from 10.3 percent to 15.2 percent of total shipments in the years 1972-76. These non-owner shippers benefited from use of the line, at the same rates charged owners, without sharing in the losses incurred.

Why Risks Deter Independent Owned Lines. Conceptually, an independent entrepreneur could be induced to build a crude oil or refined products pipeline if the prospective rate of return were sufficiently high to compensate for the risks incurred. This, however, is not a practical alternative. If the line is an interstate common carrier, subject to regulation by the Interstate Commerce Commission, now the Federal Energy Regulatory Commission, it will be limited to the rate of return allowed by such regulation.[23] If it is not subject to such regulation, it will still face the threat that a successful line, attempting to charge rates commensurate with the risks which were incurred but which did not happen to materialize, will invite the building of a competing line.

It has been argued that independent entrepreneurs could be induced to build pipelines, and could finance them, if the prospective users of the lines were willing to provide throughput or deficiency guarantees sufficient to assure an adequate return on the investment. Typically, however, the potential users have not been willing to propose such a one-sided arrangement. From the viewpoint of their stockholders they would be in the position of guaranteeing the investment in the line without the chance to earn even the return permitted by government regulation. Furthermore, they would be underwriting the line for the benefit of any shippers who may choose to take advantage of the resulting reduction in transportation cost but without participating in a throughput agreement. Persuading prospective users to share the risks on these terms would be difficult.

Transaction Cost Savings and Planning Advantages of Vertical Integration. Another reason for vertical integration is the advantage of coordinating the size, the route and the completion date of a pipeline with the requirements of the refinery or refineries it is intended to serve. Frequently this can be done more efficiently by management decisions within an integrated firm than by arms length negotiations with an independent entrepreneur.

The prospective user or users of a pipeline, either a new line or an expansion or extension of an existing line, are in the best position to determine their requirements and therefore the potential traffic. This

---

[23] This point can be illustrated by data already cited. Standard's wholly-owned Service Pipe Line charged the parent, and all other shippers, 50 cents to move a barrel of crude oil from Wyoming to Chicago. The difference between this and the alternative of rail freight at $1.55 per barrel is a rough measure of the magnitude of the saving to Standard as a producer and refiner, and therefore its incentive to see that a line was built. Since the tariff rate had to cover all of Service Pipe Line's costs, the net return on the pipeline investment was a small fraction of the 50 cents.

information is closely related to their short and long run plans for crude oil supplies, for refinery location and expansion, and their evolving strategies for marketing the refined products. Information and judgments developed in this planning process also put them in the best position to evaluate the investment risks discussed supra.

Providing all of this information, in a sufficiently detailed and convincing manner, to one or more independent entrepreneurs in the hope of persuading one of them to undertake the project, involves an added cost that is largely avoided by vertical integration. This problem can be compounded by any attempt to find and interest enough potential investors to assure some measure of competition among them. There is also the delay while the independent entrepreneur studies the project and reaches his independent judgment as to its viability. Having concluded that the pipeline is needed, the prospective user finds it simpler to just go ahead and build it.

The independent entrepreneur must recognize that the bulk of the traffic on the line will come from frequently only one or two users and rarely more than a few. To the extent that he relies on representations made by the prospective users, he must be concerned that these can be colored by their obvious interest in seeing that the line is built. He must in turn convince the prospective purchasers of the funded debt that the pipeline is a sound investment. Since he does not have the prospective user's urgent need that the pipeline get built, he is likely to be cautious in his appraisal of the risks.

We have here what some economists have termed "the small numbers bargaining problem." Where the potential participants on both sides of a bargain are quite limited, the added economic cost of an arms length transaction may more than offset the assumed advantages of a competitive market.[24] It is noteworthy that the examples of products lines that are exceptions to the general rule of vertical integration, cited later in this section, have relatively large numbers of potential shippers compared with the investment required.

One example of the greater efficiency of vertical integration is the refinery built by Standard Oil Co. (Indiana) at Mandan, North Dakota. This plant was completed in 1954 with an initial capacity of 30,000 b/d. It was designed to process crude oil from recently developed fields in northwestern North Dakota but there was no preexisting pipeline to transport this crude. Thus there was urgent need for a line running

---

[24] More specifically, these economists relate the small numbers condition, the costs of transferring or acquiring the pertinent market information and the problems of appraising risks. See Oliver E. Williamson, *Markets and Hierarchies: Analysis and Antitrust Implications,* and David J. Teece, *Vertical Integration and Vertical Divestiture in the U.S. Oil Industry.*

from this source of crude to the refinery, sized to correspond with both the refinery requirements and the available crude oil supply, and to be completed neither earlier nor later than the refinery completion.

The obvious market for the bulk of the output of this refinery was eastward into Minnesota, particularly the area around Minneapolis-St. Paul, where Standard already had a strong market position. This called for a refined products pipeline to the Twin Cities, large enough to handle the expected output of the refinery but not substantially larger than necessary, and to be completed at approximately the date the refinery was completed.

The detailed planning of crude oil lines, refinery construction and the products line, and the related arrangements for the necessary crude oil purchases and refined products marketing, were all part of a single coordinated project. The decisions on construction of the pipelines were as integral parts of this project as the decisions as to the types and sizes of the individual processing units at the refinery or the size of the onsite crude oil and products storage facilities.

Both crude oil and products lines are utterly dependent on the fortunes of this one refinery. Their location is such that they would be virtually worthless to any other shipper. This fact, plus the difficulty of coordinating plans via arms length negotiations rather than internal management decisions, explains why Standard made no attempt to find an independent investor in the lines.

The Mandan situation is only one of numerous examples of pipelines that are integral parts of one overall investment encompassing the processing units at a refinery, the crude oil supply for that plant and the arrangements for marketing the refinery products. Thus the plans for pipeline construction or expansion can best be coordinated with the corresponding plans for refining and marketing facilities by decisions within a single management. Furthermore this dependence on a single shipper tends to discourage an independent investor in the pipeline. Some examples of such "plant facilities," that are owned by smaller firms, have been noted in Section II and are described in Appendix A.

The planning advantages of vertical integration are of course most evident in the frequent instances where a line is built to serve only one refinery. Even when it is planned to serve two or more refineries, however, the principal potential users are in the best position to determine the requirements in terms of size of pipe, crude oil sources or refined products markets to be reached, and desired time of completion. There are similar planning advantages associated with a major expansion of an existing refinery, or a major change in the sources of crude oil or in the markets to be served.

337

In a somewhat less obvious fashion than at Mandan, all of the extensive products pipeline system wholly owned by Amoco Oil Co. can be considered plant facilities, designed to meet Standard's peculiar requirements. They run from Standard's refineries to the markets they serve, sometimes in directions quite contrary to the general flow of products from other refineries. They are used to move materials from one refinery to another, or from a refinery to a chemical plant. Sometimes the flow of products in a section of line has been reversed in order to balance the capacity at each of Standard's refineries with Standard's then current marketing requirements in different areas. Such investment and operating decisions can be made more easily within an integrated firm than by negotiations with an independent investor.[25]

Reasons for the Exceptions to Vertical Integration. The description of vertical integration in Section II noted a few exceptions to the general rule that refined products pipelines are built by those who expect to use them. It may be helpful to consider these exceptions in the light of the analysis, in this Section, of the reasons for vertical integration.

One notable exception is Southern Pacific Pipe Lines. Its aggregate of over 2,400 miles of refined products lines consists of four distinct lines or groups of lines. The largest of these, over 1,300 miles, runs, from connections with refineries in the Los Angeles area, *east* to Phoenix and Tucson, and also from El Paso, Texas *west* to Phoenix. Another group radiates, from refineries and tanker terminals in the San Francisco area, as far as Chico on the north, San Jose on the south and east to Reno, Nevada. Shorter lines run from tanker terminals at Portland to Eugene, Oregon, and from refineries at Bakersfield to Fresno, California. There are several reasons why these lines are an exception to the rule.

As a holding company with total assets of over $4 billion, Southern Pacific had resources not available to an entrepreneur whose sole interest is a pipeline project. Because the investment in each of these lines was small relative to its total assets, it could incur risks that might well have deterred an entrepreneur without similar ability to spread the risks. It could, and did, finance the line out of its total resources, rather than a

---

[25] In this instance they are facilitated because, contrary to other large products lines, Standard's system does not purport to be a common carrier. Thus there is no need to consider the requirements of other shippers. The same situation obviates all questions about possible discrimination against non-owner shippers, except the effect of their exclusion. The exclusive use of these lines to handle only Standard's products is no significant handicap to other shippers, both because in general the location and direction of flow in these lines make them of little use to others and because of the availability of more advantageous common carrier lines.

lien on the pipeline property, which would otherwise have been the sole recourse of an entrepreneur. Thus it did not have the problem of persuading prospective purchasers of the funded debt that the pipeline was a sound investment.[26]

The Southern Pacific lines serve a relatively large number of refiners and flow to relatively thin markets, i.e., the number of potential users is large relative to the required investment. For example, there are ten small refineries in the Bakersfield area with aggregate crude running capacity of 169,000 b/d. Only a fraction of their aggregate output of light products can be expected to move north to Fresno. The economies of scale dictate that there should be only one line, and the Southern Pacific line is only eight inches in diameter. Apparently no one or two of the refiners in Bakersfield felt that their need for low cost transportation to Fresno was sufficient to warrant their building a line.

A somewhat similar situation prevails with respect to the much longer line running east from Los Angeles. There are 19 refiners in the Los Angeles area, both large and small, with aggregate capacity of over one million b/d, but only a very small fraction of their light products moves east into Arizona. Here again there was the possibility of recognizing the opportunities for a pipeline investment that did not depend on the needs of a very few potential users.

One motive for building these lines was to regain the refined products traffic that Southern Pacific had once enjoyed but had lost as tank car shipments were displaced by trucks. Southern Pacific had the advantage that it could lay the pipe in the railroad right-of-way, whenever it was advantageous to do so, thus reducing the required investment.

A parallel situation is the San Diego Pipeline, jointly owned by the Santa Fe and Southern Pacific railroads, and running from Los Angeles to San Diego. Again there was a large number of potential users relative to their aggregate volume of traffic, the aggregate resources of the two builders were large relative to the pipeline investment, and there was the possibility of using railroad right-of-way.

There is also the Calnev Pipeline, running, from a connection with the Southern Pacific line at Colton, California, to Las Vegas, Nevada. This project was put together by an individual named J. B. Harshman, but with an engineering affiliate of Williams Pipe Line providing the expertise and the Union Pacific railroad providing the bulk of the capital. The latter subsequently bought out the minority interest. Again

---

[26] In these respects Southern Pacific is similar to a large integrated oil company, or a smaller one undertaking a pipeline project that is small relative to its total resources. This advantage would also, of course, be true of any large diversified company, or a firm, such as Buckeye, which is already a large pipeline operator.

the conditions were quite similar. This line is now refiner owned, because Union Pacific purchased Champlin Petroleum Co., but the pipeline antedates the building of Champlin's Los Angeles refinery and thus is an exception to the general rule that pipelines are built by those that expect to use them.

Another exception is the Kaneb Pipe Line, which initially ran from connections with seven refineries in southern Kansas to a terminal at Fairmont, Nebraska. In this instance the number of potential shippers, relative to the rather modest investment required, and the importance to these refiners of the market to which the line provided low cost transportation, minimized the inherent risks.[27] The early success of this segment made possible the financing of subsequent additions to the line.

Reasons for Joint Ventures. Having considered the reasons why oil pipelines are usually built by those that expect to use them, there is the further question why this vertical integration frequently takes the form of joint ventures. Where two or more prospective shippers foresee the need for major investment in transportation facilities, the economies of scale encourage the building of one large pipeline rather than two or several smaller ones. Ceteris paribus, the capacity of a crude oil line increases almost as the cube of the diameter of the pipe. There is no such increase in the required investment. For example, the required investment per barrel of capacity of a twenty-four-inch line is only a little over one third of that in a twelve-inch line and about one sixth of that in a six-inch line. There are equally great economies in the out-of-pocket expenses of operating the line. There are similar advantages in large refined products lines wherever warranted by the expected traffic.

These economies of scale have not always been achieved. Differences of judgment among the prospective shippers as to the anticipated volume of traffic, as to the route to be followed, as to how the ownership should be divided, and even conflicts of personalities, have sometimes resulted in two or more lines where one would have been more efficient. One example is the Amoco, Arapahoe and Platte lines running from crude oil producing areas in the Rockies to refineries in the Middle West. The economies of scale, however, explain why so many shipper owned pipelines built since World War II have been joint ventures.

---

[27] There were two lines, separately financed but built at the same time and subsequently merged. Augusta Pipe Line Co. ran forty-three miles from the Anderson-Prichard refinery at Arkansas City to Augusta, Kansas. The 246-mile Kaneb line, running north from Augusta, was directly connected to the other six refineries. The aggregate funded debt of the two lines was less than $7 million. The initial capacity of the ten-inch main line was only 250,000 b/d, which compares with Colonial's initial capacity of 800,000 b/d.

340

IV. The Limited Possibilities for Discrimination

This paper does not purport to examine the evidence as to whether shipper owners do or do not discriminate against actual or potential non-owner shippers in the use of shipper owned lines.[28] Merely from industry structure, however, it is possible to draw conclusions about what refineries might conceivably use, or like to use, a competitor owned refined products pipeline, and hence the potential for discrimination. In this context, "industry structure" refers only to the location of each refinery relative to existing pipelines.[29] The following analysis is limited to refineries owned by other than the sixteen largest refining companies, on the assumption that no one is greatly concerned about any possible discrimination suffered by the so-called "majors." [30]

The Bureau of Mines reported that on January 1, 1977, there were 174 operating refineries, with aggregate operating crude running capacity of 4,217,000 b/d, that were not owned by the sixteen largest refining companies. These refineries can be grouped according to the ownership of the products pipeline, or lines, if any, which by their locations they might conceivably use or like to use. The classification is with respect to the specific refinery, without regard to any other refinery that may be owned by the same firm.

The grouping of these refineries in Appendix B is not infallible, both because of the difficulty of getting precise information about some situations and because of the problem of fitting them into only five discreet

---

[28] The reader will appreciate that adequate treatment of all of this evidence would be a major undertaking, far beyond the scope of the present discussion, and that a great deal has been and will be said by others on this subject.

[29] The same analysis is not feasible for crude oil lines. A refiner might buy crude oil from producing properties located on any number of crude oil lines not directly connected to its refinery. The refinery could be supplied either through interconnecting lines or by exchanges for oil more conveniently located. Thus the location of the refinery relative to specific pipelines does not indicate the potential for discrimination. For similar reasons, refinery location is illuminating only with respect to the products of a specific refinery. If the refinery owner chooses to obtain products by purchase from or exchange with another refinery, the potential for discrimination will depend on the location of that refinery. Similarly, this analysis tells us very little about potential discrimination against non-refiner marketers.

[30] This frequently used term is admittedly ambiguous. Specifically, it is used here to mean the sixteen largest refiners, in terms of their crude oil distillation capacity in the United States in January 1977, excluding Puerto Rico and the Virgin Islands. They were Exxon Company, Standard Oil Company (California), Standard Oil Company (Indiana), Shell Oil Company, Texaco, Mobil Company, Gulf Oil Corporation, Atlantic Richfield Company, Marathon Oil Company, Union Oil Company, Sun Company, Inc., Standard Oil Company (Ohio), Continental Oil Company, Ashland Oil & Refining Company, Phillips Petroleum Company, and Cities Service Company.

categories. It is sufficiently accurate, however, to provide the general picture. It will be seen that only a few of the 174 refineries are in a position to suffer any discrimination.

Appendix Table 1 lists eighty-four refineries, accounting for 16.8 percent of the total capacity of the 174 owned by non-majors, that are not in a position to use any existing products pipeline, and hence could not suffer discrimination. They are distant from any products line, except for lines running *to* rather than *from* the vicinity of the refinery, or are too small to warrant even a short connection. The feasibility of connection is judged considering both the distance from the refinery to the nearest line and the volume of the refinery's output of products transportable by pipeline.

Another thirty-nine refineries, accounting for 18.6 percent of the aggregate non-major capacity, were situated where the only available products line is non-refiner owned. This group, in Appendix Table 2, includes a number of refineries that are obviously too small to warrant even a short connection (and thus might have been included in Table 1 rather than Table 2), and some that are not so small may not be connected to the apparently available line. The non-refiner owners of these lines, however, would have no reason to discriminate against any shipper.[31]

Another twenty-eight refineries, accounting for 42.3 percent of the aggregate capacity of the non-majors, were owned by firms that owned, or were part owners of, products lines serving these specific refineries. Presumably they did not suffer any discrimination in the use of these lines, even if only partly owned. This group can be broken into the three subgroups shown in Appendix Table 3. With respect to the nine refineries in subgroup A, accounting for 7.5 percent of the aggregate capacity of other than major owned refineries on January 1, 1977, the owned lines were the only ones by which products could be shipped from these refineries.

The ten refineries in subgroup B, accounting for 12.7 percent of the total non-major capacity, use owned products pipelines and also use, or conceivably might use, non-refiner owned lines. For several of these refineries, the non-owned line is reached via the owned line, but the owned line is more than a short connection.

The nine refineries in subgroup C, 22.1 percent of total non-major capacity, use owned products pipelines but also use, or might conceivably use, refiner owned pipelines, including such lines to which the

---

[31] Conceivably, a non-refiner owned pipeline, but one owned by an important refined products marketer, might discriminate against competitors in the use of the line. There are, however, no such lines.

owned line is connected. It is only those in this group that might possibly suffer discrimination.[32]

In the two remaining groups, the owner of the refinery does not own a products line serving this particular refinery. One group of six refineries, 7.3 percent of total non-major capacity, use, or conceivably could use, both refiner owned and non-refiner owned products lines. The potential for discrimination is thus limited to situations where the refinery might find it more advantageous to use a refiner owned line.

There are also, however, seventeen refineries, 15.0 percent of total non-major capacity, for which the only available products line, or lines, or the one most likely to be used, is owned by other refiners. This group, listed in Appendix Table 5, is probably an overstatement because it includes several refineries that may well find it advantageous to sell their total output in local markets reached by truck, and therefore may have no interest in pipeline transportation.

In summary, only about twenty-six out of 174 non-major refineries (Tables 3c and 5), constituting 37 percent of the non-major refining capacity, are located so that they might conceivably suffer discrimination in the use of a competitor owned refined products pipeline.

This same analysis of refinery location with respect to existing products pipelines also indicates what refiner owned lines are most important when considering the evidence on whether they do or do not discriminate against non-owners. For this purpose consideration can be conveniently limited to the refineries in Appendix Table 3, Part C and in Table 5: those that use, or might conceivably use, a refiner owned line and are not also in a position to use a non-refiner owned line. Of these twenty-six refineries, eighteen, accounting for 78.5 percent of the capacity in this limited group, use or conceivably might use, the Colonial and Plantation lines from the Gulf to the East Coast. The eighteen include several that are also in a position to use other refiner owned lines, and several that may have no interest in other than local markets for their products. Nevertheless it is apparent that any evidence as to whether the owners of the Colonial and Plantation lines do or do not discriminate against non-owners is of particular importance.

Among the other eight refineries in this group, three small ones might conceivably make some use of the Chevron line running northwest from Salt Lake City. Two might use the Wyco line from Wyoming to Denver. The Clark refinery at Hartford, Illinois, is in a position to use several competitor owned lines as well as water transportation on the

---

[32] There may be some overlap of subgroups B and C. In a few instances it is not clear, without much more intensive research, whether a refinery's needs could be met best by a refiner owned or a non-refiner owned line, or conceivably it might use both.

Mississippi and connecting rivers. The Clark refinery at Blue Island, near Chicago, depends primarily on partly owned lines running to Detroit and into Wisconsin but also might find it advantageous to use both non-refiner and competitor owned lines. The large Getty refinery at Delaware City has an owned pipeline into western Pennsylvania, and also depends on water transportation along the East Coast, but conceivably could use competitor owned lines. In these last two instances, the potential for discrimination is quite limited.

To reiterate, however, this analysis has to do only with what refineries might conceivably suffer discrimination in the use of competitor owned lines. There is no attempt to examine the evidence as to whether there is in fact any discrimination.

V. Alleged Restriction of Pipeline Throughput or Capacity

On several occasions the U.S. Department of Justice has advanced a theory that is summarized briefly in the following quotation:

> By restricting the throughput or capacity of a pipeline, its owners can force some oil to flow to the final market by more costly, less efficient means of transportation. As the price of crude or product downstream is generally set by the cost of the incremental barrel, throughput restriction has the effect of raising the downstream profits. Vertically integrated firms can thus capture at other unregulated states some of the potential monopoly profit on pipeline transportation denied by regulation.[33]

This section starts by pointing out a flaw in the logic which invalidates the theory. It goes on to demonstrate that other means of crude oil transportation are used only where pipelines are obviously impractical, or where the alternative mode is presumably cheaper. Data are not available for a similar analysis of refined products transportation. The facts of industry structure, however, virtually preclude pipeline owners from doing what the Justice Department says they can do. The final portion of Section V examines the theory as applied to the Colonial and Plantation pipelines, which are the two largest products lines, and concludes that it is refuted by both logic and experience.

The Flaw in the Theory. The full statement from which the above quotation is taken, and other statements by the Department of Justice,

---

[33] Hearings before the Interstate Commerce Commission on Ex Parte 308: *Valuation of Common Carrier Pipelines,* Statement of Views and Arguments by the United States Department of Justice, footnote 44, p. 22. (Hearings now under FERC, RM 78-2.)

including John H. Shenefield's testimony on June 28, 1978, before the Senate Subcommittee on Antitrust and Monopoly, make it clear that this theory is presumed to hold even if pipeline regulation is "ideal," that is, free access and no discrimination against non-owner shippers, and tariff rates provide no more than adequate return on the investment, and even if there is adequate competition at all other industry levels.

The fatal flaw in the theory is the argument that, even in the absence of any discrimination by pipeline owners against non-owner shippers, prices at the delivery point are determined by the incremental or marginal cost of shipping part of the supply at a higher cost than the bulk of the movement. The fallacy in this argument can be demonstrated by a hypothetical example using the following greatly simplified assumptions.

> That at the going market prices there is demand for 1,000,000 barrels of products that could have been shipped from the Gulf to the East Coast by pipeline but pipeline throughput is limited to 900,000 barrels.
>
> That there are ten shippers of equal size, including owners and non-owners, and that pipeline prorationing is such that each of these suppliers moves 90 percent of its shipments by pipeline and 10 percent by tanker. (Different assumptions as to the number of shippers or their relative size would not affect the conclusions.)
>
> That the 900,000 barrels are shipped by pipeline at a cost of 50 cents per barrel while the other 100,000 barrels move by tanker at a cost of $1.00 per barrel.

Under these assumptions, and according to the Justice Department's theory, undersizing the pipeline gives each shipper an excess profit of $45,000.

Cost of moving 100,000 barrels by tanker at $1.00 per barrel, which according to the Justice Department's theory will be reflected in the market price		$100,000
Actual transportation cost:		
90,000 barrels by pipeline at 50 cents per barrel	$45,000	
10,000 barrels by tanker at $1.00 per barrel	10,000	55,000
Excess Profit		$ 45,000

But each supplier would have a strong inducement to increase its share of the shipments, and hence its excess profits. Assume, for example, that one firm increased its share to 20 percent, and that the others took correspondingly smaller shares of both the shipments and the excess profits.

345

Market price reflects the cost of moving 200,000 barrels at $1.00 per barrel		$200,000
Actual transportation cost, assuming non-discriminatory prorationing:		
180,000 barrels by pipeline at 50 cents per barrel	$90,000	
20,000 barrels by tanker at $1.00 per barrel	20,000	110,000
Excess Profit		$ 90,000

Since each supplier has the same inducement to increase its share, competition among them will force the price at the delivery point down to where it covers only the average transportation cost of 55 cents per barrel. Because of this competition, the $45,000 excess profit calculated for each shipper would never occur.

If (contrary to the assumptions of adequate competition at all levels of the oil industry except the pipeline, made by the Justice Department for the purpose of expounding the theory), there were collusion among the suppliers to prevent this erosion of the theoretical excess profit, there could also be collusion to exact a monopoly profit even if all shipments went by pipeline. Thus there is a flaw in the logic which invalidates the theory.

The Evidence against Restriction. Furthermore, it can be demonstrated that there has been no restriction of crude oil pipeline traffic causing shippers to use more costly means of transportation, and that refined products pipelines are virtually precluded from doing what the Justice Department says they can do.

a. Crude oil lines. The Annual Petroleum Statement, published by the Department of Energy for the year 1976, shows receipts of crude oil by U.S. refineries broken down by mode of transportation to the refinery. The detailed report can be summarized as follows:

	Refinery Receipts in 1976	
	Thousands of Barrels	*Percent of Total*
By pipe lines	3,085,764	62.7
By tankers and barges		
Foreign oil	1,487,350	30.3
Domestic oil	270,649	5.5
By tank cars and trucks	70,428	1.5
Total	4,920,191	100.0

Analysis of these data discloses that virtually all of the deliveries by other than pipeline occurred where these alternatives presumably were more rather than less efficient.

Of the total receipts, 30.3 percent was foreign oil delivered by tanker or barge. Almost all of this went to coastal refineries, where it was obviously the only feasible mode of transportation.[34] Less than 1 percent of the total receipts was foreign crude delivered to refineries by inland waterway. The Delta refinery at Memphis has already been cited as one example of a plant that has found it more economical to rely on barge shipments up the Mississippi river rather than connect to a pipeline.

Of the total refinery receipts of crude oil, 5.5 percent was domestic oil delivered by tanker or barge. A breakdown of the 5.5 percent supports the presumption that this also was the most economical mode of transportation.

	Thousand Barrels per Year	Percent of All Refinery Receipts of Crude Oil
Deliveries to Gulf Coast refineries	151,097	3.07
Deliveries to West Coast refineries	80,099	1.63
Deliveries to East Coast refineries	21,162	.43
Deliveries by inland waterway	18,291	.37
Total receipts of domestic oil by tanker or barge	270,649	5.50

The deliveries to Gulf Coast refineries could reflect use of barges because of temporary delays in making pipeline connections to offshore production. There are numerous situations, however, where barging small quantities of oil from near shore or offshore production may be more economical than building a pipeline.

Over half of the tanker or barge deliveries of domestic oil to West Coast refineries was Alaskan crude, which obviously could not be delivered by pipeline. There is also considerable California crude oil production offshore or near shore. Here again, there presumably are

---

[34] There are substantial savings in tanker transportation cost to be achieved by offshore unloading facilities for ultra large crude carriers, with short pipeline connections to onshore refineries. Delays in achieving these economies, however, have been due to government restrictions and they are not related to any shortage of domestic pipeline capacity.

situations where it is more economical to barge the oil than to construct a pipeline.

The lower laid down cost of imported oil makes use of domestic crude in East Cost refineries uneconomic except under special circumstances.[35] The very small quantities of domestic crude delivered to the East Coast by tanker or barge, less than 60,000 b/d and only 0.43 percent of all refinery receipts, obviously did not warrant construction of a pipeline all the way from the Gulf Coast producing area. There were also small quantities of domestic crude delivered to inland refineries by inland waterway. The Delta refinery is again an example where this was the most economical mode of transportation.

Of the total receipts of crude oil by U.S. refineries in 1976, only 1.55 percent was delivered by tank car or truck. A state-by-state breakdown of this 1.55 percent indicates that the transportation cost thus incurred could not possibly have affected crude oil or refined products prices to the benefit of pipeline owner shippers.

| | Deliveries by Tank Car or Truck | |
Location of Refinery	Percent of Total Receipts by U.S. Refineries	Percent of Total Receipts by Refineries in This State
California	0.312	2.52
Texas	0.256	0.97
Louisiana	0.148	1.20
Western Pennsylvania and West Virginia	0.140	23.50
Colorado, Utah, Montana and Wyoming	0.277	8.43
Oklahoma and Kansas	0.139	2.05
New Mexico	.094	1.54
Other states	.184	

California is the only state for which there was reported any significant volume of interstate deliveries by tank car or truck. Such interstate shipments, only 0.5 percent of total receipts by California

[35] Under the complex government regulations affecting domestic crude oil prices, some domestic oil is priced far below imported oil. Under the "entitlements" program, however, this does not warrant the cost of transporting the low priced oil to the East Coast. Aggregate imports are far more than the capacity of East Coast refineries. Obviously it would be uneconomic to ship domestic oil to the East Coast rather than refining it on the Gulf. This would require an increase in the volume of foreign oil incurring the additional cost of transportation to the Gulf plus the cost of shipping the products back to the East Coast.

refineries, apparently came from Utah and did not reflect any shortage of pipeline capacity.[36] There is no readily available explanation for even this nominal amount. It may come from fields in Utah that are not adjacent to the Four Corners line and therefore have no pipeline connection to California. If so, it does not conform to the Justice Department's theory because the alternative of pipeline movement does not exist.

Intrastate shipments of crude oil by tank car or truck to California refineries amount to 2.0 percent of total California receipts, equal to the requirements of one relatively small refinery with a capacity of 34,000 b/d. Many of the approximately forty refineries in California are small ones located very close to local crude oil production. It is hardly surprising that small quantities of oil can sometimes be transported to the refinery more economically by truck because the volumes are not sufficient to warrant investment in a pipeline. Furthermore, these local truck movements are not at higher cost than incurred by pipeline owners. On the contrary, use of local crude means that in general the cost of transporting crude oil to these refineries is actually lower than that incurred by the larger refineries that receive their crude by pipeline from much greater distances. Furthermore, such small refineries tend to sell their output in local markets where they have a products transportation cost advantage over more distant refineries. Indeed these transportation cost advantages are a primary reason why such small refineries can compete with much larger, and therefore more efficient, plants.

The same reasoning applies to tank car or truck deliveries to refineries in Texas, which amounted to less than 1.0 percent of all deliveries to Texas refineries. Almost all of this was Texas crude, equivalent to the requirements of one relatively small refinery. Here, as in California, such deliveries can be explained by the fact that many of the almost fifty refineries in Texas are very small plants located close to crude oil production.

---

[36] Another table in the same report shows total interstate shipments of crude oil to California, by all means of transportation, as follows:

District V (Alaska)	38,273,000 barrels
Utah	5,792,000 barrels
Colorado	140,000 barrels
Total	44,205,000 barrels

If the two sets of data are precisely comparable, over half of the shipments from Utah and Colorado were by tank car or truck. This could not have been due to any shortage of pipeline capacity. The only pipeline from these states to California is, or was, the Four Corners Pipe Line, which was taken out of service in September 1976, because traffic had dropped to so low a level of capacity that operation was no longer economic.

A similar analysis of each of the other states would be merely repetitive because it reflects the same situations. Indeed there are a number of very small refineries located near limited amounts of crude oil production and at a distance from any crude oil line. They receive their oil by truck because neither they nor anyone else has thought that the volumes warranted investment in a pipeline. Therefore they could not possibly be adversely affected by any restrictions imposed by pipeline owners.

The higher percentage of crude oil receipts by refineries in Western Pennsylvania and West Virginia that was delivered by tank car or truck deserves special comment. The Pennsylvania type crude oil from which these refineries make lubricants comes from "stripper" wells that frequently produce less than one barrel per day. It is hardly surprising that gathering some of this oil by tank truck may be more economical than constructing a pipeline.

In summary, this analysis demonstrates that, barring unidentified but unimportant exceptions, such as delays in connecting lines to newly discovered production, there is no restriction on crude oil pipeline usage forcing oil shippers to resort to more costly modes of transportation. On the contrary, those most likely to use the alternate mode have a transportation cost advantage which is one of the reasons for their ability to compete with more efficient refineries.[37]

In evaluating the Justice Department's theory we have so far assumed that other "means of transportation" refers to other than pipelines.[38] Perhaps one should also consider that the proponents might argue that restricting the capacity or throughput of certain crude oil lines could "force some oil to flow to the final market through other more costly, less efficient" pipelines.

Such an argument cannot be rebutted by data such as have been cited on the use of pipelines versus alternative modes of transportation. There is no way of measuring whether, or to what extent, restricted throughput in certain crude oil lines forces use of other higher cost lines.

---

[37] Recently there has been some movement of Alaskan crude by tanker via the Panama Canal to the Gulf, and thence to refineries in the Midwest, at a higher cost than if there were a large diameter pipeline from the West Coast. It is noteworthy, however, that the only proposals for such a line have been initiated by the expected users who are interested as either producers or refiners. Furthermore, the Justice Department has stated that Sohio's proposed line from Long Beach to Midland, Texas, would not have the adverse effects assumed to result from the generalized application of its theory. (Statement by John H. Shenefield presented to the Senate Subcommittee on Antitrust Policy on June 28, 1978.)

[38] The hypothetical example used by the Justice Department assumes that restriction of pipeline throughput forces some shipments by higher cost tankers. Ibid.

It can be pointed out, however, that in addition to the logical flaw in the theory, it involves a series of questionable assumptions.

It assumes that the owners would deliberately undersize a line, thereby forfeiting some of the economies of scale and increasing the transportation cost to them as the principal users, and that they would do this with the intent of affecting crude oil or refined products prices. To the extent that the resulting capacity was less than the potential traffic would warrant, the owners would have to prorate all shippers, or attempt to restrict non-owners in defiance of government regulations to the contrary. Under proration they and their competitors would be equally affected by any higher cost of alternative transportation.

Alternatively, it assumes that the owners would, in some unspecified manner, restrict the throughput of an adequately sized line, again with the intent of influencing crude oil or refined products prices. The owners would be forfeiting additional pipeline revenues at the very low incremental cost associated with unused capacity.

In either instance there are the further assumptions that the difference in cost between the specified pipeline and the next cheapest line, and the volume of traffic diverted to the more costly route, would both be sufficient to have any significant effect on crude oil prices, and that this effect could be calculated by the pipeline owners with sufficient assurance to cause them to act accordingly.

Considering all of these necessary assumptions, it seems that even this attenuation of the Justice Department's theory calls for convincing evidence that the presumed restriction of capacity or throughput has actually occurred, and has had the alleged effect.

b. Refined products lines. Data such as were cited for crude oil pipelines are not available for a similar analysis of refined products pipelines. The facts of industry structure, however, provide some basis for questioning the Justice Department's theory.

About 30 percent of the miles of products lines, perhaps 20 percent of the barrel miles of capacity (see footnote a, Table 1), are owned by firms that are neither refiners nor refined products marketers. Inclusion of Texas Eastern Transmission Co., which owns a very small refinery but is primarily a transporter, would increase this to about 25 percent of capacity. The owners of such lines would have no reason to restrict throughput in an attempt to influence refined products prices. Furthermore they frequently compete with shipper owned lines reaching the same markets. They are free to increase capacity, or extend the lines into new markets, to take advantage of any restriction of throughput by shipper owned lines.

There are also the numerous lines, some of which are described in Appendix A, that exist solely to transport products from an affiliated refinery to the markets it serves, and which would be of no use to any other shipper. There would be no incentive for the owners to restrict the throughput of such plant facilities, thus forcing the owners to resort to higher cost transportation.

Frequently there are two or more products lines reaching the same market, sometimes from quite different refinery locations. One example is the four refiner owned lines to Denver: two from refineries in Wyoming, one from the Texas Panhandle and one from Kansas. There are several lines, both refiner and non-refiner owned, running from refineries in Kansas, Oklahoma and inland Texas to markets in the Upper Middle West, and also several lines from Gulf Coast refineries to these same markets. Products lines reach into Ohio from both the west and the east. Even if all of the competitive factors already noted were absent, restriction of the throughput of one line, in an attempt to influence refined products prices, would be ineffective so long as the owners of the other line or lines were free to increase their low cost transportation to such markets. Given the differing attributes and corporate strategies of the owner shippers, any concerted action would require an overt conspiracy that would be difficult to organize and even more difficult to conceal.

Furthermore, the more important refined products markets are supplied in part by refineries located at or near these markets, which refineries are supplied with crude oil by low cost pipeline or water transportation, and in part by refineries located nearer crude oil supplies, which reach these markets by products pipelines. For example, most markets in the whole upper middle west can be supplied either by refineries in that area or by mid-continent or gulf coast refineries. Any restriction of products line transportation, in an attempt to influence refined products prices, would invite increased crude line throughput and expansion of the refineries near these markets.

Admittedly, the facts summarized in the preceding two paragraphs are not in themselves a complete answer to the Justice Department's theory because of the possibility that use of competing pipelines might involve some net additional cost. Even this attenuated application of the theory, however, is invalidated because, as demonstrated at the beginning of Section V, there is no reason to expect that delivered prices will reflect the incremental cost of transporting part of the supply at a higher cost than the balance of the shipments.

Colonial and Plantation Pipelines. The Colonial and Plantation lines, running from refineries on the Gulf to southeastern states and the East

Coast, bulk so large in the total pipeline movement of refined products that they deserve special consideration. Furthermore, they are the outstanding examples of situations to which the Justice Department's allegations appear to be targeted. The following comments are in part an elaboration of the foregoing discussion.

1. The Colonial line was built in spite of doubts as to the soundness of this very large investment, and in the face of considerable opposition. It is at least open to question whether there would be a pipeline from the Gulf to the East Coast today if the large oil companies that expected to be the principal users had been barred from participating in the project. If Colonial's tariff rate is less than the alternative of tanker transportation, an important element in any such difference has been its ability to finance the project at relatively low interest cost because of the combined credit of these owners.

2. As previously noted, pipeline owners have strong incentives to expand capacity whenever this appears warranted by the expected traffic. Under prorationing, owners and non-owners are restricted to the same percentage of their desired throughput. Since the owners include the larger shippers, they have more to lose by being forced to use more costly modes of transportation. There is also the expectation that any needed increase in capacity will earn additional profits for the owners.

The owners also have strong incentive to encourage nonowner shippers and thus increase throughput up to the capacity of the line. Because most of the cost of owning and operating the line is fixed by the size of the investment, the incremental cost of moving the additional barrels is very small.

3. There is no evidence of any intent to restrict capacity in these lines. On the contrary there has been repeated expansion of capacity in an attempt to catch up with demand. Construction of the Colonial line was begun in 1963 and completed in February 1965. The initial design capacity was about 800,000 b/d. By the end of 1965 the seasonal peak of traffic was already close to capacity. The first expansion of capacity was announced in February, 1966 and completed in that year. The installation of additional pumping stations had been allowed for in the original design but it had not been anticipated that they would be needed until the mid seventies. A second expansion was announced in November 1966, and a third and fourth were completed by the end of 1968. These also involved additional pumping stations, increasing capacity to 1,152,000 b/d. In 1971, the fifth expansion, consisting of looping 461 miles of thirty-six inch pipe and adding 128 miles of spur lines, increased capacity to 1,464,000 b/d. A sixth expansion, in 1976, looped 183 miles with forty inch pipe bringing capacity to 1,620,000

b/d. A seventh expansion, currently underway, will add more forty inch pipe to bring capacity up to 2,100,000 b/d.

Neither is there any indication of an attempt to restrict throughput for the purpose of influencing destination prices. Because of seasonal fluctuations in traffic, and because of surplus capacity to some areas but not others, average daily throughput has necessarily been substantially less than total capacity. Colonial's throughput, however, has grown as follows:

	Thousand Barrels per Day
1965	637
1967	970
1969	1150
1971	1170
1973	1420
1975	1460
1977	1594

The number of shippers in the line grew from only the ten owners in 1965 to thirty-five in 1977 and forty-two in 1978. Included in the latter figures are not only other Gulf Coast refiners but also marketers obtaining products by purchase or exchange with Gulf Coast refiners. Deliveries for non-owners grew from zero in 1965 to 28 percent in 1976 and 32 percent in 1977. This does not suggest any attempt to limit throughput, or to restrict use of the line by non-owners.

Part of Colonial's traffic was apparently diverted from the pre-existing Plantation line, which suffered a 31 percent decline in traffic between 1963 and 1964, presumably because the larger pipe used by Colonial enabled it to charge lower rates. Subsequently, however, Plantation also expanded its capacity and extended the line so that by 1975 it was handling a third more traffic than in 1963.

4. The continued movement of refined products to the East Coast by tanker or barge is in large part in situations where this is the lowest cost mode of transportation. Spurs of the Colonial line reach to about 80 miles from the coast at Selma and Fayetteville, North Carolina, and to about 100 miles from the coast of Augusta, Georgia. Otherwise, both Colonial and Plantation run roughly 150 miles from the coast of North and South Carolina and Georgia, and they do not extend into Florida. Thus there would be a long truck haul from the nearest pipeline terminal in competition with tanker terminals at coastal ports. Understandably the bulk of the pipeline traffic goes to inland markets in the southeastern states, and the tanker movement to markets near the coast.

Similarly, the Plantation line stops at Washingon, D.C., and the Colonial line goes only to Linden, N.J. Thus neither is a lower cost alternative to tanker or barge shipments to New England. Obviously limitation of pipeline capacity or throughput could have no bearing on prices in areas not reached by these lines.

5. As demonstrated at the beginning of Section V, the owners of the Colonial and Plantation lines have had no reason to expect that, by restricting throughput, they could influence prices at delivery points to their advantage.

6. Even if there were not this flaw in the theory, any effect on prices in the Middle Atlantic States, where the lines do reach the coast at some points, would depend on whether, all factors considered, the pipeline transportation cost is, or was, less than the cost by tanker at the time of shipment. The variables are such that this cannot be calculated either easily or precisely. It appears that at times tankers have been the cheaper mode of transportation, and that rarely has the added cost by tanker been as much as a major fraction of a cent per gallon. More importantly, however, the uncertainties are such that it would be virtually impossible for the pipeline owners to calculate the incremental cost of tanker transportation, as it is alleged to affect destination prices, and come to a common agreement to restrict pipeline movement in order to achieve this effect.[39]

---

[39] The pipeline transportation cost for this comparison includes a number of factors other than the then current pipeline tariff rate. There is the cost of getting the products from a refinery into the Colonial or Plantation line, which can vary from nominal to substantial, depending on the distance from the refinery to Colonial or Plantation. There is also the substantial cost of truck transportation from the pipeline to the final market, which can vary depending on the location of that market relative to the nearest pipeline terminal.

There are similar variables in the cost of getting products from a refinery on board a tanker and from the nearest tanker terminal to the local market. More importantly, the tanker rates fluctuate sharply, and the only readily available data are monthly averages of the reported rates for single voyage charters. Such spot rates are what an economist would call opportunity cost. The cost as calculated by the shipper, however, may be based on long term charters, or on the cost of owning and operating the vessel.

For the above reasons, the following comparison for limited periods is an oversimplification:

Period	Tariff Rate Pasadena, Texas to Carteret, N.J. Colonial Pipe Line (cents per barrel)	Tanker Rate Gulf Coast to North of Hatteras (cents per barrel)	
		Monthly averages	Period averages
7/1/75– 3/14/76	48.25	50.7–97.3	63.3
3/15/76–10/31/76	51.65	50.3–97.3	63.6

Tanker rates are weighted monthly averages for single voyages, of clean vessels suitable for carrying gasoline, and over 30,000 DTW, reported by *Platt's*

7. Furthermore, shipments from the Gulf to the East Coast must compete with products from East Coast refineries, from refineries in Puerto Rico and the Virgin Islands, and with imports of products. Any attempt to influence products prices by restricting pipeline movement from the Gulf would invite increased competition from these sources. Even if the Justice Department's theory were valid, the net effect on prices would be difficult to calculate because of all of the uncertainties as to the availability and the relative cost of these alternatives. Again, the ten owners of Colonial and the three owners of Plantation would have to reach similar conclusions as to this effect, and would have to agree on the desirability of restricting pipeline movement in spite of these alternatives.

In summary, the theory advanced by the Justice Department will not stand the test of logic, and of empirical analysis where the data for such analysis are available.

VI. Ease of Entry, Concentration and Competition

In considering the structure of any industry, and its bearing on the effectiveness of competition, economists commonly consider both the ease with which newcomers can enter the business and the degree to which ownership of the industry is concentrated in a few firms. This section will comment briefly on these criteria, as applied to oil pipelines, and will also consider the degree to which there is competition among pipelines.

Ease of Entry. Ease of entry is considered to be important because the potential competition of those that might enter, and the actual competition of those that do, are major restraints against any attempts of existing firms to try to exact a more than competitive rate of return. Ease of entry is, of course, a matter of degree, and depends on particular circumstances. Building the Trans-Alaska Pipeline was a far different project than laying a short gathering line to a refinery from nearby production. Furthermore, the effect on both actual and potential competition can be quite different for oil pipelines, given the pervasiveness of vertical integration, than in industries where this condition does not prevail.

Oil Price Handbook and Oilmanac but here converted to cents per barrel at the rate of 8.648 barrels per long ton. There is the further qualification that north of Cape Hatteras would include voyages as far north as Maine. Thus the reported average may overstate the cost to Middle Atlantic ports.

According to this oversimplified comparison, the *average* tanker rate in these periods exceeded the pipeline tariff by about one third of a cent per gallon. Considering the fluctuations in the tanker rate, and considering all of the other variables, it would be difficult to demonstrate that there was any significant volume of tanker shipments at a total transportation cost significantly higher than by pipeline.

It should be obvious from much that was covered in Sections II and III, however, that entry into pipeline transportation has been far easier than it would have been if the expected users of a line had been prohibited from participating in its construction.

Concentration. The most commonly used measure of concentration is the percentage of an industry that is in the hands of the four, eight, or some other number of the largest firms. Such a national concentration ratio is of little significance here because, in common with a number of other industries, all pipelines do not compete with all other pipelines.[40] Because pipelines frequently run to or from a single refinery, and usually serve only a very few, and because the economies of scale deter the building of two or more lines when one larger one would serve, there are not a large number of competitors for the same traffic flow. Whether there is effective competition among pipelines cannot be answered by citing concentration ratios.

---

[40] The ten largest crude oil lines, as reported to the ICC, are:

	Millions of Barrel Miles Transported in 1975, as Reported to the ICC[a]	Percent of ICC Total
Lakehead Pipe Line	297.7	19.3
Amoco Pipe Line	181.2	11.7
Shell Pipe Line	122.7	7.9
Mid Valley Pipe Line	97.6	6.3
Mobil Pipe Line	91.9	5.9
Texas Pipe Line	82.2	5.3
Arco Pipe Line	72.2	4.7
Exxon Pipe Line	61.2	4.0
Ashland Pipe Line	57.1	3.7
Marathon Pipe Line	56.7	3.7

[a] Includes throughput of the named lines plus that of any other lines operated by the named line but owned jointly with other owners.

By far the largest, in terms of barrel miles transported, is the Lakehead Pipe Line, which is a subsidiary of the Interprovincial Pipe Line and exists primarily to transport crude oil from western Canada to refineries in eastern Canada. Particularly since the drastic curtailment of crude oil imports from Canada to the United States, this line is hardly a part of the U.S. industry. If we eliminate Lakehead, and also make a rough estimate of the total to include lines not regulated by the ICC, the concentration ratios are:

	Percent of Industry
four largest crude oil lines	36
eight largest crude oil lines	56

Thus, by the standard applied to other industries, ownership of crude oil pipelines is not highly concentrated.

It is also worth noting that the concentration ratios shown in footnote 40 would not be greatly changed if user ownership were prohibited. Colonial's large share of products pipeline transportation, for example, would not be changed. If it were somehow replaced with two or more smaller lines, this would involve considerable loss in economic efficiency.

Competition among Pipelines. It has already been noted that many small lines, and some not so small, exist solely to transport crude oil to an affiliated refinery, or products from that refinery to the markets it serves. The location of other lines is such that they are of little use to others than the two or three owners. Such lines do not compete with other pipelines. The competition is in the markets for the refined products, and the pipeline, along with the cost of the crude oil and the cost of operating the refinery, is merely one of the costs of getting the products to market.

The ten largest refined products lines, as reported to the ICC, are:

	Millions of Barrel Miles Transported in 1975, as Reported to the ICC[a]	Percent of ICC Total
Colonial Pipe Line	582,050	43.6
Plantation Pipe Line	94,901	7.1
Texas Eastern Transmission Corp.	60,020	4.6
Williams Pipe Line	59,054	4.4
Mapco[b]	41,903	3.1
Explorer Pipe Line	41,086	3.1
Phillips Pipe Line	37,554	2.8
Gulf Refining Co.	37,272	2.8
Buckeye Pipe Line	29,953	2.2
Southern Pacific Pipe Line	24,826	1.9
Four Largest		59.7
Eight Largest		71.5

[a] Includes throughput of the named line plus that of any other lines operated by the named line but owned jointly with other owners.
[b] Transports primarily products of gas processing plants rather than produces of refineries.

The two largest account for over half of the total barrel miles reported by the ICC, perhaps 45 percent of a larger total including non-ICC lines. The concentration reflects the large volume of products moving from Gulf Coast refineries to markets in the southeastern states and the Atlantic Coast, the long distance of this movement, and its restriction to only two lines because of economies of scale achieved by a large diameter line relative to several small ones.

With this exception, products pipeline ownership is decidedly not concentrated. The eight next largest lines account for less than 25 percent of the barrel miles reported by the ICC, and an even smaller share of the total including non ICC lines. It is also noteworthy that five of these eight lines are not owned by refiners (ignoring the very small refinery owned by Texas Eastern).

Even the largest crude oil lines have some of the plant facility characteristics described for some of the smaller lines in Appendix A. They were built to connect the refineries of the owners with the crude oil producing areas expected to be their principal sources of supply. Each of the larger lines, however, runs from a major producing area, or areas, to major refining centers. Interconnections with other pipelines provide considerable flexibility in the ultimate destination of oil from any producing property on these lines, and also in the route to that destination. Even in these generalized flows of crude oil there are not, and because of the economies of scale there could not be, the large numbers of competitors that, in economic theory, are alleged to guarantee effective competition. There is, however, considerable competition for this traffic.[41]

For example, the section of the Amoco Pipe Line running from producing areas in Wyoming to a junction near Kansas City, and thence on to its refinery near Chicago, is paralleled by the Platte Pipe Line from producing areas in Wyoming to a junction near St. Louis with other lines into Chicago. A third alternative is the Arapahoe Pipe Line which connects with other lines serving producing areas in Wyoming and runs east to a junction with several other lines reaching Chicago. Thus the three are competitors for the crude oil traffic from the Rocky Mountains to any refinery in the Midwest.[42]

Ten pipeline companies have been identified as possible competitors for the shipment of crude oil from the tanker terminal at St. James, Louisiana, to Toledo. Other examples that might be cited are the flows from West Texas and New Mexico to both Chicago and the Gulf, and the movement of imported crude from the Gulf to the upper Middle West. The statement prepared by Howrey & Simon for the Association of Oil Pipe Lines, referred to earlier, lists thirteen pipelines participating in the traffic from West Texas to the Texas Gulf Coast refining area. This includes five pipeline companies that are participants in the Rancho Pipe Line System, in which each sets its own tariff rates.

The number of competitors faced by a refined products line depends greatly on its route. Among those listed in footnote 40, Colonial and Plantation reach roughly the same markets, and the interconnections with other lines are such that most Gulf Coast refiners could use either

---

[41] Government regulation of pipelines is of course designed to produce conditions similar to those that could be expected to result from rivalry among a large number of competitors. Indeed, economists would judge the effectiveness of the regulation by the degree to which it achieves this objective.

[42] The proposed abondonment of Arapahoe as a crude oil line, because of insufficient traffic, would reduce this number to two.

one. Texas Eastern and Explorer provide alternative means for shipping products from the Gulf to the Upper Middle West. Each faces numerous other competitors, the number varying with different segments of their lines. This last statement also holds for the Williams, Phillips, Gulf, and Buckeye lines.[43]

At the other extreme the non-refiner owned Southern Pacific Pipe Line has no pipeline competition over much of its length. There are also the numerous examples, some cited supra and in Appendix A, of lines that have no direct competition because they exist solely to transport products from a single refinery to the markets it serves. Competition among pipelines that are plant facilities is a meaningless concept.

Competition is not limited to crude oil lines versus crude oil lines, or products lines versus products lines. For example, the losses incurred by Explorer Pipe Line, noted supra, were due in part to the alternative of shipping crude oil north by pipeline from the Gulf to refineries in the Upper Middle West, rather than refining the oil at Gulf Coast refineries and shipping the products into the Upper Middle West, and complex changes in the availability and relative cost of this alternative. There can also be the alternative of shipping crude or products by ocean tanker or inland waterway to many of the same areas reached by pipeline.

VII. Summary

Extent of Vertical Integration. The most important characteristic of oil pipeline industry structure is the pervasiveness of vertical integration. Approximately sixty-five refining companies, accounting for over 96 percent of U.S. refining capacity, own crude oil pipelines. Approximately fifty refining companies, accounting for over 94 percent of refining capacity, own refined products pipelines. The exceptions, and also those refiners that have relatively little investment in pipelines, are generally those that have little need for pipeline transportation.

About 88 percent of the U.S. crude oil pipeline mileage on January 1, 1974, was owned by domestic refiners. Most of the other 12 percent is not an exception to the generalization that the lines are usually built by those that expect to use them. It includes lines built by Canadian interests to supply Canadian refineries, lines built by crude oil producers or gatherers, and lines built by refiners but subsequently sold.

---

[43] In his statement before the ICC in Ex Parte 308, Vernon T. Jones, President of Williams Pipe Line, lists ten products lines with which his line must compete.

About 70 percent of the miles of refined products pipelines on January 1, 1974, was owned by refiners. The larger part of the other 30 percent is also not an exception to the generalization that the lines are usually built by those that expect to use them. It includes lines, such as Buckeye and Williams, that were built by refiners but later became independent. Another major segment transports products of gas processing plants, rather than refineries, and many of these lines are owned by the producers or consumers of those products.

Pipeline ownership of the larger oil companies is not out of line with their size as refiners. The eighteen largest on January 1, 1974, had 81 percent of U.S. refining capacity. At that time they owned approximately 79 percent of the *refiner owned* crude oil pipeline mileage and approximately 81 percent of the *refiner owned* refined products mileage. If allowance could be made for differences in pipe diameter, these shares would be somewhat higher. Their shares of the totals, including non-refiner owned lines, however, would be smaller.

This vertical integration calls for explanation. Why are pipelines usually built by those that expect to use them, and only rarely by independent entrepreneurs? The fact that this shipper ownership is true of small firms as well as large suggests that it does not occur because of any attempt to monopolize. Its pervasiveness suggests that there are economic efficiencies that would be lost if shippers were prohibited from owning the lines they use.

An interesting and significant aspect of this vertical integration is the large number of refiner owned pipelines that exist to serve only one refinery. They were built, frequently by relatively small refiners, to provide low cost transportation of crude oil to that refinery, or low cost transportation of products to the markets it serves, and are so situated that they are of little or no use to any other shipper. Fifteen examples are described in Appendix A.

Such situations suggest that refiners have found that their needs for transportation facilities could be met more efficiently within the firm than by arms length market transactions. It seems reasonably clear that they were not built to avoid discrimination in the use of a competitor owned line.

Forcing the divestiture of such lines, or prohibiting users from building additional ones, would serve no useful purpose. To the extent that they are situated so that they can serve only an affiliated refinery, they cannot discriminate against non-owner shippers, either in access or in the rates charged.

It is also worth noting that competition among such pipelines is a meaningless concept. The competition is in the markets for the refined

361

products, and the pipeline transportation is merely one of the costs of the supply for the refinery, or one of the costs of getting the refined products to market.

Another aspect of vertical integration is the number of large joint venture pipelines. Where two or more prospective shippers see the need for transportation facilities over roughly the same route, the major economies of scale encourage the prospective users to agree on the building of one large pipeline rather than two or several small ones.

Reasons for Vertical Integration. The cost savings achieved by pipeline transportation, as compared with the alternatives of rail or truck, are so great that refiners, or alternatively crude oil producers, must see that the pipelines get built when and where needed. These savings, where the volume of traffic warrants investment in a pipeline, are far greater than the potential return on that investment under government regulation. Indeed the saving, as compared with rail or truck transportation, can be several times the total tariff rate charged by the pipeline.

The risks incurred in the construction of either crude oil or refined products pipelines can be so great as to deter any investor whose sole interest is the potential return on the pipeline investment. These risks are inherent in the peculiarities of pipeline investment. That these risks are not merely hypothetical has been demonstrated by a series of examples of lines that got into financial difficulties.

Conceptually, an independent entrepreneur could be induced to build a pipeline if the prospective rate of return were sufficiently high to compensate for the risks incurred. The user could afford to pay this high rate because of the saving relative to rail or truck transportation cost. Given the government regulation of tariff rates, however, this is not a practical alternative.

Independent entrepreneurs might be induced to build pipelines, and could finance them, in spite of the inherent risks, if the prospective users of the lines were willing to provide throughput or deficiency guarantees sufficient to assure an adequate return on the investment. Typically, however, the potential users have not been willing to propose such a one-sided arrangement. From the viewpoint of their stockholders they would be in the position of guaranteeing the investment in the line without the chance to earn even the return permitted by government regulation. Furthermore, they would be underwriting the line for the benefit of any shippers who might choose to take advantage of the resulting reduction in transportation cost but were not willing to participate in a throughput agreement. Persuading prospective users to share the risks on these terms would be difficult.

362

Another important reason for vertical integration is the advantage of coordinating the size, route and completion date of a pipeline, with the requirements of the refinery or refineries it is intended to serve, by management decisions within a firm rather than by arms length negotiations with an independent entrepreneur.

The prospective user or users of a pipeline are in the best position to determine their requirements and therefore the potential traffic. This information is closely related to their short and long run plans for crude oil supplies and their evolving strategies for marketing the refined products. Information and judgments developed in this planning process also put them in the best position to evaluate the investment risks.

Providing all of this information, in a sufficiently detailed and convincing manner, to an independent entrepreneur in the hope of persuading him to undertake the project, involves an added cost that is largely avoided by vertical integration. There is also the delay while the independent entrepreneur studies the project and reaches his independent judgment as to its viability. These problems would be compounded by any attempt to assure effective competition among several prospective builders.

The independent entrepreneur, or entrepreneurs if there is to be competition among them, must either rely on information and projections supplied by the prospective user, or users, or else duplicate the studies they have already made. He must recognize that the bulk of the traffic on the line will come from frequently only one or two users and rarely more than a few. To the extent that he relies on representations made by the prospective users, he must be concerned that these can be colored by their obvious interest in seeing that the line is built. He must in turn convince the prospective purchasers of the funded debt that the pipeline is a sound investment. Since he does not have the prospective user's urgent need that the pipeline get built, he is likely to be more cautious in his appraisal of the risks.

We have here what some economists have termed "the small numbers bargaining problem." Where the potential participants on both sides of a bargain are quite limited, the added economic cost of an arms length transaction may more than offset the assumed advantages of such a transaction. Having decided that the pipeline is needed, the prospective user finds it simpler to just go ahead and build it. Since the line can be financed on the basis of the total resources and credit rating of the user, the problem of placing the funded debt is also reduced.

It is interesting in this connection to consider four notable exceptions to the generalization that refined products pipelines are built by those that expect to use them. In each instance the number of shippers was large, relative to the required investment, and no one or two predominated. This both reduced the risk incurred and inhibited any possible agreement among the users to build their own line. Furthermore, three of the lines were built by railroads whose total assets were far larger than the pipeline investment. Added to this spreading of the risk, and the ease of financing, there was the further advantage that some of the pipeline could be laid on railroad right of way.

Limited Possibilities for Discrimination. This paper has made no attempt to examine the evidence as to whether shipper owned pipelines do or do not discriminate against non-owner shippers. Merely from their location relative to existing lines, however, it is possible to draw conclusions about what refineries might conceivably use, or like to use, a competitor owned refined products line, and hence the potential for discrimination.

Appendix A groups those 173 operating refineries on January 1, 1977, that were not owned by the eighteen largest refining companies, in five categories. The largest group, eighty-four refineries and 16.8 percent of the capacity of all 173, are not in a position to use any existing products line, and hence could not suffer discrimination.

Another thirty-nine refineries, comprising 18.0 percent of the non-major refining capacity, were situated where the only available products line is owned by a firm that is neither a refiner nor a refined products marketer. Such lines would have no reason to discriminate against any shipper.

A third group of twenty-eight refineries, accounting for 42.3 percent of the aggregate capacity of the non-majors, were owned by firms that owned, or were part owners of, products lines serving these specific refineries. Presumably they did not suffer discrimination in the use of these lines, even if only partly owned. For nine of these refineries the owned line is the only one that could be used. Ten of them might also conceivably use a non-refiner owned line. The remaining nine might also conceivably use a competitor owned line. It is only this third subgroup that might possibly suffer discrimination.

Six refineries, 7.3 percent of total non-major capacity, use, or conceivably might use, both refiner owned and non-refiner owned lines. The potential for discrimination is thus limited to situations where the refinery might find it more advantageous to use a competitor owned line.

There are also, however, seventeen refineries, comprising 15.0 percent of the non-major refining capacity, for which the only available products line, or the one most likely to be used, is owned by other refiners.

Combining these seventeen with the nine in subgroup 3C provides some insight into what refiner owned pipelines are most important in considering the evidence on whether they do or do not discriminate against non-owner shippers. Eighteen of the twenty-six, accounting for 78.5 percent of the capacity of this limited group, use, or conceivably might use, the Colonial or Plantation lines from the Gulf to the East Coast.

It is apparent that any evidence as to whether the Colonial and Plantation lines do or do not discriminate against non-owners is of particular importance. To reiterate, however, this paper does not attempt to examine such evidence.

Alleged Restrictions of Pipeline Capacity or Throughput. On several occasions the U.S. Department of Justice has advanced the theory that, by restricting the capacity or throughput of a pipeline, its owners can force some crude oil or refined products to flow by more costly means of transportation, that crude oil prices at the well or refined products at destination are determined by the higher cost of this incremental traffic, and that integrated firms can thus capture a monopoly profit.

The basic flaw in this theory is the allegation that prices at delivery points are determined by the assumed higher cost of the alternative mode of transportation. But, given non-discriminatory prorationing of pipeline capacity, which the Justice Department admits for purpose of expounding the theory, the incremental barrels of supply do not all move at the higher cost of the alternative mode of transportation. On the contrary, the proportion of the incremental barrels moving by pipeline will be the same as for the rest of the shipments. Thus the transportation cost for the incremental barrels is the same as the average cost for all shipments, and the pipeline owners have nothing to gain by restricting pipeline traffic. This point was illustrated by a simple hypothetical example.

Since the exposition of this fatal flaw in the logic of the theory hinges on the possibly questionable assumption of non-discriminatory access to the pipeline, it is important to examine the evidence as to whether pipeline owners do or could act as is alleged. Detailed analysis of actual receipts of crude oil by U.S. refineries discloses that virtually all of the deliveries by other than pipeline occur because pipelines are impractical or because the alternative mode is cheaper.

Thus there is no evidence of restriction of pipeline traffic that could have the alleged effect.

Data are not available for a similar analysis of refined products transportation. The facts of industry structure, however, raise doubts about the relevance of the theory. About 25 percent of the refined products pipeline capacity is owned by firms that are neither refiners nor refined products marketers. They would have no reason to restrict throughput in an attempt to influence products prices. Furthermore they are important competitors of shipper owned lines, and are free to increase their throughput to take advantage of any restriction by shipper owned lines.

There are also numerous shipper owned lines that exist solely to transport products from an affiliated refinery to the markets it serves, and which would be of no use to any other shipper. There would be no incentive to restrict the throughput of such lines, thus forcing the owners to resort to higher cost transportation.

Frequently there are two or more products lines reaching the same market, sometimes from quite different refinery locations. Restricting the throughput of one line, in an attempt to influence refined products prices, would be ineffective so long as the owners of the other line or lines were free to increase their low cost transportation to such markets. Given the differing refinery locations, and the differing attributes and corporate strategies of the different owners, any concerted action would require an overt conspiracy that would be difficult to organize and even more difficult to conceal.

Furthermore, the more important refined products markets are supplied in part by refineries located at or near these markets, which are supplied with crude oil by low cost pipeline or water transportation, and in part by refineries located nearer crude oil supplies, which reach these markets by products pipelines. Any restriction of products line transportation, in an attempt to influence refined products prices, would invite increased crude line throughput and expansion of the refineries near these markets.

It can also be pointed out that application of the Justice Department's theory to products pipelines involves a series of questionable assumptions, and that the pipelines do not appear to have acted in the alleged manner. In this connection particular attention has been paid to the Colonial and Plantation lines because of their size and importance.

It is at least open to question whether there would now be a pipeline from the Gulf to the East Coast if, as proposed by the Justice Department, the prospective users had been barred from participating in this project. Furthermore, the low tariff rate on the Colonial line, thus raising the possibility that any incremental movement by tanker might be at higher cost, was made possible by financing the bulk of the investment at low interest rates because of the combined resources and credit of the shipper owners.

Any restriction of capacity or throughput, for the alleged purpose, would be contrary to the self interest of the owners. Any needed increase in capacity is likely to result in a disproportionate increase in pipeline profits. Conversely, the owners, as the principal shippers, have the most to lose, under prorationing, if adequate capacity forces them to use more costly modes of transportation. The owners also have a strong incentive to encourage non-owner shippers, up to the capacity of the line, because the incremental volume is highly profitable.

Repeated expansion projects, the seventh and latest bringing Colonial to more than two and one half times its initial capacity, do not indicate any attempt at restriction for ulterior motives. The equally great twelve-year growth in throughput, and the increase in shippers from only the ten owners in 1965 to forty-two in 1978, do not suggest any attempt to limit use of the line by non-owners.

Even if there were not the basic flaw in the Justice Department's theory, the variables making up the total cost of moving products from a specific Gulf Coast refinery to a specific East Coast market by pipeline, and the alternative total cost by tanker, are such that it would be virtually impossible to calculate how much of the total supply moves at a higher cost than by pipeline, and the amount of the cost difference. Certainly the larger part of the tanker shipments to the East Coast has been in areas not reached by pipeline or in situations where tankers were the lower cost mode. Rarely has the difference in favor of pipelines been as much as a major fraction of a cent per gallon. Considering that this cost difference varies by supply origin and destination point, and that it fluctuates with the wide fluctuations in tanker rates, it is difficult to see the pipeline owners reaching, and acting on, a common understanding as to its effect.

Shipments from the Gulf to the East Coast must compete with products from East Coast refineries using imported crude, from refineries in Puerto Rico and the Virgin Islands, and with imports of products. Any attempt to influence products prices by restricting pipeline movement from the Gulf would invite increased competition from these sources. Here again, the net effect on prices would be difficult to calculate because of all of the uncertainties as to the availability and the relative cost of these alternatives. Again, the ten owners of Colonial and the three owners of Plantation would have to reach similar conclusions as to this effect, and would have to agree on the desirability of restricting pipeline movement in spite of this possibility.

In short, this theory does not appear to be a valid basis for government policy.

Appendix A: Pipelines as Plant Facilities

The following examples illustrate that vertical integration frequently exists because a refiner owned pipeline was built to serve only one affiliated refinery. It was built by the refiner to assure low cost transportation of crude oil to that refinery, or low cost transportation to the markets for its refined products. The examples also include a few instances where a line serves two refineries and is owned jointly by the refinery owners. Frequently the location of these lines is necessarily such that they are of little or no use to any other shipper. This will be apparent by reference to the accompanying exhibits. The owner might welcome additional traffic, with the cost savings resulting from larger volume, but this is not possible. Because such lines are as much a part of the refinery complex as the individual processing units, or the storage tanks at the site, it is convenient to refer to them as plant facilities. This term, however, does not connote any attempt to exclude other shippers.

These examples shed some light on the reasons for and effects of vertical integration. They suggest that a refinery's needs for transportation facilities may, at least at times, be met more efficiently within the firm than by arms length transactions with an independent pipeline company. Furthermore, it seems reasonably clear that a pipeline serving a single refinery, where there is no alternative line that might have been used, was not built as a defensive move to avoid discrimination in the use of a competitor owned line.

It is equally true that, if a pipeline is so located that it can serve only an affiliated refinery, it cannot be guilty of discriminating against non-owner shippers. For the same reason, there is no reason to question whether what the refiner charges itself for transportation is more or less than adequate. This is equally true if the pipeline is located to serve only two or three refineries owned by the joint owners of this line.

It will also be noted that competition among such pipelines is a meaningless concept. The competition is in the markets for the refined products, and the pipeline is merely one of the costs of the supply for the refinery, or one of the costs of getting the refined products to market.

Holly Corp. (Navajo Refining). Holly Corp. and its wholly-owned subsidiary, Navajo Corp., are the two general partners of Navajo Refining Co., which operates a refinery at Artesia, N.M., with a capacity of 29,930 barrels per day. Average throughput in the 1977 fiscal year was 27,600 barrels per day.

The principal sources of crude oil for this refinery are in southeastern New Mexico, within approximately fifty miles of Artesia. The company owns a 423-mile crude oil gathering system connecting this area with the refinery. It is also a 50 percent owner of a fleet of tank trucks gathering oil from wells not connected by pipelines. The total oil thus gathered is sufficient to meet the refinery requirements.

A wholly owned refined products pipeline runs from the refinery about forty miles north to Roswell, N.M., and about 150 miles west to El Paso, Texas. It has a capacity of 14,000 barrels per day. Most of the products not shipped via this line are delivered from the refinery by tank truck.

Exhibits 1 and 2 are maps of these pipelines, which were built and are operated exclusively to serve this refinery. There is no other refinery near enough so that it could possibly use them. (Maps supplied by Holly Corp. All other information from its Form 10K report to the Securities and Exchange Commission for the fiscal years ending July 31, 1974 and 1977 and its *Annual Report* for 1977.)

Union Pacific Corp. (Champlin Petroleum Co.). Union Pacific's petroleum operations are owned through its subsidiary, Champlin Petroleum Co. Champlin owns three refineries at Corpus Christi, Texas, Enid, Oklahoma, and Wilmington, California. Certain pipelines owned in connection with the Corpus Christi refinery could be counted as plant facilities, but the Enid refinery provides the clearest example.

Champlin owns and operates approximately 500 miles of crude oil gathering lines in Oklahoma serving the Enid refinery. It also owns and operates 600 miles of products lines running from Enid north through Kansas and Nebraska into Iowa and South Dakota. This is a plant facility, built and used solely to transport products of this refinery. In a famous legal case, Champlin successfully defended its right to exclude all other shippers. (*Champlin Refining Co. vs. United States*, 379 U.S. 29)

Apco Oil Corp. Until recently, Apco had two refineries. In 1977, that at Arkansas City, Kansas, had throughput of 37,100 barrels per day and that at Cyril, Oklahoma, 12,800 barrels per day. Apco owned two crude oil lines and one products line that can be counted as plant

EXHIBIT 1

Navajo Crude Oil Lines

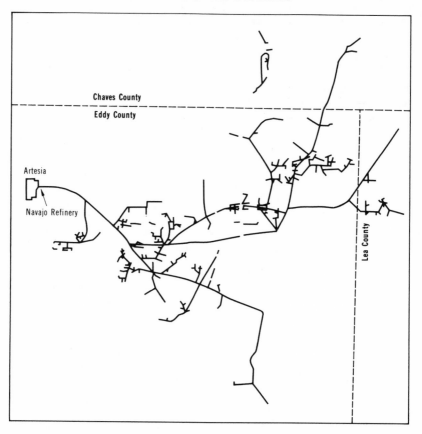

facilities, and was also part owner of several other crude and products lines that it used.

Apco Pipe Line, serving the Arkansas City refinery, owns and operates 468 miles of crude oil gathering and trunk lines. This line, with a capacity of 30,000 barrels per day, gathers oil from fields in Kansas and Oklahoma near the refinery. That refinery and this pipeline, plus interests in certain other lines, were recently sold to Total Petroleum (NA) as part of the planned liquidation of all of Apco's assets. Apparently the pipeline was considered to be an integral part of the refinery complex.

The wholly owned Anderson-Prichard Pipe Line owns and operates 120 miles of pipeline, with a capacity of 15,000 barrels per day,

EXHIBIT 2

NAVAJO PRODUCTS PIPELINE

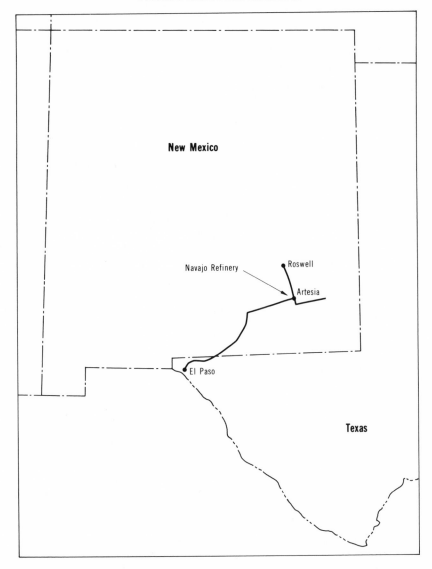

which gathers oil in Oklahoma for transportation to the Cyril refinery. This subsidiary also owns and operates forty-seven miles of refined products pipeline extending from the refinery at Cyril to Sunray Village, Oklahoma, where it connects with the Sun pipeline running east to Memphis and, indirectly, with the Williams and other lines reaching the upper Middle West. Both lines are clearly plant facilities. (All data from Apco's 10K report to the SEC for 1977.)

Mid-Continent Petroleum Corp. Exhibit 3 is a map of Mid-Continent Pipeline taken from the parent company's 1949 Annual Report. It shows that this network of crude oil lines was built to connect the refinery at West Tulsa with the producing areas from which that refinery was supplied. Mid-Continent Petroleum Corp. was merged with Sunray Oil Co., and later with Sun Oil Co. The pipelines are now part of Sun's whole pipeline system, thus making it difficult to present a more recent map of those that could be considered plant facilities.

Pasco, Inc. The following excerpt is from Pasco's 10K report for 1973. At that time it owned a 40,000 barrel per day refinery at Sinclair, Wyoming.

> *Pipeline Operations*—Registrant presently owns approximately 270 miles of crude oil gathering and truck lines serving its producing fields into the Sinclair Refinery. Registrant's petroleum products pipeline interests consist of the Medicine Bow Pipe Line System, "Medicine Bow System", which is wholly-owned by Registrant, and a 20 percent stock interest in Pioneer Pipe Line Co., "Pioneer". The remaining 80 percent stock interest in Pioneer is owned by Continental Pipe Line Co., "Continental," which increased its interest from 65 percent by the purchase on March 1, 1973, of a 15 percent interest from Registrant for $1,150,000. The Medicine Bow System extends from the Sinclair Refinery to a terminal at Denver, Colorado, which is also wholly-owned by Reg-

---

NOTES TO EXHIBIT 3

NOTE: Pipeline runs set another new high record in 1949, being 20,734,340 barrels as compared with 18,398,725 barrels in 1948. During the year 167 miles of new line were laid. By laying a larger line between Beggs and Tulsa, operating efficiency and capacity have been improved materially. Main lines were looped at several strategic points, and additional tankage was installed at six pump stations, including floating roof tanks for more efficient handling of natural gasoline which is being used in ever-increasing quantities at the West Tulsa refinery.

During the year the Company acquired, through lease and purchase, gathering systems in Lincoln and Creek Counties, Oklahoma, involving approximately 2,500 barrels a day of additional crude connections. Total leases connected at the end of the year were 1,492 compared with 1,248 last year.

The total number of miles of line in operation at the close of the year was 1,651 as compared with 1,530 at the end of 1948.

SOURCE: Mid-Continent Petroleum Corp. Annual Report, 1949.

EXHIBIT 3

Map of Mid-Continent Pipeline

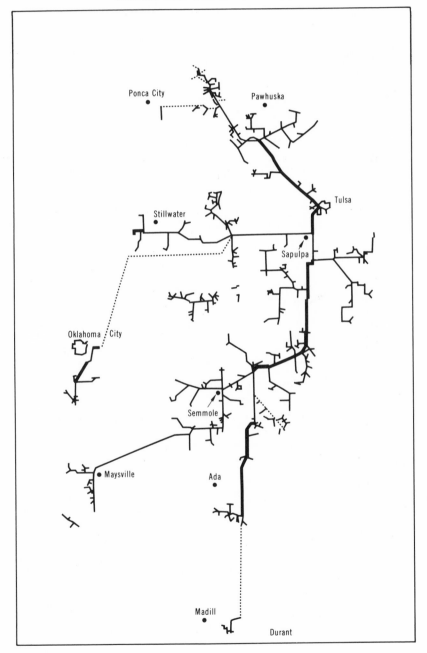

istrant. Registrant increased its interest in the Medicine Bow System and the Denver Terminal from 50 percent to 100 percent by the acquisition on May 1, 1973, of the 50 percent undivided interest of Skelly Pipe Line System for $1,750,000. Pioneer is managed by Continental and owns a pipeline system which extends from the Sinclair Refinery to a terminal at Salt Lake City, Utah.

Registrant has undertaken construction, at an estimated cost of $6,640,000, of a new ten inch crude oil pipeline extending from Casper to Sinclair, Wyoming, which it hopes to complete during the summer of 1974. The initial capacity of the pipeline will be 36,000 b/d and it is designed for an ultimate capacity of 55,000 b/d. As part of a pipeline modernization program, the capacity of the Medicine Bow System is being increased from its present capacity of approximately 13,000 b/d to 20,000 b/d.

Pasco has since sold all of its properties and liquidated the enterprise. The refinery and pipelines were purchased in one transaction by Little America Refining Co., thus emphasizing that the pipelines were considered as plant facilities.

Indiana Farm Bureau Cooperative Association. This farmer cooperative has a refinery at Mount Vernon, Indiana, with a reported capacity of 20,000 barrels per day. Subsidiaries own approximately 200 miles of crude oil gathering and trunk lines in adjacent areas of southern Indiana, southeastern Illinois and western Kentucky. The association also owns approximately 230 miles of refined products line running north from Mount Vernon to Peru, Indiana. This is clearly a plant facility, built and used to transport products from the refinery to the markets it serves. The location of the line makes it of little or no value to any other shipper.

OKC Corp. This company owns a refinery at Okmulgee, Oklahoma, with a 1976 throughput of approximately 21,000 barrels per day. In connection with the purchase of this refinery, OKC acquired a crude oil trunk line running approximately ninety miles from Oklahoma City to Okmulgee "which is used to transport substantially all of the crude oil requirements of the refinery" (10K report for 1974). There is no other refinery near Okmulgee and hence no other user of this line. At the same time, OKC also acquired crude oil gathering lines in the vicinity of Oklahoma City.

United Refining Co. United's wholly-owned Kiantone Pipeline, transporting crude oil seventy-eight miles south from a connection with the

Lakehead (Interprovincial) pipeline near Buffalo to its refinery at Warren, Pennsylvania, is a clear example of a plant facility. This line was completed in 1971 in response to the notice that the 290 mile line from Cygnet, Ohio, to Warren, owned by Buckeye Pipe Line, was to be abandoned as obsolete and unsafe. Buckeye reported that the volume handled was not sufficient to warrant modernization, and it was not interested in building a new line to serve this one refinery.

Kiantone was built when United expected to depend largely, if not entirely, on Canadian crude oil transported via the Interprovincial line crossing Canada from Sarnia, Ontario, to Buffalo, New York. See Exhibit 4. With the sharp reduction in the availability of Canadian crude, United was forced to turn to other sources, notably overseas oil. It acquired a 7 percent interest in Texoma Pipeline, completed in 1975 and running north from the Gulf of Mexico into Oklahoma. From thence connecting pipelines provide a circuitous route to a connection with Interprovincial and thus with Kiantone. Kiantone is an ICC (now FERC) regulated common carrier, and would welcome other shippers, but there are none. Other refineries in western Pennsylvania process local Pennsylvania type crude oil primarily into lubricants.

EXHIBIT 4

KIANTONE PIPELINE CORPORATION: DIVISION OF
UNITED REFINING COMPANY

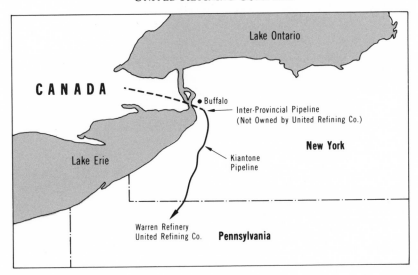

Farmers Union Central Exchange. Farmers Union built and operates the 443 mile Cenex Pipe Line extending from its refinery at Laurel, Montana, to Minot, North Dakota. Exxon's refinery at Billings, Montana is also connected to this products line and is the only shipper other than Farmers Union.

In an exchange arrangement, Continental Oil Co. delivers products from its Billings refinery to Farmers Union along the Yellowstone Pipe Line extending west into Washington and Farmers Union delivers products to Conoco along the Cenex line.

Exhibit 5 is a map of the Cenex line. These three refineries are the only ones that could possibly use it.

Allied Materials Corp. This is one of a number of very small refiners owning pipelines that can be counted as plant facilities. The *Pipe Line News Directory* for 1977 reports that Allied Materials has thirty-nine miles of crude oil gathering lines serving its refinery at Stroud, Oklahoma.

Diamond Shamrock Corporation. The pipelines owned by Diamond Shamrock Corporation are another example of lines that are essentially plant facilities, although certain of these lines are, or could be, used to

EXHIBIT 5

MAP OF CENEX PIPELINE

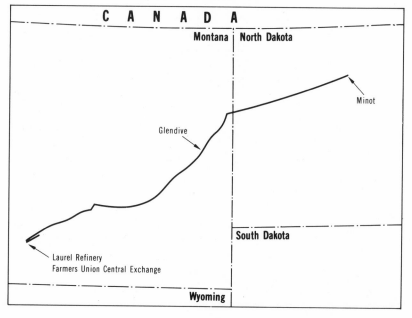

serve two other refineries. The following information has been confirmed by Diamond Shamrock, which also furnished the two maps in
Exhibits 6 and 7. The numbers on the maps correspond to the following
numbered paragraphs.

1. The wholly-owned The Shamrock Pipe Line Corporation owns
and operates over 885 miles of crude oil pipelines from fields in
western Oklahoma, southeast Colorado, southwest Kansas and the
Texas Panhandle to Shamrock's only refinery which is located at
Sunray (McKee), Texas. This is a common carrier, reporting to
and regulated by the FERC, but its location is such that it is of little
use to any other refinery.

2. Diamond Shamrock also owns about 207 miles of crude oil
gathering and trunk lines in two counties adjacent to its refinery.

3. The jointly-owned D-S Pipe Line Corporation is operated by
Diamond Shamrock and consists of 544 miles of crude oil gathering
and trunk line extending from the Abilene, Texas, area to Borger,
Texas, where it delivers oil into the Diamond Shamrock crude oil
pipeline mentioned in paragraph 2. This is a common carrier and
delivers relatively small volumes of oil to the other two refineries
in the Texas Panhandle, the Phillips plant at Borger and the
Texaco Plant at Amarillo.

4. The Shamrock Pipe Line Corporation operates the 170-mile
SAAL Products System running south from Sunray past Amarillo
to Lubbock, Texas. Shamrock shares the ownership of different
segments of this line in varying proportions with Phillips and
Texaco, in line with their expected usage. Its share in the jointly-
owned section is 33⅓ percent. This line is a common carrier,
reporting to and regulated by the FERC, but the only refineries in
a position to use it are those of the three owners at Sunray, Borger,
and Amarillo.

5. Shamrock's wholly-owned Emerald Pipe Line Corporation
owns a 114-mile refined products pipeline from Sunray northeast
into Oklahoma. This also is a regulated common carrier but the
only other refinery that could possibly use it is the Phillips Plant
at Borger.

6. Shamrock's wholly-owned West Emerald Pipe Line Corporation
operates approximately 297 miles of products line running west
from Amarillo, Texas into New Mexico. The pipeline is known
as the ATA Products System which is equally owned by Shamrock,
Phillips and Texaco. This also is a regulated common carrier but
again the only refineries that might use it are the three owners at
Sunray, Borger and Amarillo.

EXHIBIT 6

DIAMOND SHAMROCK'S CRUDE OIL LINES

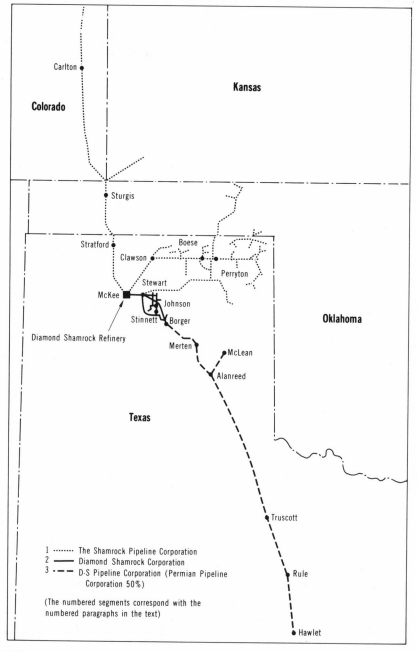

EXHIBIT 7

DIAMOND SHAMROCK'S REFINED PRODUCTS PIPELINES

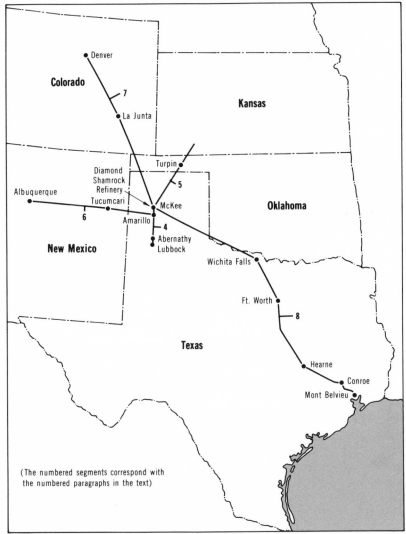

(The numbered segments correspond with
the numbered paragraphs in the text)

7. A refined products pipeline running 318 miles from Sunray to
Denver is owned jointly with Phillips and operated by Phillips. It is
not in a position to serve other than the two refineries.

8. Diamond Shamrock also has a 621-mile line from Sunray to
Houston. According to ICC (now FERC) definition, this is classed

as a refined products line. It exists, however, to transport natural gas liquids from Shamrock's gas processing plant near Sunray and from Phillips' facilities near Borger.

In summary, Diamond Shamrock is the sole or part owner of over 1,500 miles of refined products pipelines radiating in five directions from its only refinery. It is also the sole or part owner of over 1,600 miles of crude oil lines supplying that refinery. These qualify as plant facilities even though certain of the lines serve two other refineries.

Esmark, Inc. (Vickers Petroleum Corp.). Esmark's petroleum operations are owned through its subsidiary, Vickers Petroleum Corp. The approximately 700 miles of crude oil lines owned by Vickers in southern Oklahoma and northern Texas are all connected to its refinery at Ardmore, Oklahoma, and all of the crude gathered goes to this refinery. Connections with other lines provide some additional crude. Only the main lines are shown on the map in Exhibit 8, but the areas in which the crude is gathered are also indicated.

The line running north from Ardmore to Wynnewood was built by Vickers to connect with the Seaway Pipeline, thus enabling it to receive imported oil shipped north to Oklahoma from the Gulf.

Crude oil lines still owned by Vickers in Kansas were built to supply its refinery at Potwin, and were thus a plant facility, but this refinery was shut down in 1964. (As in the other examples, both the map and the above information were obtained from Vickers.)

EXHIBIT 8

VICKERS PETROLEUM CORPORATION: OWNED MAIN CRUDE OIL LINES
SERVING THE ARDMORE REFINERIES

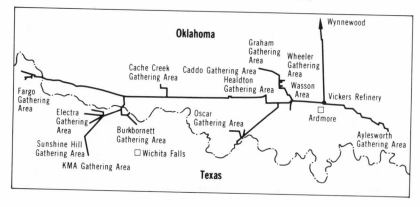

National Cooperative Refinery Association. Exhibit 9 is a simplified map of the pipelines owned by the National Cooperative Refinery Association, just as it was supplied by NCRA. It shows the main lines but not the network of small diameter crude oil gathering lines feeding

EXHIBIT 9

MAIN CRUDE OIL AND REFINED PRODUCTS LINES: OWNED BY NATIONAL
COOPERATIVE REFINERY ASSOCIATION

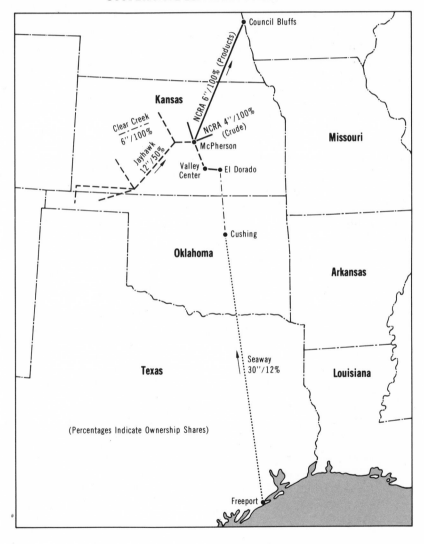

into those main lines. The total mileage of the Jayhawk line, for example, is about double that shown on the map.

The 242-mile refined products pipeline running from the NCRA refinery at McPherson, Kansas, north through Nebraska to Council Bluffs, is a clear example of a plant facility. It was built to transport products of this refinery and there are no other shippers.

NCRA, including its subsidiary, Clear Creek, Inc. owns an estimated 260 miles of crude oil lines in Kansas. The *Midwest Oil Register* for 1975 reports 190 miles for NCRA and 93 miles for Clear Creek. The latter, however, is not directly connected to the refinery and thus does not qualify as a plant facility.

NCRA is also a 50 percent owner of Jayhawk Pipe Line which has 229 miles of gathering lines in western Kansas and adjacent Oklahoma and 513 miles of trunk line from there to the refineries of NCRA at McPherson and Derby Refining Co. (Coastal States Gas Co.) at Wichita. While Jayhawk was built to serve the refineries of the two owners, it is a common carrier used by other shippers.

Derby Refining Co. (Coastal States Gas Co.). Derby's 50 percent interest in Jayhawk Pipe Line is described in the previous item. It is the sole owner of approximately 150 miles of crude oil lines in south central Kansas near its refinery at Wichita.

It also owns a 55-mile refined products line running from Wichita to McPherson, where it connects with other products lines.

Total Petroleum. Total's 273 miles of crude oil lines in Michigan are owned through its subsidiary, Michigan-Ohio Pipeline. This is an ICC (now FERC) regulated common carrier, and it is connected to the very small refinery (6,200 b/d) of Crystal Refining Company at Carson City, Michigan, but it exists primarily to supply Total's 41,000 b/d refinery at Alma, Michigan.

One section of this line runs roughly 150 miles to Alma from a point near Toledo where it connects with both Buckeye and Marathon lines. This section also connects, at Stockbridge, with the southern leg of the Lakehead Pipe Line running between Chicago and Sarnia. From these connections Total receives crude oil produced in southern Michigan, western Canada, western United States, and imported oil from the Gulf.

Another section runs about 60 miles to Alma from near Bay City, Michigan, where it connects with the northern leg of the Lakehead Pipe Line. From this connection Total receives oil produced in northern Michigan and Canada.

A third section runs south about 50 miles from Freemont to Alma and delivers oil gathered in north central Michigan.

Total owns two refined products lines, totalling 91 miles, running from Alma to company-owned terminals at Lansing and Bay City, Michigan. The line to Bay City is owned by Michigan-Ohio, and therefore an FERC regulated common carrier, but both lines were built to serve this refinery and are of no use to any other shipper. It also owns an LPG pipeline running from Alma about 40 miles east to a connection with a Shell LPG line running from Kalkaska to Marysville, Michigan.

These owned lines, with the exception of that handling LPG, are shown in Exhibit 10.

Many other examples could be cited but these fifteen should suffice. They emphasize that pipelines are usually built by those that expect to use them, and that frequently the lines are designed to serve specific refineries and are of little or no use to other shippers.

Appendix B: Five Tables Classifying All Refineries, Other Than Those Owned by the Sixteen Largest Refining Companies, According to Their Location with Respect to Existing Refined Products Pipelines.

Of the total of 273 operating refineries in the United States on January 1, 1977, as reported by the Bureau of Mines, 174 were not owned by the sixteen largest refining companies. The following tables group these 174 in five categories according to their location with respect to existing refined products pipelines. Each refinery is listed under the name used in the Bureau of Mines report, without identification of any parent company, and the classification of that refinery is without regard for the corresponding situation with respect to any other refinery that may be owned by the same firm.

This grouping of these refineries is not infallible, both because of the difficulty of getting precise information about some situations and because of the problem of fitting all of them into only five discreet categories. This qualification, however, is not sufficient to distort the general picture. The classification provides useful insight into which refineries are in a position where they might conceivably suffer discrimination in the use of competitor owned lines. The conclusions supported by these tables were reached in Section IV. As noted there, this paper does not purport to examine the evidence as to whether in fact there is any such discrimination.

EXHIBIT 10

MAIN CRUDE OIL AND REFINED PRODUCT LINES IN MICHIGAN OWNED BY TOTAL PRODUCTION

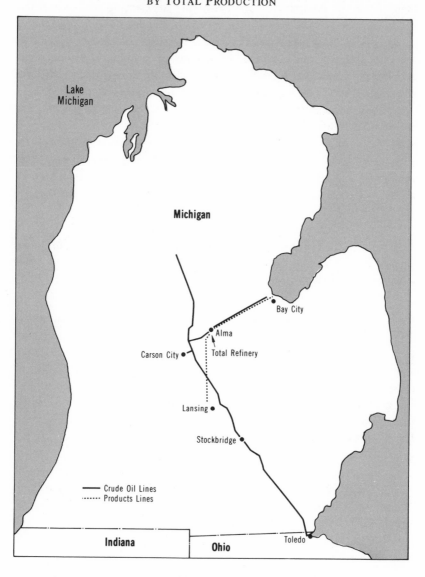

APPENDIX TABLE 1

REFINERIES DISTANT FROM ANY PRODUCTS PIPELINE OR TOO SMALL TO WARRANT CONNECTIONS

Name	Location	Operating Capacity in Barrels per Calendar Day
Louisiana Land & Exploration Co.	Mobile, Ala.	38,432
Marion Corp.	Theodore, Ala.	19,200
Vulcan Asphalt Refining Co.	Cordova, Ala.	5,000
Warrior Asphalt Corp.	Holt, Ala.	3,000
Tesoro-Alaskan Petroleum Corp.	Kenai, Alaska	38,000
Arizona Fuels	Fredonia, Ariz.	5,000
Beacon Oil Co.	Hanford, Calif	12,400
Edgington-Oxnard Refinery	Oxnard, Calif.	2,500
U.S.A. Petroleum Corp.	Ventura, Calif.	15,000
Asamera Oil Co.	Denver, Colorado	21,500
Seminole Asphalt & Rfg. Co.	St. Marks, Fla.	6,000
Young Refining Co.	Douglasville, Ga.	5,000
Bi-Petro, Inc.	Pana, Ill.	1,000
Richards, M. T.	Crossville, Ill.	100
Wireback Oil Co.	Plymouth, Ill.	1,800
Yetter Oil Co.	Colmar, Ill.	1,000
Gladieux Refinery	Fort Wayne, Ind.	12,200
Laketon Asphalt Rfg. Co.	Laketon, Ind.	8,500
E-Z Serve	Shallow Water, Kansas	10,000
Mid-America Refining Co.	Chanute, Kansas	3,100
Kentucky Oil & Refining Co.	Betsy Lane, Ky.	3,000
Somerset Refinery	Somerset, Ky.	5,000
Canal Refining Co.	Church Point, La.	4,800
Evangeline Refining Co.	Jennings, La.	5,000
Hill Petroleum	Krotz Springs, La.	5,000
Consumers Power Co.	Marysville, Mich.	34,097
Crystal Refining Co.	Carson City, Mich.	6,200
Dow Chemical Co.	Bay City, Mich.	17,000
Texas American Petrochemical	West Branch, Mich.	12,500
Lakeside Refining	Kalamazoo, Mich.	5,600
Southland Oil Co.	Lumberton, Miss.	5,800
Southland Oil Co.	Sandersville, Miss.	11,000
Southland Oil Co.	Yazoo City, Miss.	4,200
Big West Oil Co.	Kevin, Mont.	5,123
Wesco Refining Co.	Cut Bank, Mont.	4,658
Tesoro Petroleum Co.	Wolf Point, Mont.	2,500
CRA, Inc.	Scottsbluff, Nebr.	5,000

APPENDIX TABLE 1 (continued)

Name	Location	Operating Capacity in Barrels per Calendar Day
A. Johnson & Co.	Newington, N.H.	11,700
TransOcean Petroleum	Wilmington, N.C.	10,000
Caribou Four Corners	Farmington, N.M.[a]	2,500
Giant Industries	Farmington, N.M.[a]	8,800
Plateau, Inc.	Bloomfield, N.M.[a]	8,689
Thriftway Oil Co.	Bloomfield, N.M.[a]	7,500
Northland Oil & Refining Co.	Dickinson, N.D.	5,250
Westland Oil Co.	Williston, N.D.	4,658
Allied Materials Co.	Stroud, Okla.	5,500
Tonkawa Refining Co.	Arnett, Okla.	3,954
Pennzoil Co.	Rouseville, Pa.	10,000
Quaker State Oil Refining Co.	Emlenton, Pa.	3,320
Witco Chemical Co.	Bradford, Pa.	9,000
Quaker State Oil Refining Co.	Smethport, Pa.	6,500
Wolfs Head Oil Refining	Reno, Pa.	2,100
Delta Refining Co.	Memphis, Tenn.	43,900
Adobe Refining Co.	La Blanca, Tex.	5,000
Flint Chemical Co.	San Antonio, Tex.	1,500
Gulf States Oil & Rfg. Co.	Quitman, Tex.	4,500
J & W Refining Co.	Palestine, Tex.	4,000
Petrolite Corp.	Kilgore, Tex.	1,000
Pioneer Refining Co.	Nixon, Tex.	2,200
Tesoro Petroleum Corp.	Carrizo Springs, Tex.	26,100
Texas Asphalt & Refining Co.	Euless, Tex.	6,000
Thriftway Oil Co.	Graham, Tex.	1,000
Wickett Refining Co.	Wickett, Tex.	8,000
Winston Refining Co.	Fort Worth, Tex.	20,000
Eddy Refining Co.	Houston, Tex.	3,088
Mid-Texas Refinery	Hearne, Tex.	900
Monsanto Chemical Co.	Alvin, Tex.	8,500
Howell Hydrocarbons	San Antonio, Tex.	3,500
Morrison Petroleum Co.	Woods Cross, Utah	2,500
Plateau, Inc.	Roosevelt, Utah	7,500
Sound Refining Co.	Tacoma, Wash.	4,500
United Independent Oil Co.	Tacoma, Wash.	1,000
U.S. Oil & Refining Co.	Tacoma, Wash.	21,400
Elk Refining Co.	Falling Rock, W. Va.	5,500
Quaker State Oil Refining Co.	Newell, W. Va.	9,700
Quaker State Oil Refining Co.	St. Marys, W. Va.	5,000
Murphy Oil Corp.	Superior, Wisc.	45,400

APPENDIX TABLE 1 (continued)

Name	Location	Operating Capacity in Barrels per Calendar Day
C&H Refinery	Lusk, Wyo.	190
Husky Oil Co.	Cody, Wyo.	10,800
Mountaineer Refining Co.	La Barge, Wyo.	300
Sage Creek Refining Co.	Cowley, Wyo.	1,200
Southwestern Refining Co.	La Barge, Wyo.	500
Tesoro Petroleum Co.	New Castle, Wyo.	10,500
V-1	Glenrock, Wyo.	100
Total of 84 refineries		709,959

a These four small refineries are all in northwestern New Mexico. A Mapco line (non-refiner owned) running from that vicinity southeast into Texas transports natural gas liquids. Conceivably it might move refinery products, if these were available in sufficient quantity to permit batching, but we have no indication that it does so. El Paso Natural Gas Co. owns a line running from Gallup, N.M., *to* rather than *from* northwestern New Mexico.

NOTE: Connection to the nearest line is believed to be uneconomic considering both the distance and the refinery output of products transportable by pipeline. Includes a few instances where a pipeline flows *to* but not *from* the vicinity of the refinery.

APPENDIX TABLE 2

REFINERIES FOR WHICH THE ONLY AVAILABLE
PRODUCTS LINE IS NON-REFINER OWNED

Name	Location	Operating Capacity in Barrels per Calendar Day
Kern County Refinery	Bakersfield, Calif.[a]	15,900
Lion Oil Co.	Bakersfield, Calif.[a]	40,000
Mohawk Petroleum Co.	Bakersfield, Calif.[a]	22,100
Road Oil Sales	Bakersfield, Calif.[a]	500
Sabre Oil & Refining Co.	Bakersfield, Calif.[a]	3,500
San Joaquin Oil Co.	Bakersfield, Calif.[a]	20,000
Sunland Refining Co.	Bakersfield, Calif.[a]	15,000
West Coast Oil Co.	Bakersfield, Calif.[a]	15,000
Witco Chemical Co.	Bakersfield, Calif.[a]	11,000
Basin Petroleum	Long Beach, Calif.[b]	3,000
Demenno Resources	Compton, Calif.[b]	2,500
Champlin Petroleum Co.[c]	Wilmington, Calif.[b]	30,700
ECO Petroleum	Long Beach, Calif.[b]	3,000
Edgington Oil Co.	Long Beach, Calif.[b]	29,500
Fletcher Oil & Rfg. Co.	Carson, Calif.[b]	20,100
Golden Eagle Rfg. Co.	Carson, Calif.[b]	13,000
McMillan Ring Free Oil Co.	Long Beach, Calif.[b]	12,200
Lunday-Thagard Oil Co.	Southgate, Calif.[b]	3,222
Newhall Refining Co.	Newhall, Calif.[b]	11,500
Powerine Oil Co.	Sante Fe Springs [b]	44,120
Lion Oil Company	Avon, Calif.[d]	126,000
Pacific Refining Co.	Hercules, Calif.[d]	53,300
Berry Petroleum Co.	Stephens, Ark.[e]	3,500
Cross Oil & Refining Co.	Smackover, Ark.[e]	4,200
Lion Oil Co.	El Dorado, Ark.[e]	47,000
McMillan Ring Free Oil Co.	Norphlet, Ark.[e]	4,500
Atlas Processing Co.	Shreveport, La.[e]	45,000
Bayou State Oil Co.	Hosston, La.[e]	2,000
Bayou State Oil Co.	Hosston, La.[e]	2,000
Calumet Refining Co.	Princeton, La.[e]	2,400
Cotton Valley Solvents	Cotton Valley, La.[e]	7,000
Claiborne Gasoline Co.	Lisbon, La.[e]	6,500
Kerr McGee Refining Co.	Dubach, La.[e]	11,000
CRA, Inc.	Phillipsburg, Kansas	20,126

APPENDIX TABLE 2 (continued)

Name	Location	Operating Capacity in Barrels per Calendar Day
OKC Refining Co.	Okmulgee, Okla.	22,902
Vickers Petroleum Corp.	Ardmore, Okla.	62,500
Southern Union Refining Co.	Lovington, N.M.	36,100
Southern Union Refining Co.	Monument, N.M.	5,000
Longview Refining Co.	Longview, Tex.	9,000
Total of 39 refineries		785,870

a Southern Pacific Pipe Line runs from Bakersfield to Fresno.

b These are all near Los Angeles. Southern Pacific Pipe Line runs from Los Angeles to Tucson and Phoenix. San Diego Pipeline, running from Los Angeles to San Diego, is jointly owned by the Sante Fe and Southern Pacific railroads. Short refiner owned pipelines in the Los Angeles area would be of little use to non-owners.

c Calnev Pipe Line, owned by Champlin, runs from a connection with Southern Pacific Pipe Line, at Colton, California, to Las Vegas, Nevada. Champlin is included here, rather than in Table 3, because this line does not start at its refinery.

d These are near San Francisco. Southern Pacific Pipe Lines has a network of lines in the San Francisco area and extending north into the Sacramento valley, south into the San Joaquin valley and east into Nevada.

There are short refiner owned pipelines in the immediate San Francisco area but these appear to reach markets also reached by Southern Pacific Pipe Lines. (Whether either of these refineries could use a refiner owned line, thus putting it in Table 4 rather than Table 2, is not readily ascertainable.)

e These are all in southern Arkansas, and extreme northwestern Louisiana. The only near products line is Texas Eastern Transmission Co., which is here counted as non-refiner owned, although it owns a small refinery, because its refinery output is nominal relative to its pipeline traffic.

NOTE: Includes some plants whose output of products transportable by pipeline is probably too small to warrant even a short connection. These might have been included in Table 1.

APPENDIX TABLE 3

Refineries Whose Owners Own or Are Part Owners of One or More Products Lines Serving These Refineries

Name	Location	Operating Capacity in Barrels per Calendar Day
A. *Owned lines are the only ones available*		
Indiana Farm Bureau Cooperative	Mt. Vernon, Ind.	21,200
Total Petroleum	Alma, Mich.	41,000
Farmers Union Central Exchange	Laurel, Mont.	33,650
Navajo Refining Co.	Artesia, New Mexico	29,900
United Refining Co.	Warren, Pa.	52,000
Diamond Shamrock Co.	Sunray, Tex.	51,500
Sigmor Corp.	Three Rivers, Tex.	10,000
Sinclair Corp.	Sinclair, Wyo.	49,000
American Petrofina Co.	Mount Pleasant, Tex.	26,000
Total of 9 refineries		314,250
B. *Refinery also uses, or might conceivably use, non-refiner owned lines*		
Apco Oil Co.	Arkansas City, Kans.	38,020
Derby Refining Co.	Wichita, Kans.	27,982
National Cooperative Refinery Assn.	McPherson, Kans.	54,150
Skelly Oil Co.	El Dorado, Kans.	78,731
Koch Refining Co.	Rosemont, Minn.	127,300
Apco Oil Co.	Cyril, Okla.	14,000
Champlin Petroleum Co.	Enid, Okla.	53,800
Kerr McGee Refinery	Wynnewood, Okla.	50,000
La Gloria Oil & Gas Co.	Tyler, Tex.	24,400
Cosden Oil & Chemical Co.	Big Springs, Tex.	65,000
Total of 10 refineries		533,383
C. *Refinery also uses, or might conceivably use, refiner owned lines, including such lines to which the owned line is connected*		
Getty Oil Co.	Delaware City, Del.	140,000
Clark Oil & Refining Co.	Blue Island, Ill.	70,000

APPENDIX TABLE 3 (continued)

Name	Location	Operating Capacity in Barrels per Calendar Day
Murphy Oil Co.	Meraux, La.	92,500
Tenneco Oil Co.	Chalmette, La.	86,000
Champlin Petroleum Co.	Corpus Christi, Tex.	125,000
American Petrofina Co.	Port Arthur, Tex.	110,000
Coastal States Petrochemicals	Corpus Christi, Tex.	185,000
Crown Central Petroleum Co.	Pasadena, Tex.	100,000
Husky Oil Co.	Cheyenne, Wyo.	23,600
Total of 9 refineries		932,100

NOTE: The twenty-eight refineries in Table 3 are owned by twenty-three companies. These twenty-three plus the sixteen largest refining companies, all of which own products lines, compare with the fifty refining companies reported in the introduction section as owning products lines. Lines owned by the other eleven refiners cannot be identified with a specific refinery. These include a number of lines that are reported, according to ICC (now FERC) definition, as refined products lines but transport products of gas processing plants rather than crude oil refineries. Two refiners, Pennzoil Co. and Witco Chemical Co., are the owners of National Transit Co., which is primarily a crude oil line but does report five miles of products line which may be connected to Pennzoil's refinery at Reno, Pennsylvania. Amerada Hess is, or was until recently, half owner of a short connecting line in New Jersey but its refinery in that state is now closed.

APPENDIX TABLE 4

REFINERY OWNER DOES NOT OWN A PRODUCTS LINE
SERVING THIS REFINERY. IT USES, OR CONCEIVABLY COULD USE,
BOTH REFINER AND NON-REFINER OWNED LINES.

Name	Location	Operating Capacity in Barrels per Calendar Day
Clark Oil & Refining Co.	Hartford, Ill.	55,000
Energy Cooperative	East Chicago, Ind.	126,000
Rock Island Refining Corp.	Indianapolis, Ind.	43,200
CRA, Inc.	Coffeyville, Kans.	39,531
American Petrofina Co.	El Dorado, Kans.	25,000
Midland Cooperative	Cushing, Okla.	19,600
Total of 6 refineries		308,331

APPENDIX TABLE 5

REFINERIES FOR WHICH THE ONLY AVAILABLE
PRODUCTS LINE IS OWNED BY OTHER REFINERS

Hunt Oil Co.	Tuscaloosa, Ala.	30,000
Good Hope Refineries	Good Hope, La.	70,000
La Jet, Inc.	St. James, La.	15,000
Placid Oil Co.	Port Allen, La.	36,000
Amerada Hess	Purvis, Miss.	30,000
Pride Refining	Abilene, Tex.	36,500
Charter International	Houston, Tex.	70,000
Independent Refining Corp.	Winnie, Tex.	13,490
Quintana-Howell Joint Venture	Corpus Christi, Tex.	44,559
Sabre Refining Co.	Corpus Christi, Tex.	9,950
South Hampton Co.	Silsbee, Tex.	18,100
Southwestern Refining Co.	Corpus Christi, Tex.	120,000
Texas City Refining Co.	Texas City, Tex.	74,500
Caribou Four Corners Oil Co.	Woods Cross, Utah	7,500
Husky Oil Co.	North Salt Lake, Utah	23,000
Western Refining Co.	Woods Cross, Utah	10,000
Little America Refining Co.	Casper, Wyo.	24,500
Total of 17 refineries		633,099